**The Role of Fine-Scale Magnetic Fields
on the Structure of the Solar Atmosphere**

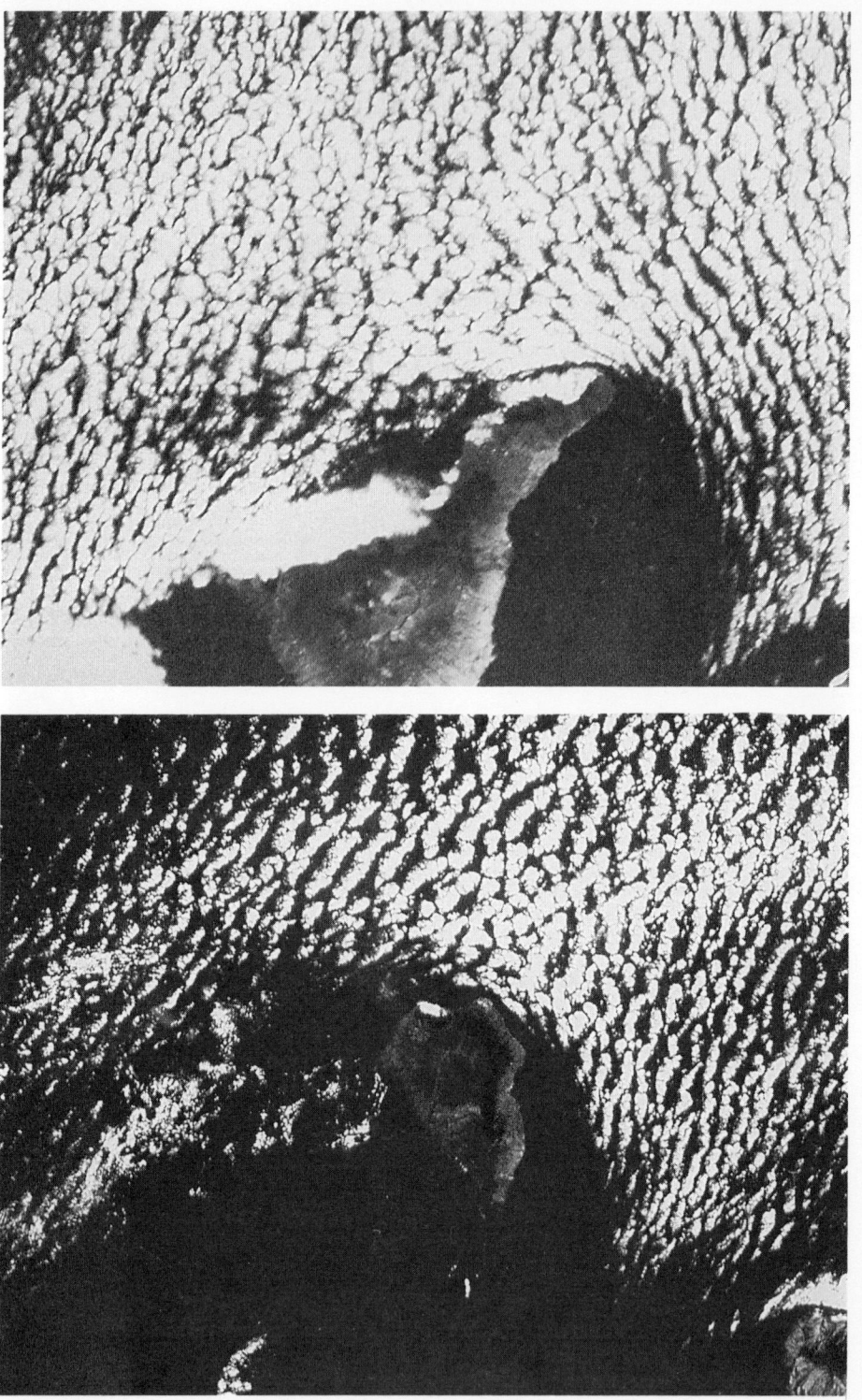

Landsat MSS photographs of La Palma and Tenerife (ESA-EARTHNET program). (Estación Espacial de Maspalomas INTA-Espana).

The Role of Fine-Scale Magnetic Fields on the Structure of the Solar Atmosphere

Proceedings of the Inaugural Workshop
and Round Table Discussion for the
D-E-S Telescope Installations on the Canary Islands
La Laguna, Tenerife, Spain 6–12 October 1986

Editors
E.-H. Schröter, M. Vázquez, and A.A. Wyller

Sponsored jointly by

- Ⓓ Deutsche Forschungsgemeinschaft, Bonn
- Ⓔ Instituto de Astrofisica de Canarias, La Laguna
- Ⓢ The Royal Swedish Academy of Sciences, Stockholm

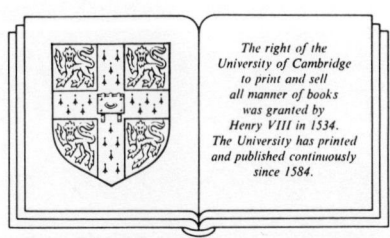

Cambridge University Press

Cambridge
New York   New Rochelle
Melbourne   Sydney

Published by the Press Syndicate of the University of Cambridge
The Pitt Building, Trumpington Street, Cambridge CB2 1RP
32 East 57th Street, New York, NY 10022, USA
10 Stamford Road, Oakleigh, Melbourne 3166, Australia

© Cambridge University Press 1987

First published 1987

Printed in Great Britain at the University Press, Cambridge

British Library Cataloguing in Publication Data

D.E.S. Inaugural Workshop and Round Table Discussion
(1986: Tenerife)
The role of fine-scale magnetic fields on the structure
of the solar atmosphere: proceedings of the D.E.S.
Inaugural Workshop and Round Table Discussions, held
in Tenerife, Spain, October 6–12, 1986.
1. Solar atmosphere. 2. Magnetic fields (Cosmic
physics)
I. Title II. Deutsche Forschungsgemeinschaft
III. Instituto de Astrofisica de Canarias
IV. Vetenskapsakademien V. Schroter, Egon-Horst
VI. Vazquez, Manuel VII. Wyller, Arne
523.7 QP528

ISBN 0 521 34281 3 hard covers

Contents

Preface	x
Editorial Note	xi

SESSION 1 Chairman: *T. Roca Cortés*
Invited Reviews

Atmospheric Structure and the Activity Cycle *H. Holweger*	1
Line Asymmetry and the Activity Cycle *W. Livingston*	14

Oral Presentations

Meridional Flows and Latitudinal Dependence of the Convection *F. Cavallini, G. Ceppatelli, A. Righini*	21
The Limb Shift Effect and its Variation with the Solar Cycle *M. Anguera, P. L. Pallé, C. Régulo, T. Roca Cortés, G. R. Isaak,* *C. P. McLeod, H. B. van der Raay*	24

Posters

On the Differences between Line Bisectors in Quiet and Active Sun *C. Marmolino, G. Roberti, G. Severino*	30
Center-to-Limb Variation of the Asymmetries of the K 7699 Å line in Solar Quiet and Active Regions *J. A. Bonet, I. Marquez, M. Vázquez, H. Wöhl*	32

SESSION 2 Chairman: *R. Rosner*
Invited Review

Interaction between Magnetic Fields and Convection *N. E. Hurlburt, N. O. Weiss*	35

Oral Presentations

A 2D Study of Compressible Granular Flow and Predicted Spectroscopic Properties *M. Steffen*	47
On the Correlation of the C-Shape of the FeI-Line 5576 Å with the Brightness of Ca^+ K-Faculae *S. Immerschitt, E.-H. Schröter*	53
Joint Discussion on Topics of Sessions 1 and 2 *T. Roca Cortés, R. Rosner*	61

SESSION 3 Chairman: *J. O. Stenflo*
Invited Review
  Structure of Magnetic Flux tubes as derived from Observations with
  Moderate Spatial Resolution   67
    S. K. Solanki
Oral Presentations
  FTS Measurements of Solar Line Asymmetries in Quiet and Active Regions   82
    P. N. Brandt, S. K. Solanki
  Numerical Simulations of Umbral Stokes-V Profiles Considering
  Influences of Unresolved Dots   88
    K.-D. Pahlke

SESSION 4 Chairman: *W. Mattig*
Invited Review
  Needs and Limits of Magnetic and Velocity Field Measurements with
  Sub-Arcsecond Resolution   93
    E. Wiehr
Oral Presentations
  Properties of a Concentrated Magnetic Field Region   103
    S. Koutchmy, G. Stellmacher
  Drift Velocities in Flux Tubes Inferred by Spatially Averaged Line
  Bisectors   110
    F. Cavallini, G. Ceppatelli, A. Righini
Joint Discussion on Topics of Sessions 3 and 4   116
  J. O. Stenflo, W. Mattig

SESSION 5 Chairman: *C. Zwaan*
Invited Review
  Types of Magnetic Flux Emergence   118
    A. Title
Oral Presentations
  Observations of the Magnetic Fine Structure of a Facula
    J. C. del Toro Iniesta, M. Semel, M. Collados
  Asymmetry of Stokes Profiles across a Sunspot; Measurements   131
    K. S. Balasubramaniam
Posters
  Sunspot Proper Motions as a Tool for the Study of the Behaviour of
  Rising Magnetic Flux Tubes   136
    F. Mazzucconi, G. Godoli

SESSION 6 Chairman: *P. Maltby*
Invited Review
  Established and Non-Established Properties of Umbral and Penumbral
  Fine Structures   140
    J. I. García de la Rosa
Oral Presentation
  Photospheric Fine Structure Close to a Sunspot   156
    O. von der Lühe
Poster
  High Resolution Spectroscopy of Sunspot Penumbrae   162
    E. Wiehr, M. Knölker, H. Grosser, G. Stellmacher
Joint Discussion on Topics of Sessions 5 and 6   165
  C. Zwaan and P. Maltby

SESSION 7 Chairman: *H. U. Schmidt*
Invited Review
  The Subsurface Structure of Sunspots and the Origin of Solar Active
  Regions   167
    F. Moreno-Insertis
Oral Presentations
  The Wilson Effect in Sunspots   183
    M. Collados, J. C. del Toro Iniesta, M. Vázquez
  A New Mechanism for the Evershed Effect   187
    F. H. Busse
Poster
  Decay Rates of Sunspot Groups from 1874 to 1939   196
    F. Moreno-Insertis, M. Vázquez

SESSION 8 Chairman: *A. Righini*
Invited Review
  How is the Penumbra formed?   199
    H. C. Spruit
Oral Presentations
  Nonlinear Compressible Convection in Regions of Intense Magnetic
  Fields   210
    N. E. Hurlburt
  The Intensity Distribution in Sunspot Penumbras   214
    M. Collados, J. C. del Toro Iniesta, M. Vázquez
Joint Discussion on Topics of Sessions 7 and 8
  H. U. Schmidt

SESSION 9 Chairman: *P. Mein*
Invited Review
  Structure and Dynamics of Small Magnetic Flux Concentrations:
  Observation versus Theory   223
    M. Schüssler
Oral Presentations
  Adiabatic Longitudinal-Transverse Magnetohydrodynamic
  Tube Waves   243
    Kurt Zähringer, Peter Ulmschneider
  Stability of Magnetic Arcades   248
    U. Anzer, A. W. Hood
Poster
  Thermal Dissipation in Slender Flux Tubes and Structured Media   250
    P. Edwin, B. Roberts

SESSION 10 Chairman: *B. Roberts*
Invited Review
  Wave and Thermal Instabilities in Flux Tubes: Their Role for the
  Structure of the Chromosphere and Transition Region   255
    R. Hammer
Oral Presentations
  Non-Linear Alfvén-Waves and the Fine Structure of Magnetic Fields in
  the Photosphere   274
    E. Jensen
  Poloidal Mode Coupling of Alfvén Continuum Modes in
  2D Coronal Loops   272

    S. Poedts, M. Goossens
  Radiative Relaxation in Small Scale Structures   281
    J. Trujillo-Bueno, F. Kneer
Poster
  Some Statistical Properties of the Magnetic Field in Prominences   289
    I. Kim, S. Koutchmy, G. Stellmacher, A. I. Stepanow
Joint Discussion on Topics of Sessions 9 and 10   292
  P. Mein, B. Roberts

SESSION 11   Chairman: *W. Deinzer*
Invited Review
  Appearance and Disappearance of Magnetic Flux at the Solar surface   297
    E. R. Priest
Oral Presentations
  Transition Zone Flows in Sunspots   317
    O. Kjeldseth-Moe, N. Brynildsen, O. Engvold, P. Maltby,
    J.-D. F. Bartoe, G. E. Brueckner
Overshoot of the Solar Granulation   322
  A. Nesis
Joint Discussion on Topics of Session 11   328
  W. Deinzer

SESSION 12   Chairman: *J. Zirker*

Solar Instrumentation
Oral Presentations
  Cross Correlation of Images of Solar Fine Structure and Possible
  Applications – a Progress Report   330
    T. A. Darvann
  The Swedish Fabry-Perot Echelle Scanner as a Two-Dimensional
  Universal Filter and Stokesmeter   335
    A. A. Wyller
Poster
  Imaging Interferometry with Non-Redundant Arrays   342
    J. B. Zirker
  Magnetic Field Measurements with the Swedish Solar Telescope on
  La Palma   343
    H. Lundstedt
Joint Discussion on Topics of Session 12   348
  J. B. Zirker

Presentations
The Swedish 50 cm Vacuum Solar Telescope: Concepts and Auxiliary
Instrumentation   349
  G. B. Scharmer
Present and Future Facilities for the Vacuum Gregory Coudé
Telescope at Izaña   354
  E. Wiehr
Present and Future Observational Facilities of the German Vacuum
Tower Telescope   362
  D. Soltau

The French Polarization-Free Telescope THEMIS *J. Rayrole*	367
Other Solar Facilities at the Observatorio del Teide *M. Vázquez*	370

Round Table Discussion
Chairmen: *E.-H. Schröter, M. Vázquez, A. Wyller*

What can be done best, what with restrictions, and what can not be done with the presently foreseen facilities on the Canary Islands?
What extensions of these facilities are desirable?
What are the candidate problems for future integrated observational experiments, for which all available solar observing facilities could be used simultaneously?

Summary of the Round Table Discussion *R. B. Dunn*	374
List of Participants	377

Preface

Last year, at the end of June 1985, were solemnly inaugurated the international observatories of Roque de los Muchachos (La Palma) and Teide (Tenerife) and the Instituto de Astrofisica de Canarias of La Laguna (Tenerife).

This stylish and elegant affair was attended by 13 royalties and excellencies and some 600 guests from 8 different nations of the scientific communities of the countries involved.

The natural idea arose of organizing also a scientific inauguration in the following year, when powerful new telescopes would be operational in both islands. As a result emerged this jointly organized Inaugural Workshop and Round Table Discussion with almost 100 registered participants from 15 different countries.

We, the organizers, had several goals in mind when structuring this workshop. Since these telescopes will be operational well into the next millennium, we wanted as many *young* solar physicists as possible to participate. They will be active with these telescopes far beyond the allotted time of some of the members of the organizing committee! We wanted them stimulated by the workshop itself and by physically experiencing the observational facilities. As such we individually endeavoured to provide travel stipends, and we were delighted to see that so many young people indeed have participated in the conference. This hopefully augurs well for the future recruitment in solar physics.

We also wanted to bring together experienced observers and articulate provocative theoreticians to interact creatively. We are very grateful for the large turnout from both these groups as witnessed by the many excellent invited reviews and papers and the very spirited discussions which followed.

However, we have also aimed beyond the goals of traditional workshops by arranging a Round Table Discussion in the hope of specifically linking the superb solar climate on the Canary Islands with the extraordinary powerful battery of new solar telescopes on these two islands. It is our conviction that they will open a new era in Sub-Arc-second Solar Physics, and thus much careful thought should be given to structure and formulate research avenues which make the most of these opportunities. Also the use of these telescopes in joint orchestration on certain research problems will need the full measure of international collaboration, and hopefully many new workshops will arise as a result of this.

Freiburg/Tenerife/La Palma December 1986
E.-H.S. M.V. A.W.

Editorial Note

A major goal of the workshop has been the stimulation of a discussion between observers and theoreticians on certain current topics of solar physics. We devoted three sessions to each such topic. In session A the field was reviewed from the observational point of view. In session B the state of the art of theory was presented. Finally in session C a comprehensive discussion of both aspects was organized and guided by the chairpersons of the two previous sessions.

A summary of these discussions has been prepared jointly by both chairpersons for publication in the present proceedings. Perhaps owing to our failure to issue strict instructions regarding their structure and content the summaries which were received proved to be more heteromorphic than we would have liked.

Nevertheless, in order to achieve a timely publication of the proceedings we made no effort to bring these summaries in more homogeneous form and publish them without any alteration.

ATMOSPHERIC STRUCTURE AND THE ACTIVITY CYCLE

H. Holweger
Institut für Theoretische Physik und Sternwarte,
Universität Kiel, Olshausenstr. 40, D-2300 Kiel, F. R. G.

1 INTRODUCTION

Is the granulation modified in active regions? Does the granulation vary with the cycle? Conflicting results have been obtained by experienced observers at excellent observing sites. As a spectroscopist, I am not the expert to decide who is right, but I find the possibility exciting - and attractive for reasons to be discussed - that those ubiquitous convective elements which carry the luminosity of the Sun to its surface are cycle modulated.

Rather than concentrating exclusively on this controversial issue, I will also try to summarize certain other <u>activity-related phenomena that are known - or suspected - to occur on a more or less global scale,</u> not restricted to the sunspot belts. In view of this, a global cycle modulation of granular properties does not appear exotic. It might be responsible, solely or in part, for the long-term variations seen in the full-disk photospheric spectrum.

2 GRANULATION
2.1 Quiet photospheres - active regions

A number of earlier studies provided evidence that the mean diameter of granules is smaller in the immediate neighbourhood of isolated sunspots or in complex spot groups (Macris 1949, 1953; Danielson 1961; Macris and Prokakis 1962; Schröter 1962). Additional support for this came from a more recent analysis by Macris (1979) of 40 selected sunspot photographs obtained at Pic-du-Midi (Table 1).

In contrast, Bray and Loughhead (1964) did not find such a difference. The same conclusion was reached by Collados et al. (1986), who studied photographs of five sunspot regions obtained with the 40 cm Vacuum Telescope at Izaña, Tenerife (Table 1). Unlike previous work, their analysis employed objective photometric criteria for identifying

Table 1: Mean diameter of granules (arc sec)

	Collados et al. (1986) Izaña	Macris (1979) Pic-du-Midi
Quiet photosphere	1.52 ± 0.16	1.78 ± 0.23
Near sunspots	1.41 ± 0.19	1.28 ± 0.25
Light bridges	1.51 ± 0.10	...

granules. They found that the mean diameter of granules in the quiet photosphere (d_{ph}), near sunspots (d_{spot}) and in light bridges, are not significantly different, though spatial variations may occur near sunspots.

In comparing these results with those of Macris (1979), Collados et al. discovered a systematic trend in Macris' data which they found difficult to explain and which may indicate a systematic bias: in a plot of d_{ph} versus d_{spot}, the individual pairs of determinations in Macris' set of data follow a linear relation, i.e. if d_{ph} is large, d_{spot} is also large, and vice versa. As will be pointed out in Section 2.2, this may in fact be a real solar phenomenon rather than the result of a systematic bias.

Clearly, further work has to be done to settle the question as to whether the granulation is modified in active regions. The application of objective photometric criteria will be essential. The study by Collados et al. (1986) is an important step forward, though I feel there still remains considerable evidence in favour of spatial variations. According to Title (1986), Spacelab 2 photographs definitely demonstrate that the granulation has different dynamical properties in magnetic and non-magnetic areas.

2.2 Granulation and the activity cycle

The possibility that properties of the granulation might vary with the solar cycle was first mentioned by Macris (1955). The first quantitative results (Macris and Elias 1955) indicated that the number of granules on the Sun increases with increasing sunspot number. However, the observational material available at that time was sparse and inhomogeneous. Birkle (1967) could not confirm the reported variability. His study was based on Potsdam routine photographs, a more homogeneous set of observations yet of comparably poor spatial resolution.

The observational situation improved substantially when the solar facilities at Pic-du-Midi and Sacramento Peak became available. Macris and Rösch (1983) confirmed the variability reported by Macris and Elias; the mean distance between granules was found to vary between 2.3 arc sec at sunspot minimum and 1.9 arc sec at an activity level where the smoothed sunspot number, R, was close to 110. Their study is based on photographs obtained at Pic-du-Midi with apertures of 38 cm (1966-1972) and 50 cm (1973-1978), plus one photograph taken in 1975 with the Sac Peak 76 cm telescope. All three sets of observations gave consistent results.

More recently, Macris et al. (1984) analyzed high-quality frames obtained exclusively with the Pic-du-Midi 50 cm refractor. The period 1973-1983 was covered. The number of granules per 10 arc sec x 10 arc sec area was found to vary from 35 at sunspot minimum (1975/1976) to about 44 around 1980-1982, with a phase lag of 2-3 years with respect to sunspot maximum. The method of analysis was still counting "by eye", but automatic reduction using photometric criteria is in progress. Preliminary results (Muller and Roudier 1984a) supported the inferred variability, at least qualitatively. Additional support came from an autocorrelation analysis of two frames obtained in 1979 and 1984, respectively (Muller 1985a).

The cycle dependence of granular properties has implications for the discussion in Section 2.1. It may be responsible for the linear relation between d_{ph} and d_{spot} in Macris' (1979) data, which Collados et al. (1986) found to be suspect. The observations used by Macris cover the period 1966-1974. The granular diameters quoted in his Table 1 show a time variation consistent with the cycle dependence (and phase lag) reported above. The linear relation would mean that both d_{ph} and d_{spot} vary and are in phase. In contrast to Macris' data, those of Collados et al. would not be expected to show such a conspicuous variation. The period covered by their observations is shorter (1980-1984) and, according to the Pic-du-Midi workers (Macris et al. 1984), this coincides with the time when the number density of granules was essentially stationary.

Another property of the granulation which should be examined for possible cycle modulation is the granular contrast. Alissandrakis et al. (1982) presented evidence for a pronounced variation with time, with (uncorrected) contrast values increasing from 1.1 to 1.3 during the ascending part of the present solar cycle. However, as these authors emphasized, there is as yet no firm observational basis for this potentially important result.

Variations in granular size (and contrast) imply that the mean thermal structure of the photosphere varies with the cycle. The basic question is: what determines, or modifies, the size of the granules? One aspect of this puzzle, which may be studied theoretically in the near future, is the lower and upper bound (if any exists) for the size of a granule that is still being driven by convection. Numerical simulations, such as those described by Steffen (1986) will be carried out in order to investigate this in more detail. In the next section, we shall briefly discuss some relevant observations.

2.3 Size distribution of granules

Most of the earlier studies were based on Stratoscope photographs which have a diffraction-limited resolution of 0.37 arc sec. Typically, a characteristic size of the granules has been inferred, but the actual value has remained controversial. Namba and Diemel (1969) found a size distribution which was peaked at 725 km, with 70% of all diameters lying in the range d = 500-1200 km. However, the numerical experiments of Musman (1969) showed drastically how sensitively the appearance of the granulation depends on spatial resolution.

At Pic-du-Midi, a resolution of 0.25 arc sec can be achieved under good conditions. From computer-processed photographs of this quality, Muller and Roudier (1984b) concluded that the granulation does not have a characteristic scale: the number of granules found in a given area increases towards smaller diameters. However, the main contribution to the total area, and radiation, comes from granules with d ≈ 1000 km.

The somewhat delicate problem of defining what is a granule can be circumvented by using two-dimensional power spectra of the spatial brightness fluctuations. From Spektrostratoskop photographs, Durrant et al. (1983) inferred that there is little evidence of distinct peaking. Rather, the power decreases exponentially with wavenumber, in accordance with speckle interferometry (Aime et al. 1978; Ricort and Aime 1979).

Expressed in terms of spatial power spectra of granular brightness fluctuations, the reported increase of the number of granules with sunspot number (Section 2.2), and also the supposed increase of granular contrast, would mean that at high levels of activity, the power is shifted towards higher wavenumbers.

Does the granulation vary with the cycle? I feel the evidence is strong, yet an independent confirmation of the Pic-du-Midi results would be highly desirable. In the next sections, I shall summarize other features of more or less global occurrence that may vary with the cycle. Other than the granulation, most or all of them are magnetic

3 EPHEMERAL ACTIVE REGIONS (ER)

The basic properties of ER were established by Harvey and Martin (1973), Harvey et al. (1975) and Martin and Harvey (1979). ER are small, short-lived bipolar magnetic regions with typical dimensions of about 10000 km and lifetimes ranging from less than one day to ten days or so. Hundreds are present on the whole Sun at any one time. Their latitude distribution is much broader than that of sunspots. Unlike the large-scale magnetic fields, the spatial orientation of ER is nearly random. They do not form at supergranule boundaries.

The number of ER varies with the solar cycle in a nontrivial way. According to Martin and Harvey (1979), the number of ER occurring at any instant of time on the whole Sun varies approximately with sunspot number R. Longer-lived ER automatically enter with higher weight into this statistics. Surprisingly, if ER with lifetimes shorter than 3 days are counted separately, a clear anticorrelation with R is found. This was recognized by Harvey and Martin (1973) in earlier records of ephemeral Ca plages. Another distinguishing property of the shorter-lived ER is their very broad latitude distribution, which is not far from being uniform.

The anticorrelation between short-lived ER and sunspot number was not easily accepted as real by its discoverers, but there is a much more conspicuous phenomenon, to be discussed in the next section, which is closely related to ER and strongly supports this anticorrelation.

4 X-RAY BRIGHT POINTS (XBP)

The properties of XBP were described by Davis (1983), Golub et al. (1977, 1979) and in earlier work. XBP are small coronal emission features visible in soft X-ray images of the Sun. Typical dimensions are 15000-20000 km, and most lifetimes are in the range 2-48 hours. About 500 XBP are estimated to occur on the whole Sun at any one time (Golub et al. 1979). Their distribution in latitude is nearly uniform.

All XBP can be associated with Ephemeral Active Regions, but approximately half of the ER identified on magnetograms do not coincide with obvious XBP. Golub et al. (1977) argue that most of the "missing" XBP are obscured by overlying coronal structures.

The frequency of occurrence of XBP shows a striking anticorrelation with sunspot number (Golub et al. 1979; Davis 1983); the available data cover the entire period 1970-1981. Davis emphasizes the apparent conflict between this result and that of Martin and Harvey (1979), mentioned in Section 3, who found that the total number of ER varies with R rather than is anticorrelated. I suspect that this lack of correspondence between the cycle dependence of XBP and ER is only superficial and that mainly the shorter-lived ER, which are anticorrelated with R, correspond to XBP. The reported lifetime of XBP seems to support this hypothesis.

From a comparison of the X-ray data with magnetograms, Golub et al. (1979) concluded that a typical XBP contains about 3×10^{19} Mx total magnetic flux. Using the relative number of XBP and active regions, they estimated that in 1970, near sunspot maximum, about 40% of the flux had emerged in the form of XBP. This figure increased to 95% in 1976 at sunspot minimum. The substantial contribution of XBP to the total flux emerging at the solar surface is consistent with an earlier estimate by Harvey and Martin (1973) of the flux emerging in short-lived ER.

In contrast, the total magnetic flux existing at any instant of time is dominated by structures other than XBP/ER (Section 6).

5 NETWORK BRIGHT POINTS (NBP)

Network bright points - also called facular points - are sub arc sec photospheric structures whose occurrence is not restricted to faculae. Their properties have been described by Mehltretter (1974), Muller (1983, 1985b), Muller and Keil (1983), Muller and Roudier (1984c) and others. NBP differ from the features discussed so far in many respects. Typical dimensions are estimated to be as small as 150 km (0.2 arc sec) and their mean lifetime is about 18 min. Their occurrence is related to both the granulation and the supergranulation: they are always located in intergranular lanes, mostly at supergranular boundaries along which they are distributed non-uniformly. Longer aggregates of these tiny structures make up what Dunn and Zirker (1973) termed "filigree" and "crinkles". In NBP, Fraunhofer lines are generally weakened ("line gaps", Sheeley 1967), a property which has been attributed to a specific vertical temperature stratification (e.g. Koutchmy and Stellmacher 1978; Stellmacher and Wiehr 1979).

NBP occur at all latitudes. In 1983, at an intermediate level of activity, the distribution was found to be non-uniform, with a spectacular increase of the surface density towards the pole (Muller and Roudier 1984c). An outstanding property of NBP also reported by Muller and Roudier is their pronounced anticorrelation with sunspot number. Near the equator, the number of NBP per 100 arc sec x 100 arc sec area increased from about 60 at sunspot maximum (1979/1980) to 300 in 1983 ($R \approx 70$).

More observations are required to prove this anticorrelation beyond doubt. Some support comes from earlier work. Waldmeier (1955) noted that polar facular points are conspicuous only at sunspot minimum. Sheeley (1964) found a clear anticorrelation of the number of polar faculae with R lasting over the entire period 1935-1963.

NBP are generally considered to be the signature of magnetic flux tubes. The reported anticorrelation then implies that the number of flux tubes in the quiet photosphere is also anticorrelated with R.

The anticorrelation with sunspot number of both types of small-scale magnetic features, X-ray bright prints (short-lived ephemeral active regions) and network bright points, suggests that the activity cycle actually consists of two complementary phenomena: large-scale flux concentrations (sunspots) confined to certain latitude zones, and small-scale features, anticorrelated in time with sunspots, occurring more or less on the whole Sun. Or, as Golub et al. (1979) expressed it in their discussion of XBP, "the solar cycle may be characterized as an oscillator in wavenumber space with relatively little variation in the average total rate of flux emergence".

In the next two sections, I shall discuss two global properties which are probably influenced by small-scale, global structures, such as those summarized above.

6 FULL-DISK TOTAL MAGNETIC FLUX

The Mount Wilson magnetograph data constitute an invaluable record of global magnetic activity. Details of observational technique and data reduction were described by Howard (1974a, 1976, and references therein). Among the quantities that can be derived from these data, the total magnetic flux

$$F^T = |F^+| + |F^-|$$

has the advantage of being independent of any possible instrumental zero drift.

The full-disk magnetic flux is obtained by summing up the fluxes measured within each surface element. Naturally, these fluxes are underestimates if mixed polarities occur within the aperture. Since 1975, all Mount Wilson full-disk magnetograms have been taken with a 12.5 x 12.5 arc sec aperture (Howard 1976). Only the longitudinal component of the field was measured, and no hidden corrections were made for field line orientation. In view of the expected complex geometry, such a correction would be highly arbitrary, even at disk center.

Mount Wilson flux data covering almost the whole cycle 21 have been published by Bruning and LaBonte (1985). These data, together with the monthly mean sunspot numbers, are shown in Figures 1 and 2. Apart from the general parallelism between F_T^H and R, there is a remarkable phase lag of F^T, the maximum of F^T having occurred about 2 years after sunspot maximum. According to Howard and LaBonte, a similar effect was seen in cycle 20. I shall come back to this point in the next section.

Another interesting difference between the cycle variation of F^T and R is apparent in the relative amplitude. Whereas R is seen to have increased from 1976 to 1979/1980 by at least a factor of 15, the magnetic flux has changed by only a factor of 3 to 4. The question is: what kind of magnetic features other than sunspot fields contribute to F^T? Obvious candidates are the small-scale structures discussed above.

The contribution of the ephemeral active regions underlying X-ray bright points can be estimated from the data given by Golub et al. (1979). They inferred that in 1973 when the activity was at a moderate level (R $\approx$ 40), typically 250 XBP were present on the visible disk at

Fig. 1. Smoothed sunspot number for solar cycle 21 (Reinsch et al. 1986).

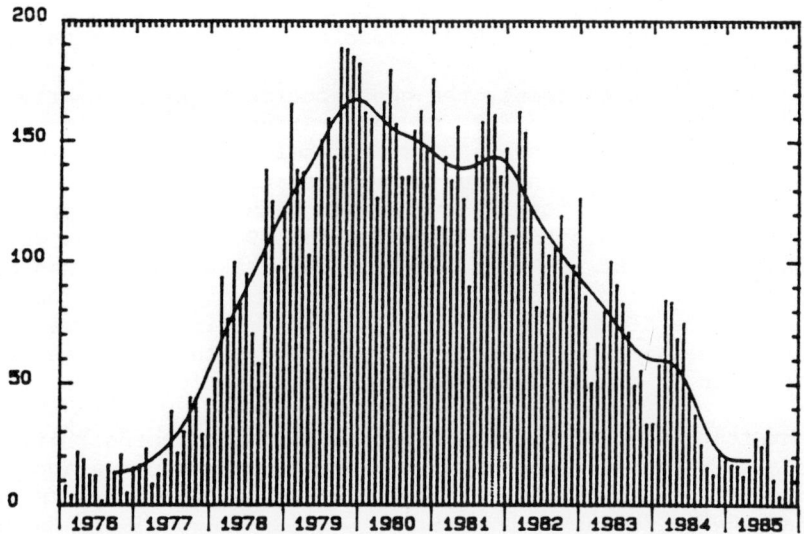

Fig. 2. Full-disk total magnetic flux (Bruning and LaBonte 1985). The position of the sunspot maximum is indicated. The three vertical markings are referred to in Section 7.

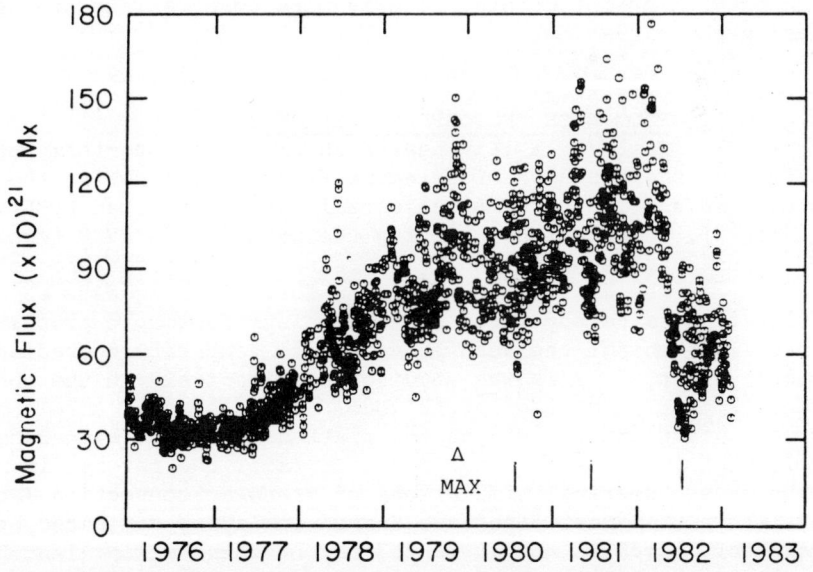

any one time, each contributing some 3×10^{19} Mx. The total contribution of XBP to the full-disk flux is thus 7.5×10^{21} Mx, which is only about 15% of the flux observed at a corresponding level of activity in cycle 21. Similarly, Howard (1974b) concluded that the influence of ephemeral regions is small.

In contrast, the contribution of network bright points to F^T appears to be substantial, at least around sunspot minimum. Using the observed distribution of NBP on the solar disk, and assuming that each NBP corresponds to one flux element of 2.5×10^{17} Mx, Muller and Roudier (1984c) derived a total contribution, due to network bright points, of 40×10^{21} Mx. This figure is valid for 1983 when R had decreased to about 70 and the number of NBP per 100×100 arc sec had correspondingly increased from 60 near sunspot maximum to about 300. Thus, almost all of the surface flux present around sunspot minimum appears to be in the form of network bright points. This may explain why the relative cycle variation of F^T is so much smaller than that of sunspot number.

Since the number of NBP varies with the cycle, their contribution to F^T is also expected to vary. In their discussion of the role of NBP, Muller and Roudier (1984c) point out an apparent conflict between this prediction and the finding of LaBonte and Howard (1982) that the quiet Sun flux near the equator shows almost no variation. However, the flux values discussed by LaBonte and Howard were defined as the lower bound of disk-center measurements covering at least 14 consecutive days. The question as to whether NBP - which are distributed non-uniformly along supergranule boundaries - are effectively omitted by this procedure deserves further examination. The same may hold for the lack of variation of K3 intensity, inferred by White and Livingston (1981) from observations of quiet regions near disk center. Interestingly, their Figure 3a seems to indicate a slight enhancement of K3 intensity around sunspot minimum, such as would be expected to arise from network bright points. Clearly more observations are needed to clarify the situation.

7 FULL-DISK PHOTOSPHERIC SPECTRUM

In 1975, W. Livingston initiated a long-term program for monitoring the strength of selected Fraunhofer lines in the spectrum of integrated sunlight. Technical details were described by Livingston et al. (1977). Observations covering the period 1976-1980 revealed that equivalent widths had steadily decreased (Livingston and Holweger 1982).

The basic theory of line formation suggests an obvious explanation as to why lines of various spectroscopic properties show a general weakening: the mean photospheric temperature gradient has become slightly flatter. This was studied quantitatively in the above-mentioned paper. Employing mixing-length model atmospheres, I have argued - as it turned out, prematurely - that a flattening of the temperature gradient implies an increase of convection. In fact, the reverse is more likely to be true. Numerical simulations of granular convection (Steffen 1986) have shown that most lines are formed in layers dominated by convective overshoot, which is not accounted for by these mixing-length models. In these - convectively stable - layers, an increase of convection generally leads to a steeper gradient, and vice versa. That is, the 1976-1980 data, if interpreted in terms of convection, would be indicative of a decrease rather than an increase.

The already mentioned variation of line strength has been tentatively associated with the marked increase of solar activity that occurred during the period 1976-1980. The link between activity and thermal structure of the photosphere has remained unidentified. Although this is still the case, the cycle-modulated features discussed in this review are candidates for further examination. An obvious question was whether the downward trend seen in line strength would reverse after sunspot maximum. An updated set of data, kindly provided by W. Livingston, is shown in Figures 3 and 4. As can be seen, after sunspot maximum, equivalent widths remained stationary for about 2 years and then indeed started to increase again. Among the lines studied, MnI 5394.67 Å shows the most conspicuous variation. This line arises from the ground state of MnI and exhibits the strongest sensitivity to changes in the thermal structure of synthetic granules studied by Steffen (1986) in his numerical simulations. Both facts taken together lend support to the view that the observed variations are due to changes in photospheric thermal structure.

The full-disk K3 intensity shows a clear anticorrelation with the strength of the manganese line. To visualize this, the K3 data have been plotted in Figures 3 and 4 with intensity increasing downward. Both time sequences in Figure 3 are seen to run closely parallel. This is true not only for the long-term trend but also for fluctuations with time scales of the order of months. Three such "events" have been marked in some of the plots to facilitate comparison. The other lines in Figure 4 also show indications of short-term variations, though less clear

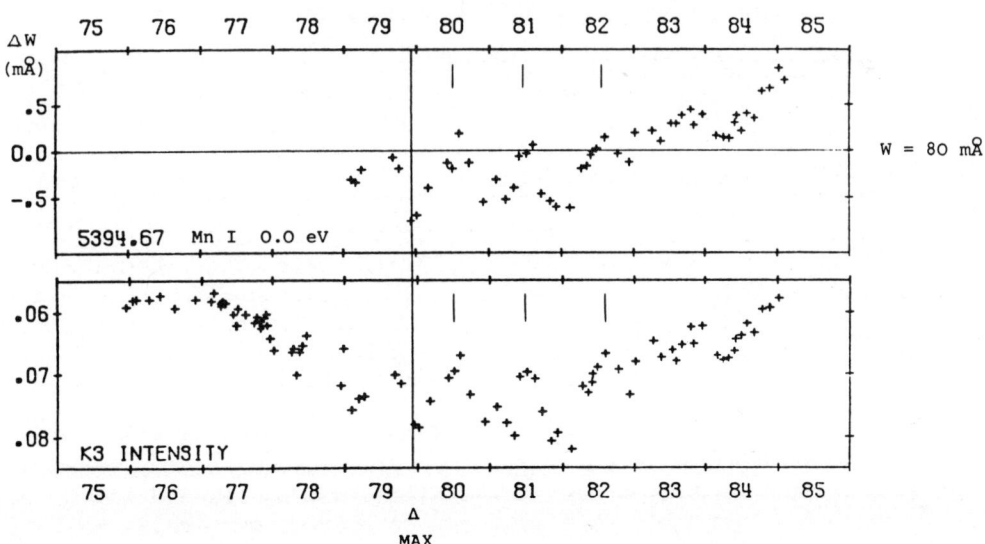

Fig. 3. Time variation of the equivalent width of MnI 5394.67 Å and of the core brightness of CaII K. Full-disk observations (Livingston 1985).

because of "noise". The variation of K3 intensity is large and cannot be due to instrumental effects (see also below). Thus, both the long-term trend and the short-term fluctuations seen in the MnI 5394.67 Å, and indicated in other photospheric lines, are most probably also real.

A close comparison of Kitt Peak full-disk K3 intensity and Mount Wilson total magnetic flux (Figure 2) reveals a striking correlation. This can be visualized by superimposing transparencies of both photos with their time scales made to register. The correlation exists on the solar-cycle time scale and also on shorter time scales. The three "events" mentioned above can also be identified in the flux data. A more detailed comparison, based on the original data is under way. This correlation also implies that the phase lag, with respect to sunspot maximum, is present in all three types of data: magnetic flux, K3 intensity and photospheric lines.

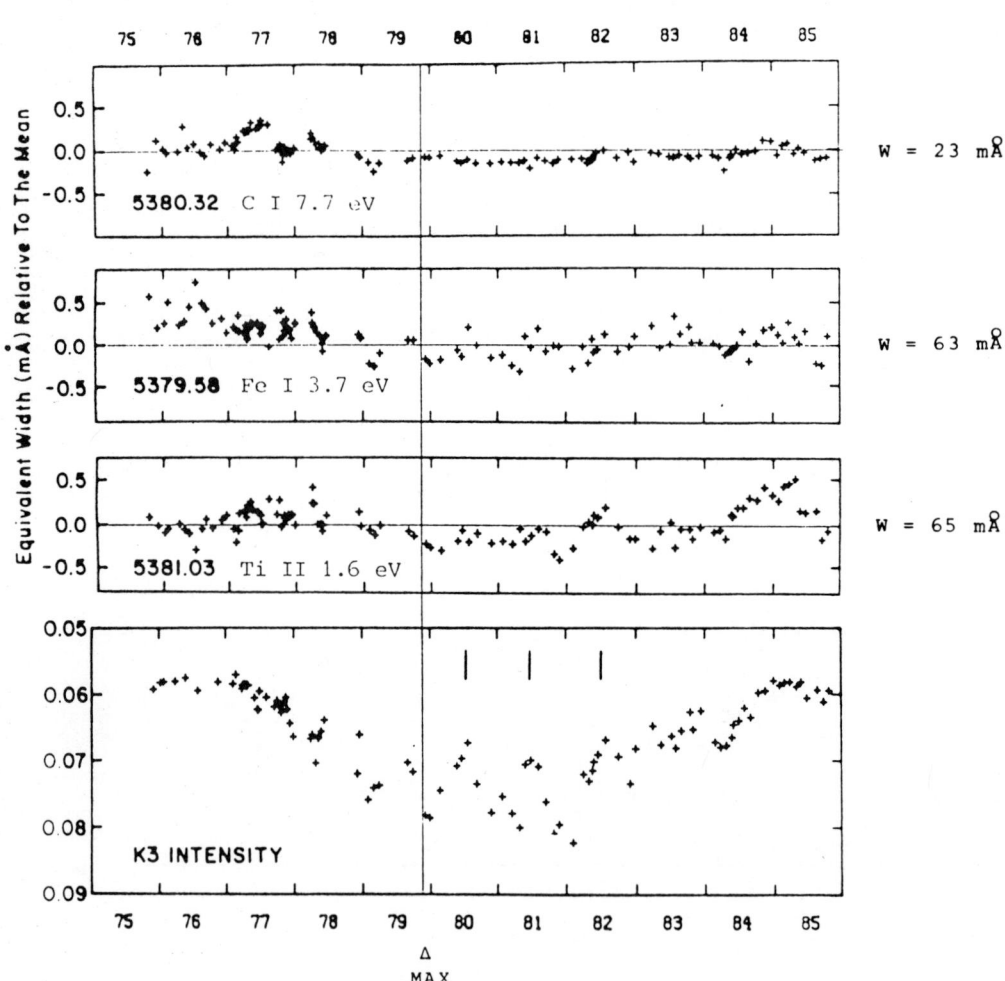

Fig. 4. Same as Figure 3, but for three of the lines studied earlier. Data from Livingston (1986).

What is the cause of the Fraunhofer line variations in the full-disk spectrum? The contribution of plage areas can be shown to be negligible because their line spectrum is barely different from that of the quiet Sun and the fractional area covered by plage is too small (Livingston and Holweger 1982; Brandt and Schröter 1984).

The K3 intensity variations have revealed a link between the strength of photospheric lines and total magnetic flux. One might speculate that the network bright points, which contribute substantially to this flux, are the cause of the equivalent width variations, particularly since these structures show a general line weakening which could be responsible for the weakening of full-disk lines with increasing activity. However, a direct contribution of network bright points to the full-disk variability by virtue of their radiation is unlikely for two reasons. First, the fractional area covered by NBP is quite small. Using the data given by Muller and Roudier (1984c), one can estimate this value to be of the order of 0.1%. Second, the number of NBP is anti-correlated with sunspot number, i.e. the line weakening which they introduce in the full-disk spectrum will be most effective at sunspot minimum, contrary to what is observed.

A more attractive hypothesis is that the Fraunhofer line variations can be attributed to the structural changes of the granulation discussed in Section 2.2. The fact that global magnetic properties are correlated with the equivalent widths would then mean that photospheric convection is to some extent controlled by those magnetic structures responsible for part of the variability in the full-disk total magnetic flux. At this point, network bright points and the associated flux tubes once again become of interest.

8 REFERENCES

Aime, C., Ricort, G. & Harvey, J. (1978). Astrophys. J., 221, 362
Alissandrakis, C.E., Macris, C.J. & Zachariadis, T.G. (1982). Solar Phys.,76, 129
Birkle, K. (1967). Z. Astrophys.,66, 252
Brandt, P.N. & Schröter, E.H. (1984). In Small-Scale Dynamical Processes in Quiet Stellar Atmospheres, ed. S.L. Keil, pp. 371. Sacramento Peak.
Bray, R.J. & Loughhead, R.E. (1964). Sunspots, pp. 65. London: Chapman and Hall.
Bruning, D.H. & LaBonte, B. (1985). Solar Phys.,97, 1.
Collados, M., Marco, E., Del Toro, J.C. & Vázquez, M. (1986). Solar Phys.,105, 17.
Danielson, R.E. (1961). Astrophys. J.,134, 287.
Davis, J.M. (1983). Solar Phys.,88, 337.
Dunn, R. & Zirker, J.B. (1973). Solar Phys. 33, 281.
Durrant, C.J., Mattig, W., Nesis, A. & Schmidt, W. (1983). Astron. Astrophys.,123, 319.
Golub, L., Davis, J.M. & Krieger, A.S. (1979, Astrophys. J. 229, L145.
Golub, L., Krieger, A.S., Silk, J.K., Timothy, A.F. & Vaiana, G.S. (1974). Astrophys. J.,189, L93.

Golub, L., Krieger, A.S., Harvey, J.W. & Vaiana, G.S. (1977) Solar Phys.,53, 111.
Harvey, K.L., Harvey, J.W. & Martin, S.F. (1975). Solar Phys.,40, 87.
Harvey, K.L. & Martin, S.F. (1973). Solar Phys.,32, 389.
Howard, R. (1974a). Solar Phys.,38, 283.
Howard, R. (1974b). Solar Phys.,38, 59.
Howard, R. (1976). Solar Phys.,47, 575.
Howard, R. & LaBonte, B.J. (1983). In Solar and Stellar Magnetic Fields: Origins and Coronal Effects, ed. J.O. Stenflo, pp. 101. Dordrecht: Reidel.
Koutchmy, S. & Stellmacher, G. (1978). Astron. Astrophys.,67, 93.
LaBonte, B.J. & Howard, R. (1982). Solar Phys.,80, 15.
Livingston, W. (1985). Private communication.
Livingston, W. (1986). Private communication.
Livingston, W. & Holweger, H. (1982). Astrophys. J.,252, 375.
Livingston, W., Holweger, H. & White, O.R. (1986). In Proc. Second Indo-US Workshop on Solar-Terrestrial Physics, eds. M.R. Kundu, B. Biswas, B.M. Reddy & S. Ramadurai, pp. 427. New Delhi: National Physical Laboratory.
Livingston, W., Milkey, R. & Slaughter, C. (1977). Astrophys. J., 211, 281.
Macris, C.J. (1949). Compt. Rend. Adac. Sci. Paris,228, 1792.
Macris, C.J. (1953). Ann. Astrophys.,16, 19.
Macris, C.J. (1955). The Observatory,75, 122.
Macris, C.J. (1979, Astron. Astrophys. 78, 186.
Macris, C.J. & Elias, D.P. (1955). Ann. Astrophys.,18, 143.
Macris, C.J., Muller, R., Rösch, J. & Roudier, T. (1984). In Small-Scale Dynamical Processes in Quiet Stellar Atmospheres, ed. S.L. Keil, pp. 265. Sacramento Peak.
Macris, C.J. & Prokakis, T.J. (1962). Compt. Rend. Acad. Sci. Paris, 255, 1862.
Macris, C.J. & Rösch, J. (1983). Compt. Rend. Acad. Sci. Paris,296, 265.
Martin, S.F. & Harvey, K.L. (1979). Solar Phys.,64, 93.
Mehltretter, J.P. (1974). Solar Phys.,38, 43.
Muller, R. (1983). Solar Phys.,85, 113.
Muller, R. (1985a). Private communication.
Muller, R. (1985b). Solar Phys.,100, 237.
Muller, R. & Keil, S.L. (1983). Solar Phys.,87, 243.
Muller, R. & Roudier, T. (1984a). 4th European Meeting on Solar Physics, pp. 239. Nordwijk.
Muller, R. & Roudier, T. (1984b). In High Resolution in Solar Physics, ed. R. Muller, pp. 242. Heidelberg: Springer Verlag.
Muller, R., & Roudier, T. (1984c). Solar Phys.,94, 33.
Musman, S. (1969). Solar Phys.,7, 178.
Namba, O. & Diemel, W.E. (1969). Solar Phys.,7, 167.
Reinsch, K., Gericke, V., Junker, E. & Schwab, M. (1986). Sterne u. Weltraum,25, 536.
Ricort, G. & Aime, C. (1979). Astron. Astrophys.,76, 324.
Schröter, E.H. (1962). Z. Astrophys.,56, 183.
Sheeley Jr., N.R. (1964). Astrophys. J.,140, 731.
Sheeley Jr., N.R. (1967). Solar Phys.,1, 171.
Steffen, M. (1986). This conference.

Stellmacher, G. & Wiehr, R. (1979). Astron. Astrophys.,$\underline{75}$, 263.
Title, A. (1986). This conference.
Waldmeier, M. (1955). Ergebnisse und Probleme der Sonnenforschung, Akademische Verlagsgesellschaft, pp. 210. Leipzig.
White, O.R. & Livingston, W. (1981). Astrophys. J.,$\underline{249}$, 798.

Fig. 3. Time variation of the equivalent width of MnI 5394.67 Å and of the core brightness of CaII K. Full-disk observations (Livingston 1985).

Line Asymmetry and the Activity Cycle.

W. Livingston
National Solar Observatory
National Optical Astronomy Observatories*
P.O. Box 26732, Tucson, AZ. 85726, U.S.A.

Abstract

Moderate strength Fraunhofer lines in spatially averaged spectra display a line asymmetry arising from the convective motions of surface granulation. Several workers have shown that line asymmetry diminishes in magnetic regions, implying an interaction between surface magnetism and granular convection. We review the evidence that the 11 year activity cycle modulates line asymmetry when the sun is viewed as a star.

Introduction

The outer layer of the sun's visible disk is in a state of vigorous convection: the granulation. Spectroscopic examination of the line-of-sight motions of granules at disk center reveals that the brightest elements are rising (blue doppler shifted) and the surrounding darker areas falling (red shifted) - see for example Kirk and Livingston, 1968. Spatially averaged, these motions impart an asymmetry to photospheric Fraunhofer lines. The line bisector, i.e. the locus of the mid points across the line profile, has a characteristic C-shape implying that convective velocities are non-uniform through the line forming region (Dravins et al., 1981).

Observed at a moderate resolution of about 3 arc-sec, line bisectors originating with a few granules display a bewildering variety of curvatures and displacements which change with time. Fig. 1 is such a temporal sequence for the bisectors of the deeply formed O 7774A line (excitation potential = 9.1v, central depth = .25). Presumably this record is a sample representing the evolution of this local group of granules.

Because Fig. 1 is hard to interpret, we find it expedient to enlarge the observed area to better determine the average convective pattern. Ultimately we have been led to observe the entire disk of the sun, in which case we are averaging ~ 10^6 granules and the C-

*Operated by the Association of Universities for Research in Astronomy, Inc., under contract with the National Science Foundation.

shape is defined to a high precision. Full disk line bisectors are reproducable from day to day and ought to be comparable between observers, discounting any differences in instrumentation.

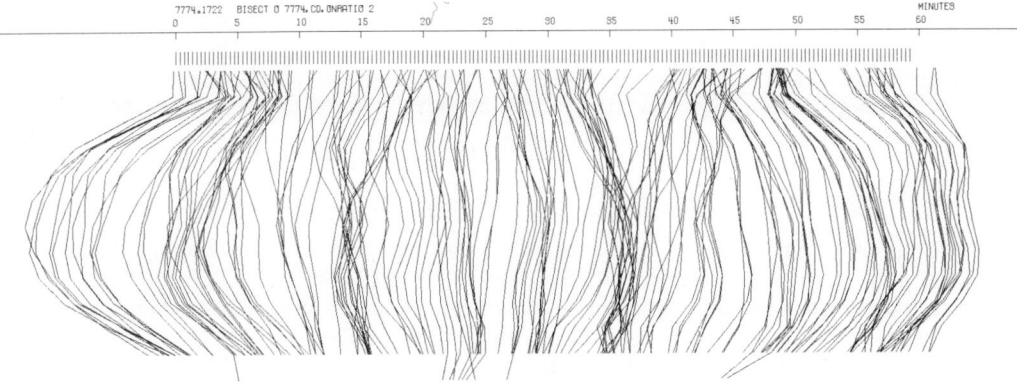

Fig 1) Temporal history of line bisector for O 7774 for a fixed position on solar disk. Time increments are 1 minute for 60 minutes.

C-shape curvature is found to be reduced in plage regions on the disk compared to the nearby quiet sun, and the inference is that magnetic fields there inhibit granular convection (Kaisig and Schröter, 1983; Brandt and Schröter, 1984; Cavallini et al 1985; Brandt and Solanki,

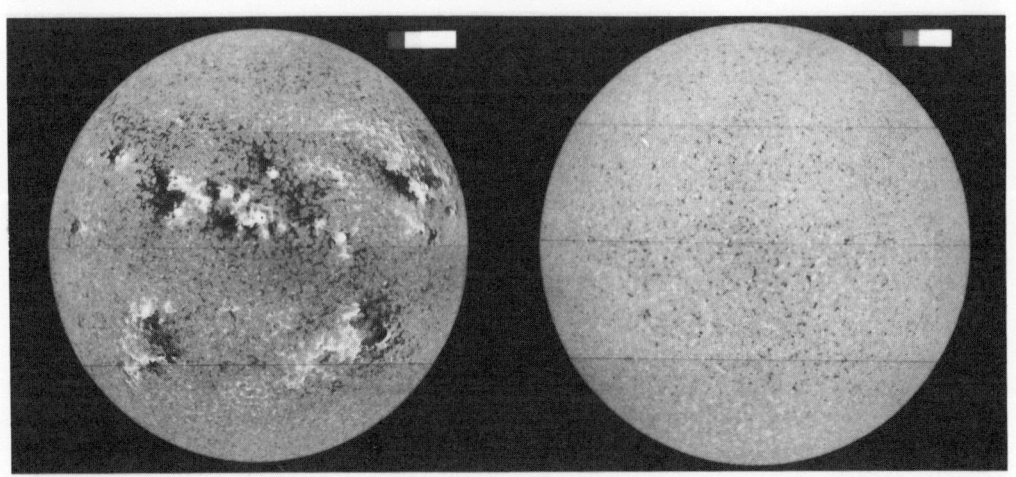

Fig 2) Full disk magnetograms representative of maximum and minimum activity.

1986; Immerschnitt and Schröter, 1986; and M. Steffan, 1986). The distribution of magnetic fields over the disk varies markedly during the 11 year activity cycle, fig. 2. Quantitatively, J. Harvey (1986) finds that the total line-of-sight magnetic flux threading the photosphere intensifies by a factor of ~8 during the period around solar maximum. <u>Hypothesis</u>: A solar cycle variation will be found in the C-shape of the line bisector of the sun-viewed-as-a-star. This variation would arise because of cyclic changes in magnetic flux over the solar disk and a presumed interaction between this flux and the granules.

Preliminary evidence for a secular change in line bisectors has been reported by Livingston (1984) and Cavallini et al (1986). In possibly related work Macris et al (1984) propose that the spacing between granules is cycle dependent. Spacelab 2 granulation movies demonstrate that granule lifetimes are longer in magnetic regions (Title, 1986) and their morphology appears different there, although this property is yet to be defined exactly. Anguera et al (1986), find that the convective blue shift for the K 7699 line is cycle dependent. On the other hand, negative results about secular changes come from Collados et al (1986) who have failed to find the expected significant difference between the diameters of granules nearby and distant from sunspots. Bruning and LaBonte (1985) found that the asymmetry of Fe 5250.2 was constant in the full disk from May 1982 to February 1983, a time during which total magnetic flux changed significantly. It may be, however, that Fe 5250.2 is too weak a line to show significant change in asymmetry; i.e. it might be subject to a wavelength shift but not much of a curvature change.

<u>Temporal Records of Full Disk Line Asymmetry.</u>

At Kitt Peak we have continued to obtain integrated light observations 2-3 days per year with the 1-m Fourier Transform Spectrometer. The FTS is employed for this work because it provides a high spectral resolution of 600,000 and the instrumental profile is symmetric. At the McMath telescope an auxilary flat mirror replaces the usual image forming concave to provide unvignetted full disk light at the FTS. Our FTS archives now span the period 1980-86, covering solar maximum to minimum, for the spectrum interval 5000-6350Å. Signal-to-noise has been enhanced by averaging the bisectors of several similar lines. After critically evaluating candidate lines for suitability of the C-shape and freedom from telluric blends, the following 13 Fe I lines were selected for study: 5137, 5198, 5217, 5225, 5250.6, 5263, 5302, 5339, 5365, 5389, 5393, 5432, and 5501Å. For each line we measured a parameter $\Delta\lambda$ = (wavelength of the bisector at a fixed intensity level near the line core) - (wavelength of maximum bisector blue shift). A mean $\Delta\lambda$ was then computed for the 13 lines in each FTS spectra and the spectra were combined to arrive at a daily mean value. These data are plotted in fig. 3. The error bars represent a formal internal variance for the day, but there may be unknown errors from day to day, or between observing periods. These errors can arise from collimation differences, irregularity in the reflectivity of FTS optical surfaces on which the sun is imaged, and differential extinction across the solar disk which is a function of sky transparency and Zenith distance.

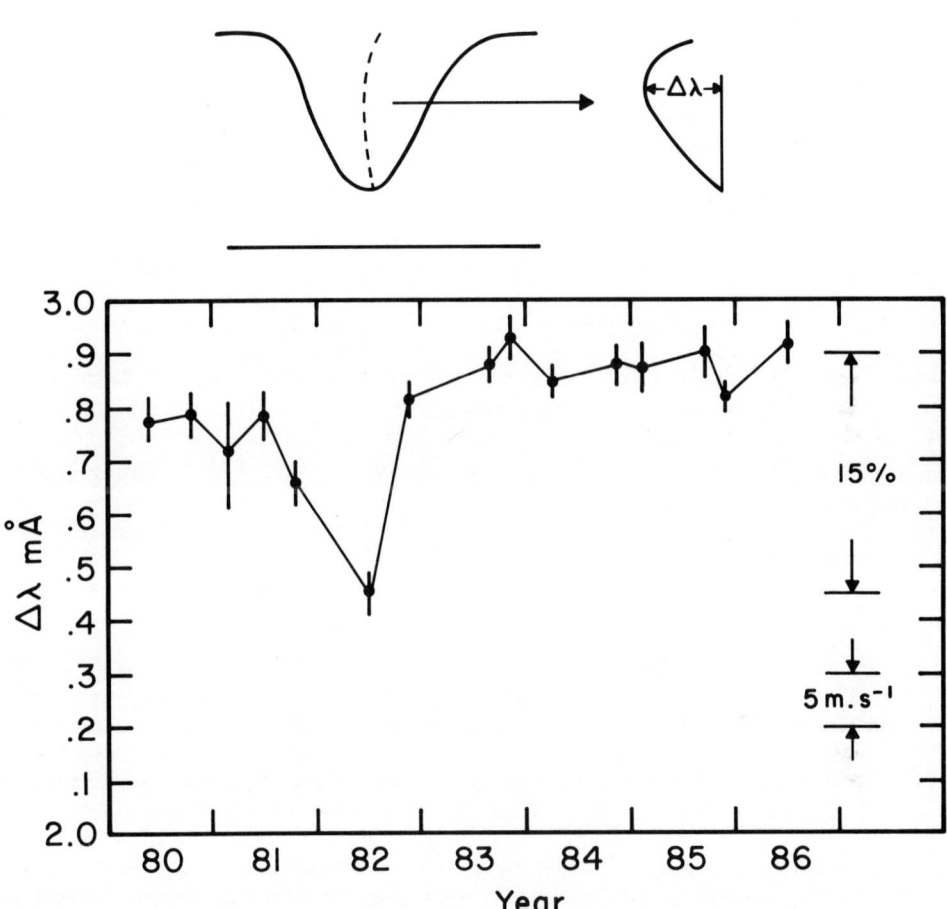

Fig 3) Mean line bisector curvature for the full solar disk.

From 1980-82, the full disk bisector curvature Δλ was observed to become less, prompting us to propose the detection of a cycle dependence (Livingston, 1982). In late 1982, abruptly, Δλ underwent a step rise and has since remained constant with a standard deviation of 2.2 m/s. This post 1982 level is distinctly higher (2.87 mÅ) than in 1980-81 (2.76 mÅ).

Reality of the 1980-82 C-shape Curvature Decline

The results displayed in fig. 3 are homogeneous as to the instrument, observing technique, and data analysis methods. For comparison, fig. 4 gives an indication of chromospheric activity over the same period. We see little or no correlation between the two figures. One is led to question, then, whether the 1980-82 trend was solar in origin or just an instrumental artifact.

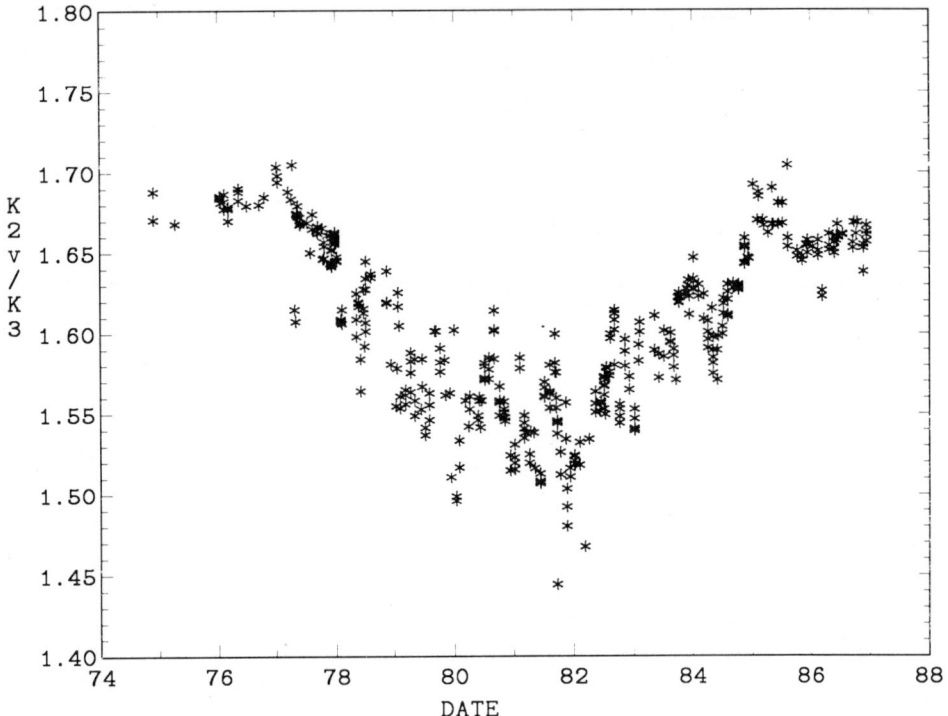

Fig 4) Chromospheric activity in Ca K 3933 for same interval as Fig. 3.

There are several clues that during 1982 the activity cycle entered a transitional state. McIntosh (1986) has noted a sudden decrease around December, 1982, of 10 cm radio flux. That decrease has not recovered. If the reader examines the Bruning and LaBonte (1985) plot of total magnetic flux during cycle 21, an earlier, sharp decline is noticed in 1982 (see also Holweger, 1986). Another piece of evidence comes from our line intensity archives (Livingston and Wallace, 1987) which indicate a distinct increase of line depth in mid-1982. One may suppose that the influence of total magnetic flux on the convective signature is non-linear and that after late in 1982 any magnetic modulation of the bisector curvature fell below our threshold of detection.

Conclusions

Previous work quoted from the literature have established that line asymmetry as described by C-shape curvature is reduced in magnetic regions. Theoretical modeling is needed to support the idea, but we may suppose this attenuation of asymmetry is a signature of some kind of interaction between magnetic fields and granular motions. Our Kitt Peak FTS observations of line asymmetry indicate a steepening decline 1980-82 followed by a rise and constant behavior since. A comparison between this full disk line asymmetry record and the total magnetic flux as measured at Mt. Wilson suggests there may be a correlation and we propose that the 1982 dip in line asymmetry is indeed real. The topic of granular convection and its dependence on surface magnetism appears ripe for exploration.

I wish to acknowledge the contribution of You-ran Huang, Lloyd Wallace, and Dick White to the preparation of this review.

References:

Anguera, M., Palle, P. L. Roca Cortes, T., Mcleod, C. P., & van der Raay, H. B., (1986), this conference.

Brandt, P. N. & Solanki, S. (1986), this conference.

Brandt, P., & Schröter, E. H. (1984), p. 371 in "Large-scale dynamical processes in quiet stellar atmospheres", Keil S., Ed., Sunspot, New Mexico.

Bruning, D. H., & LaBonte, B. (1985) Solar Phys. **97**, 1.

Cavallini, F., Cappatelli, G., & Righini, A. (1985) Astron. Astrophys. **143**, 116.

Cavallini, F., Ceppatelli, G., & Righini, A. (1986) Astron. Astrophys. **158**, 275.

Collados, M., Marco, E., DelToro, J. C., & Vazquez, M. (1986) Solar Phys. **105**, 17.

Dravins, D., Lindegren, L., & Nordlund, A (1981) Astron. Astrophys. **96**, 345..

Harvey, J. (1986) private communication.

Immerschnitt, S., & Schröter, E. H. (1986), this conference.

Kaisig, M., & Schröter, E. H. (1983) Astron. Astrophys. **117**, 305.

Kirk, J. G., & Livingston, W. C. (1968) Solar Phys. **3**, 510.

Livingston, W. (1984) p. 330 in Keil ref.

Livingston, W. & Wallace, L. (1987) Ap. J. (in press).

Livingston, W. C. (1982) Nature **297** 208-209.

Macris, C. J., Mueller, R., Rosch, J., & Roudier, T. p. 265 in Keil ref.

McIntosh, P. S. (1986) Solar Phys. (submitted).

Steffan, M. (1986), this conference.

Title, A. (1986), this conference.

MERIDIONAL FLOWS AND LATITUDINAL DEPENDENCE OF THE CONVECTION

F. Cavallini, G. Ceppatelli
Osservatorio Astrofisico di Arcetri
Largo Enrico Fermi 5 - 50125 Firenze, Italy

A. Righini
Istituto di Astronomia, Università di Firenze
Largo Enrico Fermi 5 - 50125 Firenze, Italy

1   INTRODUCTION

Theoretical studies on the differential rotation of the Sun require a surface meridional flow (e.g. Gilman, 1972, 1974; Glatzmaier and Gilman, 1982). Recent measurements of the drift of tracers suggest that this flow must be lower than 1.4 m/s (Balthasar et al., 1986), while spectroscopical measurements give contradictory results (Beckers, 1978; Howard, 1979; Duvall, 1979; LaBonte and Howard, 1982; Snodgrass, 1984). These measurements may be strongly affected by different convection between equator and poles produced by the magnetic field which locally inhibits convective effects, as shown by local changes of granulation in active regions (Schröter, 1962; Macris, 1978). In the last years several measurements of meridional motion have been performed at the Arcetri Solar Tower. In this paper we discuss from a critical point of view the obtained results.

2   MEASUREMENTS

The ensemble of the experimental data consists of spectra taken at several distances from the disk center along the polar and the equatorial diameters. The observations have been carried out on the following lines of Fe I: 5569 Å, 5576.1 Å, 6297.8 Å, 6301.5 Å, 6302.5 Å, on the Fe II 6149.2 Å and on the Ca I 6162.2 Å. All the observed positions on the solar disk were previously examined through a Zeiss H_α filter in order to ascertain the absence of active phenomena. For a further check, the quietness of the observed regions was verified by inspecting the Ca II-k spectroheliograms obtained at the Meudon Observatory. The measurements cover the period 1983-1985. Some preliminary results for 1986 are also available.

3   RESULTS

The line bisectors do not show any significant difference between equatorial and meridional diameters, as shown in Cavallini et al., 1985a, 1986a. These results suggest that the convective effects which produce the line asymmetry are not latitude dependent. However, we cannot exclude that slight variations of convective efficiency may occur. In fact, the line bisector seems to be less sensitive to convective changes than the line shifts. This is shown by active regions line asymmetry and shift measurements (Cavallini et al., 1985b, 1986b). In fact, if the line shift and asymmetry variations observed in presence of a magnetic field are attributed to a weakening of the convection, it can be seen that a small reduction of convective effects, corresponding to a weak magnetic field, produces line shifts but not line deformations.

Three of the observed lines suggest a poleward meridional flow of 50 m/s at 45° latitude (Fig. 1), while the two lines at 6297.8 Å and 6301.5 Å do not show any systematic meridional motion (Cavallini et al., 1985a). A comparison of these results with those obtained by other authors brought us to the conclusion of a possible line dependence of the observed effects and then to suppose a biasing of the results by a latitudinal dependence of the convection. This conclusion suggests that measurements performed on lines very different in the convective sensitivity should exhibit a different meridional motion. Preliminary measurements performed in the Summer 1986 on the 6149.2 Fe II and 6162.2 Ca I lines contradict this last conclusion, imposing a reexamination of the preceding observations. We have therefore performed some measurements on the three Fe I lines around 6300 Å. The paucity of the data prevents us from finding any significant result about meridional motion. However, the daily averages of the equator-meridian differential shift are very similar for all the three lines.

Figure 1. Wavelength differences between meridional and equatorial line shifts. The solid lines are the best fitting 3rd order polynomials passing through the zero point at the disk center.

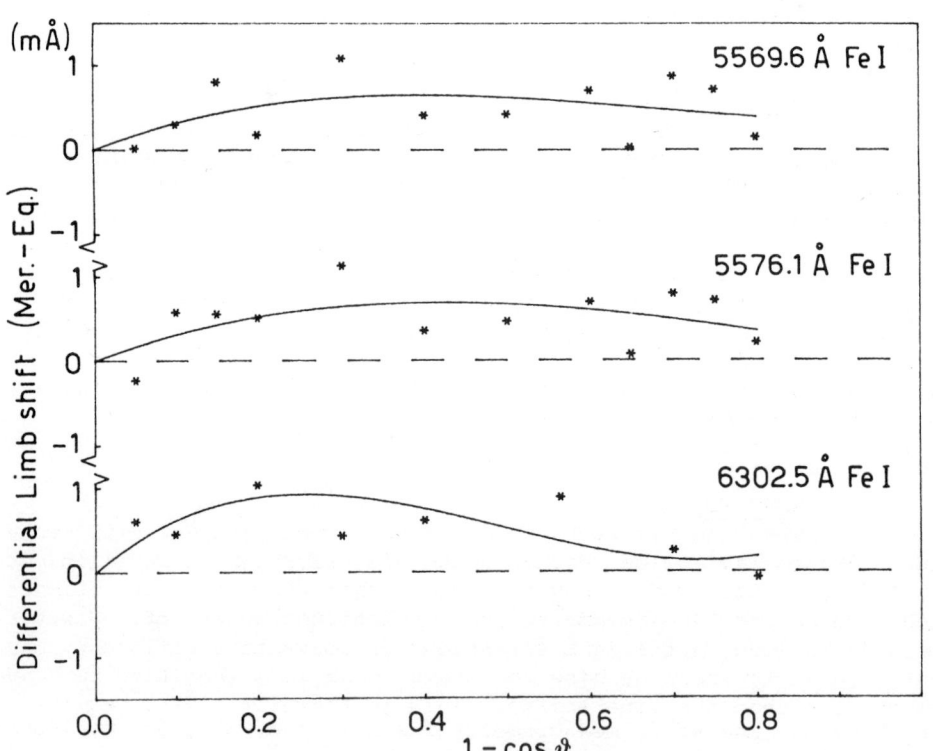

This suggests that a larger ensemble of data should show a similar meridional motion for all these lines. This fact contradicts the different meridional motions exhibited by these lines in previous observations (Cavallini et al., 1985a). The reason for this conflicting result might be found in the different measuring techniques adopted in the two cases. Former results were obtained by comparing meridional and equatorial limb shift curves obtained on different days. In this case the differential limb shift is therefore affected by the residuals arising from the solar and instrumental fluctuations of the line position observed at the disk center on different days. It may therefore occur that an additional noise, not generally equal for the three lines, is introduced in the measurement of the differential limb shift. This additional noise, not completely cancelled, might be the cause of the discordant results obtained concerning these lines.

The conclusion on meridional motions cannot yet be drawn. However, the first analysis of the 1986 data seems to show the marginal role of the convective changes on determining the observed latitudinal effects.

REFERENCES

Balthasar, H., Vazquez, M., Wöhl, H. (1986). Astron. Astrophys. 155, 87.
Beckers, J.M. (1978). In Proceedings Workshop on Solar Rotation. Univ. of Catania and Catania Astrophysical Observatory, p. 166.
Cavallini, F., Ceppatelli, G., Righini, A. (1985a). Astron. Astrophys. 150, 256.
Cavallini, F., Ceppatelli, G., Righini, A. (1985b). Astron. Astrophys. 143, 116.
Cavallini, F., Ceppatelli, G., Righini, A. (1986a). Astron. Astrophys. 163, 219.
Cavallini, F., Ceppatelli, G., Righini, A. (1986b). Astron. Astrophys. (in press).
Duvall, T.L. Jr. (1979). Solar Phys. 63, 3.
Gilman, P.A. (1972). Solar Phys. 27, 3.
Gilman, P.A. (1974). Ann. Rev. Astron. Astrophys. 12, 47.
Glatzmaier, G.A., Gilman, P.A. (1982). Astrophys. J. 256, 316.
Howard, R. (1979). Astrophys. J. Letters 228, L45.
LaBonte, B.J., Howard, R. (1982). Solar Phys. 80, 361.
Macris, C.J. (1978). Astron. Astrophys. 78, 186.
Schröter, E.H. (1962). Z. Astrophys. 56, 183.
Snodgrass, H.B. (1984). Solar Phys. 94, 13.

THE LIMB SHIFT EFFECT AND ITS VARIATION WITH THE SOLAR CYCLE

M. Anguera, P.L. Pallé, C. Régulo, T. Roca Cortés
Instituto de Astrofísica de Canarias, Universidad de La Laguna,
Tenerife, Spain

G.R. Isaak, C.P. McLeod, H.B. van der Raay
Department of Physics, University of Birmingham,
United Kingdom

Abstract. The radial velocity limb shift effect has been measured for the KI 7699A line using a resonant scattering spectrophotometer in the summer of 1982. On the other hand, using integral sunlight, the line of sight velocity has been measured during the years 1976 to 1986 and the gravitational redshift determined. This value shows a variation over those years and, when compared with the phase of the solar activity cycle, the most probable interpretation is a change of the limb shift effect with the cycle.

1 INTEGRAL SUNLIGHT MEASUREMENTS

Integral sunlight has been used to measure the line of sight velocity between the Sun and the observer, by means of a resonant scattering spectrophotometer (Brookes et al., 1978). The observations have been carried out at Izaña (Tenerife) over the years 1976 to 1986, (Jimenez et al., 1986). Briefly, solar radiation passes through an interference filter centered on the potassium resonance line (7699A), polarizer and electro-optical light modulator. Then a stable potassium vapour is used and, when placed in a longitudinal magnetic field, the two Zeeman components sample the solar line. By alternately measuring the intensities on either side of the line, the relative position of the solar and laboratory lines can be established and hence the Doppler velocity shift measured.

The relative line of sight velocity can be daily obtained as:

$$V_{obs}(t) = V_{orb} + V_{spi}(t) + V_{grs} + V_o + V_{osc}(t)$$

where V_{orb} is the line of sight orbital velocity of the Earth around the Sun; V_{spi} the spin velocity of the instrument around the Earth's axis; V_{grs} is the gravitational redshift and V_o includes other solar velocity as well as any instrumental shifts which are constant over a daily run. Finally, V_{osc} is the time dependent velocity due to solar oscillations and its value is smaller than 1 m/s (Claverie et al., 1979).

1.1 Analysis

In the present analysis we are interested in the study of the daily constant term, $(V_{grs}+V_o)$, and its variation along the years, so $V_{osc}(t)$ can be neglected. There are several ways to determine the daily value of this term; the easiest and most convenient is a null measurement. Indeed, if $V_{obs}(t=t_c)=0$ then

$$V_{grs}+ V_o = -V_{orb}(t_c) - V_{spi}(t_c)$$

So, knowing the crossing time, t_c, the value can be easily found using the Astronomical ephemeris. This is a useful technique because it is independent of line shape (if symmetrical) and sensitivity calibration. Unfortunately, because of the variation of V_{orb} during the year, such a null measurement can only be achieved at Izaña over the period July to November. With this method, the yearly mean value $(V_{grs}+V_o)$ has been calculated using always the first 15 days of August of each year, whenever available.

The results are presented in Figure 1 together with two different solar activity indices. The two main features of this figure with respect to the $(V_{grs}+V_o)$ values are: (a) The mean value over the years is some 8% smaller than the expected V_{grs} value, 636 m/s. This means that the contribution of the V_o term can not be neglected and must be explained; (b) The obtained values show a variation along the years.

Figure 1.- Yearly values of $(V_{grs}+V_o)$ with their error bars. Twice the standard error of the mean. Two solar activity indicators, the Calcium Plage Index (•) and the Zurich International Sunspot Number (☉), are also plotted for comparison.

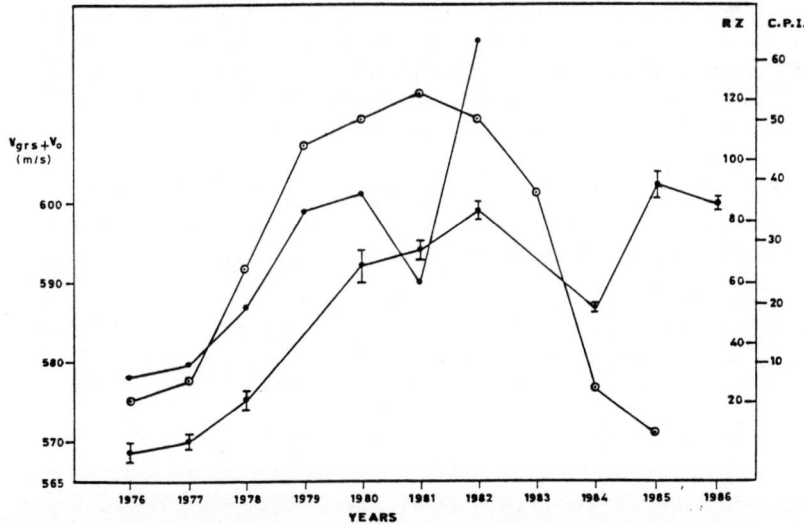

1.2 Interpretation

Although the instrument has remained basically unchanged over the last eleven years, the first aspect to be taken into consideration is whether instrumental effects can produce both the variation of $(V_{grs}+V_o)$ along the years, and the blue shift V_o. This term, V_o, accounts for solar effects, as the limb shift, and for non-solar ones as systematic instrumental errors. Let us then briefly comment on some possible non-solar effects.

a) <u>Isotopic shift</u>. Because of the existence of the isotope ^{41}K with a proportion $^{39}K/^{41}K=13.5$ a shift from 0 to -76 m/s can be produced as the optical thickness changes from zero to infinity (optically thin to thick). Of course, this can take place only if this parameter has different values in the Sun from those in the cell. The power applied to the oven to supply potassium vapour to the cell was changed by as much as a factor of two and no change in the $(V_{grs}+V_o)$ value was noticed. On the other hand, a variation with time of this parameter in the Sun is not expected to be able to explain the variation found.

b) <u>Outgassing in the cell</u>. Since the same potassium cell has been used for all years, one may believe it is becoming old and some outgassing of the cell is able to produce velocity shifts. In order to check this possibility, two new cells were used during 3 weeks in 1985, and when comparing the $(V_{grs}+V_o)$ values, gave a maximum difference of 6 m/s. Also running with a similar instrument, but with different oven and magnet design and different cell, the results found are similar and not greater than this value.

c) <u>Other instrumental effects</u>. Additional tests, changing mirror types, electro-optical light modulator, polarizers and electronics showed that none of them produces velocity shifts greater than 1 m/s. An additional, instrumental effect must be considered: if the permanent magnet used to create the longitudinal magnetic field is such that the intensity of the field is decreasing along the years, then a progressive red shift can appear as a consequence of the C shape of the line bisector of the KI 7699A line (Roca Cortés et al., 1983; Livingston, 1982 and 1984). However this would mean a change of less than 2 cm/s per gauss.

2 THE LIMB SHIFT EFFECT

If solar phenomena are considered to explain our observations, the first of them to take into account is the well known "limb shift effect". When the Sun is observed as a star, the wavelengths of the spectral lines appear to be blue shifted with respect to laboratory ones (Dravins, 1982; Schröter, 1957). This shift is different for different lines (Adam, 1948).

In order to measure the limb shift for our line, KI 7699A, in 1982 we carried out a type of observation at Izaña (Tenerife), different to the ones previously described and simultaneously with them. A similar spectrophotometer was used to measure the radial velocity in selected areas of the Sun's disk (Brookes et al., 1981). The input beam of light from the whole disk is directed to two closely spaced thin rotatable prisms; the output passes through a lens forming an image in the plane of an aperture placed in the optical axis of the undeflected beam. Any point on the image of the solar disk may be placed at the center of the aperture by rotating each prism to the desired angle. In practice, positions are pre-determined by a cycle of angular changes of each prism. The area observed of the solar disk is controlled by the aperture size (in these observations it was 2.3 arc minutes). The output from the aperture passes through a second lens to produce a parallel beam which enters a resonant scattering spectrophotometer.

2.1 Observations and analysis

The rotation axis of the sun was scanned in nine different positions. Figure 2 shows, at relative scale, the Sun's disk and the nine scanned positions along its meridian ± 1, ± 2, ± 3, ± 4 and 0 corresponding to heliocentric angles ± 16, ± 32, ± 46, ± 55 and 0 degrees respectively. These observations were extended from July to September, a total of 43 days; a daily run consists of nearly 80 points, time and velocity, for each of the nine positions, with a sampling time of 7.5 minutes.

Figure 2.- The nine selected positions along the Sun's meridian at which measurements were taken.

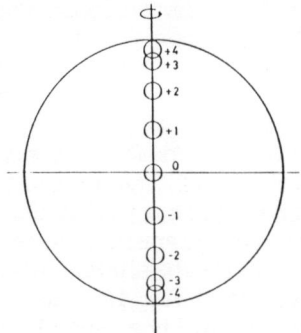

From the analysis of the radial velocity, for each of the nine positions, as in the integral light measurements the value V_o, for each position relative to the central one can be calculated every day. The mean values over the days observed are shown in Figure 3-a, where the error bars are twice the standard error in the mean values. The limb shift effect is nicely seen even in this case where measures very near the limb have not been taken.

Assuming the limb shift has a dependence $V_1 = K(1-\cos \Theta)^2$, the best fit to the observational data gives us a value $K = 100 \pm 20$ m/s. Once this effect is removed, see Figure 3-b, the data still presents some curvature which probably means that a more complex model is needed to account for the limb shift effect (Beckers and Nelson, 1978). Moreover, Figure 3-b clearly shows that if meridional currents exist, they are smaller than 10 m/s at the top photosphere level, where the potassium line is formed.

Figure 3.- (a) Mean values of V_o relative to the central position for each of the nine positions. Once the limb shift effect is removed the residual velocities are obtained (b) which still show some curvature.

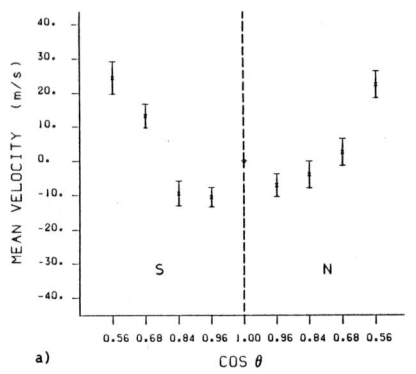
a)

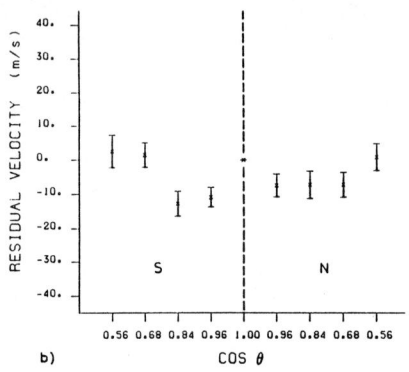
b)

2.2 Interpretation of the results

Having measured this effect, it can be introduced in a numerical model (Herrero et al., 1983) which simulates the instrument and the observations. It can be easily calculated what the contribution of the limb shift in integral light observations would be. The simulated value ($V_{grs}+V_o$) for 1982 becomes 10% lower than the observed one, showing that additional redshifts should still be present.

If one assumes that the variation of ($V_{grs}+V_o$) along the years is only due to variations on the limb shift, following the numerical models by Beckers and Nelson (1978) it implies that the granular size or its contrast, or both, is varying with the solar cycle, being smaller at maximum activity. Thus if the size of the granules is smaller then the depth of the convection zone changes and is shallower at the maximum of the cycle. It must be noticed, that if this is the only effect it implies a large variation of the physics of the sun layers along the solar cycle.

3. CONCLUSIONS

The most probable interpretation of the observations presented is based on the idea that the variation found (see Figure 1) is the combination of two effects. The first could be, either solar (related with some solar periodicity with a period larger than 12 years) or, less probable, instrumental (a progressive redshift due to a hypothetically progresive loss of magnetic field by the permanent magnet). In addition, it is believed that there exists a second effect which is a variation of the limb shift with the solar activity cycle. Under these conditions, the limb shift variation will only be some 20% of the observed effect (~ 6 m/s).

Of course, future additional tests are needed to clearly show the solar nature of this variation. In this sense it would be convenient to repeat the limb shift measurements near the phase of minimum activity in the sun (1986-87). The use of cells with isotopically pure ^{39}K will allow us to infer some properties on the optical thickness of the Sun and check our previous hypothesis. The possible change of the magnetic field of the permanent magnet has to be measured and, finally, one should wait a few more years in order to see how the parameter $(V_{grs}+V_o)$ behaves.

ACKNOWLEDGEMENTS

We would like to thank the IAC and the University of Birmingham for the finnancial and technical support provided during so many years of observation. We are also indebted to all those attached to the solar groups of these institutions. This work was partially funded by the S.E.R.C. (UK) and the C.A.I.C.Y.T. (Spain).

REFERENCES

- Adam M.G.(1948).Interferometric measurements of solar wave-lenghts and a investigation of the Einstein Gravitational Displacement. Mon. Not. Roy. Astron. Soc., **108**, 446.
- Brookes J.R.; Isaak G.R.; Van der Raay H.B.(1978). A resonant scattering spectrometer. Mon. Not. Roy. Astron. Soc., **185**, 1.
- Brookes J.R.; Isaak G.R.; Van der Raay H.B.(1981). A two-dimensional solar spectrometer. Solar Physics, **74**, 503.
- Beckers J.M.; Nelson G.D.(1978). Some comments on the limb shift of solar lines. Solar Physics, **58**, 243.
- Claverie A.; Isaak G.R.; McLeod C.P.; Van der Raay H.; Roca Cortés T.(1979). Solar structure from global studies of the 5-minute oscillation. Nature, **282**, 591.
- Dravins D.(1982). Photospheric spectrum line asymmetries and wavelenght shifts. Ann. Rev. Astron. Astrophys., **20**, 61.
- Herrero A.; Jimenez R.; Roca Cortés T.(1983). Velocity signal produced by passage of active regions compared by observations. Mem. Soc. Astro. Ital., **55**, 331.
- Jimenez A.; Pallé P.L.; Régulo C.; Roca Cortés T.; Elsworth Y.P.; Isaak G.R.; Jefferies S.M.; McLeod C.P.; New R.; Van der Raay H.B.(1986). Variations in the mean line of sight velocity of the sun: 1976-1985. IAU Symp. No. 123 "Advances in Helio and Asteroseismology". Aarhus. Denmark.
- Livingston W.(1982). Magnetic fields, convection and solar luminosity variability. Nature, **297**, 208.
- Livingston W.(1984). Secular change of full disk line asymetries. **In** Small scale dynamical processes in quiet stellar atmospheres, ed. S. L. Keil. Pub. of. Sac. Peak. Obs. U.S.A.
- Roca Cortés T.; Vazquez M.; Wöhl H.(1983). Space and time variations of the KI 7699 solar line profile. Solar Physics, **88**, 1.
- Schröter E.H.(1957). Zur deutung der rotverschiebung und der mitte-rand variation der Fraunhoferlinien. Z. Astrophys., **41**, 141.

ON THE DIFFERENCES BETWEEN LINE BISECTORS IN QUIET AND ACTIVE SUN

C. Marmolino, AFGL-NSO/SP, on leave from Napoli University

G. Roberti, Dipartimento di Fisica N.S.M.F.A., Napoli, Italia

G. Severino, Osservatorio Astron. Capodimonte, Napoli, Italia

The asymmetry and shift of solar lines show systematic variations between quiet (QR) and active (AR) regions, which have been documented by several authors (Bonet et al. 1984; Cavallini et al. 1985 and references therein). Marmolino et al. (1984, 1986) studied the effects produced by photospheric motions (waves and granulation) on the KI 7699 line in the quiet Sun. In the present paper we extend this study to the synthesis of line bisectors in plages.

In our calculations a plage is composed by small scale hot magnetic flux tubes embedded in a cool surrounding atmosphere; also the quiet Sun is a two component mixture but having a smaller density of flux tubes than a plage has. The VALC is the cool model (Fig. 1a).

The hot element is brighter than the cold one by 80% in the line core and by only 3% in the continuum. Adding the same velocity field to both the cold and the hot model, we find that the only model difference does not affect significantly the line bisectors (Fig.1b). Then their observed variations between QR and AR have to be attributed mainly to a different dynamical structure in the two atmospheric components. In the cool atmosphere we assume that velocity, temperature and pressure fluctuations are due to the combined effects of granulation and short period acoustic waves (Marmolino et al., 1986). In the hot element we assume no granulation but the presence of short period acoustic waves, since between the types of motion in a flux tube there is a longitudinal mode which can propagate likewise an acoustic pulse. In Fig.2 we compare our synthetic bisectors with those observed by Bonet at al. Each synthetic bisector is obtained by taking the liner combination of the cool and hot profiles weighted according to different values of the filling factor of flux tubes. The synthetic bisectors are less and less blueshifted when the filling factor increases, which agrees with the behaviour observed by Bonet et al. However the values of the flux tube convering fraction required to fit the bisector of the KI 7699 line in AR appear to be greater than those used by Ayres et al.(1986) in the case of the CaII K line and the CO bands. This could be an indication that the simple dynamical model we assumed in the flux tube is incomplete. Finally the "pure" hot model does not reproduce the upper redshifted tail of the bisector in AR. This tail could imply an influx into the flux tube. Some evidence of such a downdraft is reported by Labonte & Howard (1982).

Fig. 1   a) Temperature vs. mass column density in the cool and hot components; b) synthetic bisectors of the KI 7699 line in the presence of granulation (Nelson 1978): solid line refers to the cool component and dots to the hot component.

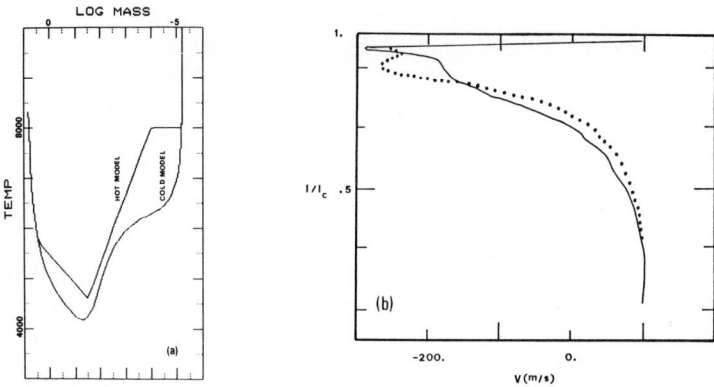

Fig. 2   Bisectors of the KI 7699 line at disk center. The symbols "x" and "+" refer to the observations in QR and AR respectively. Solid line represents the synthetic line bisectors each labelled by the values of the flux tube fraction.

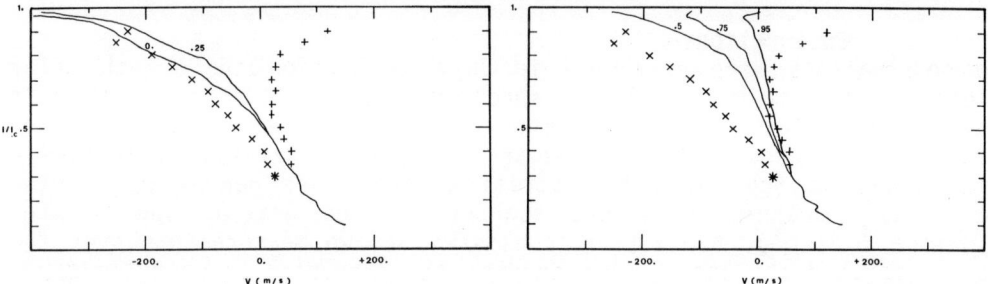

REFERENCES

Ayers T.R., Testermann L., Brault J.W.: 1986, Astrophys. J. 304, 542
Bonet J.A., Marquez I., Roca Cortes T., Vasquez M., Woehl H., Wittmann A. : 1984, in S.L. Keil, ed: "Small Scale Dynamical Processes in Quiet Stellar Atmospheres", NSO/SP, p. 323
Cavallini F.,Ceppatelli G.,Righini A.:1985,Astron.Astrophys.143, 116
Kelch W.L., Linsky J.L.: 1978, Solar Phys., 58, 37
Labonte B.L., Howard R.: 1982, Solar Phys., 80, 361
Marmolino C., Roberti G., Severino G., Vasquez M., Woehl H.: 1984, Proceed. 4th Europ, Meet. on Solar Phys., ESA SP - 220, p. 191
Marmolino C., Roberti G., Severino G.: 1986, Solar Phys., in press
Nelson G.D.: 1978, Solar Phys. 60, 5
Severino G., Roberti G.,Marmolino C.,Gomez M.T.:1986, Solar Phys,104,259

CENTER-TO-LIMB VARIATION OF THE ASYMMETRIES OF
THE K 7699 Å LINE IN SOLAR QUIET AND ACTIVE REGIONS

J.A. Bonet, I. Marquez, and M. Vázquez
Instituto de Astrofisica de Canarias
E-38071 La Laguna, Tenerife, Spain

H. Wöhl
Kiepenheuer-Institut für Sonnenphysik
Schöneckstr. 6, D-7800 Freiburg, F.R.G.

Introduction

Within the last years the profile shape variation of the K 7699 Å line has gained some interest. The main aim of this investigation is to verify and complete the results found by Roca-Cortes et al. (1983) and Bonet et al.(1984) using better resolved spectra with high temporal resolution in quiet regions and plages. In addition an improved data reduction procedure is applied. The center-to-limb variation of the main parameters defining the line and the oscillations were presented in the poster, but only the center-to-limb variations of averaged bisectors are given in this paper. The complete results, especially of the temporal behaviour, will be published elsewhere.

Observations

The observations were performed in July 1983 using the McMath Solar Telescope and its main grating spectrometer operated in double pass mode at the Kitt Peak Observatory, USA. The telescope has been described by Pierce (1964). A spectral region containing the K 7699 Å solar line and the three terrestrial oxygen lines nearby was photo-electrically scanned using the scanner of the grating spectrometer (1 step = 5.37 mÅ). Every scan consisted of two measurements over the region (forward and backward) recording their average on magnetic tape. The duration of this procedure was 14.5 s per double-scan. It was aimed to have a minimum of 250 cycles scanned, but this was - mainly due to clouds - not always possible. The apertures were 5 arcseconds squared for the quiet regions and for the plages. The plages measured were in general from different active regions. An image slicer was used to put the light in the slit of 220 microns. The scanning exit slit had a width of 200 microns.

Reduction procedures
Instrumental profile and noise effects

The instrumental profile of the spectrograph was determined using a multi-mode He-Ne-LASER for which we expect a FWHM of 10 mÅ. The measured profile of the LASER line was 21 mÅ. The transformation of the instrumental profile from the measured wavelength of the 6328 Å LASER line to the K 7699 Å line was performed by a convolution process

yielding a conversion factor of 1.076.
The intensity rms - noise of the data is of the order of about 0.1 %, nevertheless a smoothing with an optimum filter was applied. The instrumental profile corrections were only applied to the K line and not to the oxygen reference lines.

Determination of line positions and bisectors

The line positions have been determined by fitting with polynomials of the order 4 for the K line and of the order 2 for the oxygen lines. The bisector line was determined in steps of 5 % of the continuum. In order to increase the number of sampling points by a factor of 4 the Fourier interpolation technique was used.

Results

The results for the center-to-limb variations of averaged bisectors are given for the quiet regions in Fig. 1 and for the active regions in Fig. 2, respectively. The investigation performed yields mainly the same results already shown by Roca-Cortes et al. (1983) and Bonet et al. (1984). Less noise in the data makes more results significant. The main difference between quiet and active regions is, that the curvature of the bisectors is more pronounced in the latter case. The bisectors from quiet regions show in general a linear tilt from the line bottom towards the blue wing and only from regions near the solar limb a curvature of the bisector line is found.

Acknowledgements

We thank the staff of the National Solar Observatories, especially Dr. W.C. Livingston and Mr. B.Graves, for many discussions and the support obtained. H.W. thanks the German Research Foundation (DFG) for a travel grant (Wo 191/10-1). We thank the referee, Dr. P.N.Brandt, for several suggestions, which improved the paper.

References

Bonet,J.A., Marquez,I., Roca-Cortes,T., Vázquez,M., Wöhl, H., Wittmann, A. : 1984, in 'Small - scale dynamical processes in quiet stellar atmospheres', S. Keil (ed.), p.323
Pierce,A.K. : 1964, Appl.Optics **3**,1337
Roca-Cortes,T., Vázquez,M., Wöhl,H. : 1983, Solar Phys. **88**,1

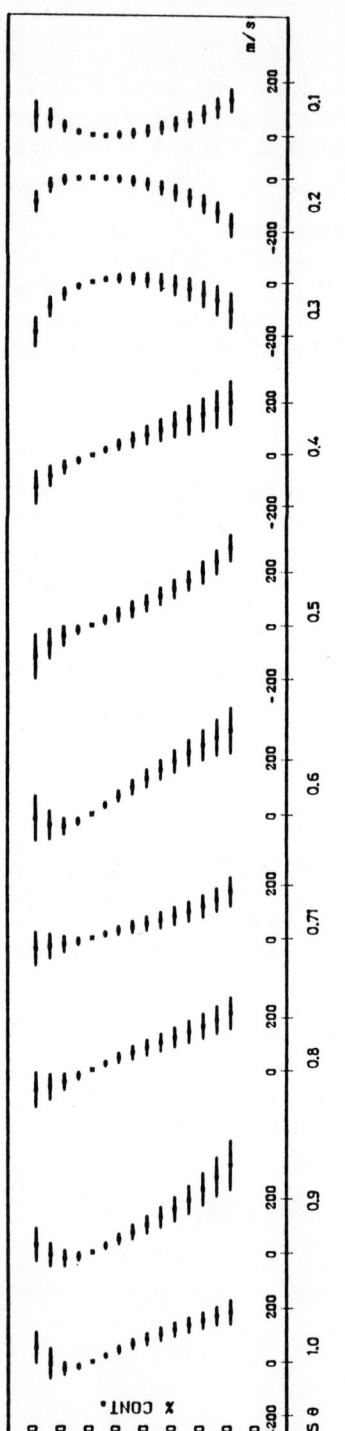

Fig. 1: Quiet regions: Center-to-limb variation of the average bisector points of the K 7699 Å line determined from 250 scans superposed at the 70% level. The bars give the r.m.s. - deviation. In the case of the C-shape at cos Θ = 0.1 another measurement resulted in a similar 'inverse' C-shape like at cos Θ = 0.3 and 0.2, respectively.

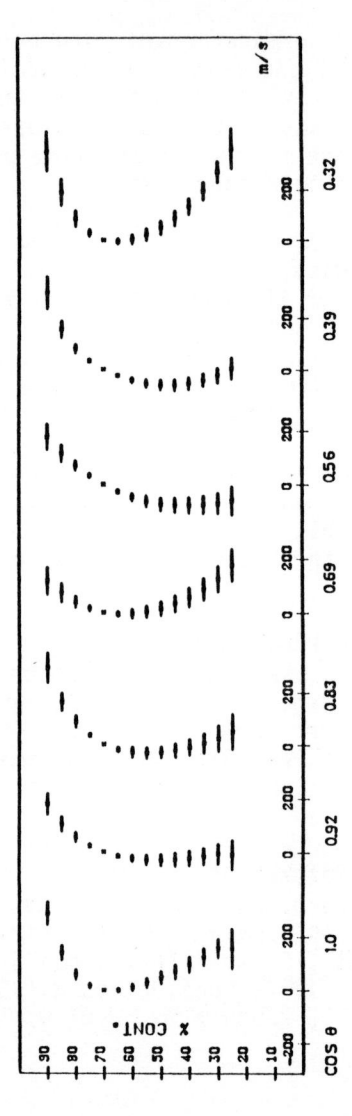

Fig. 2: Active regions: Center-to-limb variation of the average bisector points of the K 7699 Å line determined from 250 scans superposed at the 70% level. The bars give the r.m.s. - deviation. In the case of the C-shape at cos Θ = 1.0 there were only 50 scans averaged and in the case cos Θ = 0.39 only 114.

Interaction between Magnetic Fields and Convection

N. E. Hurlburt and N. O. Weiss

Department of Applied Mathematics and Theoretical Physics,
University of Cambridge, Cambridge, CB3 9EW, England

Abstract. We discuss nonlinear convection in the presence of an imposed vertical magnetic field and its influence on the fine structure of the resulting field. The emphasis in recent work has been on compressible behaviour; we contrast recent results of numerical experiments on steady and oscillatory magnetoconvection with those obtained in the Boussinesq approximation. An attempt is also made to relate idealized model calculations to the structure of observed magnetic fields in the solar photosphere.

1. Introduction

The fine structure of magnetic fields in the solar photosphere can only be understood by studying the nonlinear interaction between magnetic fields and convection. To solar physicists that may be a sufficient reason for working on the problem but it is worth emphasizing that this interaction is important in a wider context. As basic physics, it provides an example of a nonlinear dynamical process that cannot be explored in the laboratory. In astrophysics, we need to understand the structure of solar magnetic fields in order to discuss activity in other stars where detailed structures cannot be resolved. Any investigation of such a complicated nonlinear problem inevitably leads to large-scale numerical computations for which there are two different approaches. Engineers attempt to simulate turbulent processes, including as many realistic details as they can, while physicists prefer to model complicated systems by isolating individual effects. In what follows we adopt the later approach. In particular, we stress the importance of systematically varying parameters in idealized numerical experiments.

We shall not embark on another survey of magnetoconvection, which has already been dealt with in a number of reviews (Priest, 1982; Proctor & Weiss, 1982; Nordlund, 1984, 1985a,b; Weiss, 1985). Instead we shall focus on recent developments in compressible magnetoconvection. In the next section we discuss observational and theoretical constraints on modelling photospheric convection and in §3 we summarize what has been learnt from calculations using the Boussinesq approximation. Then we consider steady, compressible convection, with and without a magnetic field, in §4, followed by more complicated oscillatory behaviour in §5. This treatment will emphasize results obtained in Boulder and Cambridge by a group including D. P. Brownjohn, F. Cattaneo, M. R. E. Proctor, J. Toomre and ourselves. Finally, we attempt to relate these idealized calculations to small-scale magnetic features in the solar photosphere.

2. Modelling convection in the solar photosphere

High resolution magnetograms obtained by the Lockheed group show that small flux tubes with intense magnetic fields form between the convection cells that give rise to solar granulation (Title & Tarbell 1986). The tubes appear in regions with downward velocity but the downflow within a tube is drastically reduced if the flux is sufficiently large. The Lockheed observations also indicate the presence of significant quantities of magnetic flux at the centres of granules, though the corresponding fields are typically less than 600 G. Future observations will show how the distribution of magnetic flux responds to the dynamical evolution of granules seen in the results obtained from Spacelab 2 (Title et al. 1986). The behaviour of facular points suggests that flux tubes appear at junctions in the intergranular network and remain within the dark intergranular lanes (Muller 1983).

These observations raise an immediate question for theoreticians: how are the location and strength of these intense fields related to the pattern of ambient convection? The formation of isolated sheets or tubes of flux from which the motion was excluded had been predicted well before they were observed (Parker 1963; Weiss 1964), though the field strength was underestimated and convective collapse of isolated flux tubes was not investigated until kilogauss fields had been detected. A complete theory should describe both the fine structure of the magnetic field and its relationship to granular convection, taking account of the special properties of the solar photosphere. Throughout most of a star the energy density of any plausible magnetic field is much less than the thermal energy of the gas, so that the ratio

$$\beta = P/(B^2/2\mu_o) = \gamma V_S^2/2V_A^2,$$

(where P is the gas pressure, B the field strength, γ the ratio of specific heats, V_S the sound speed and V_A the Alfvén speed) is large; in addition the radiative diffusivity κ is much greater than the magnetic diffusivity η. In the photosphere, however, β may be of order unity or less, while ionization increases the opacity and reduces κ so that the ratio $\varsigma = \eta/\kappa$ is greater than unity for depths in the range 2,000 - 20,000 km. To model photospheric convection a fully compressible treatment is required, allowing both for local density fluctuations and for mean stratification. Processes in the deep convection zone may be adequately represented in the anelastic approximation (which suppresses sound waves and is only valid for $\beta \gg 1$) or the more restrictive Boussinesq approximation (which eliminates stratification and assumes an incompressible flow).

Nordlund (1982, 1985a) has attempted an ambitious simulation of convection in the solar photosphere, using the anelastic approximation but including the detailed chemistry of stellar atmospheres. His results show broad upwellings of hot gas, corresponding to bright granules, with magnetic flux confined to intergranular lanes, in qualitative agreement with the observations. The fine structure of the fields is, however, limited by numerical resolution. These models represent convection with adequate precision for the purpose of studying line formation but are not able to describe the magnetohydrodynamic processes responsible for the detailed structure of intergranular magnetic fields. Steffen (in these Proceedings) discusses a similar model assuming axial symmetry, which allows much greater accuracy in the computation.

Even with supercomputers, numerical experiments on convection can only be carried out for idealized configurations and it is important to contrast the effects of different assumptions. Segregation of magnetic fields from the flow can be described within the Boussinesq approximation but substantial variations in density require a treatment that is anelastic or fully compressible. Radiative transport may be represented in the diffusive approximation (valid only for regions that are optically thick), in the Eddington approximation (currently being studied by J. M. Edwards) or by solving the equations for radiative transfer (as in the work of Nordlund and of Steffen). There is a choice of possible boundary conditions: in the diffusive approximation either the temperature or the conductive flux may be fixed; the horizontal component of the field may be set to zero or the field may be matched to a potential field above the convecting layer; the lower boundary may be a fixed plane on which the vertical velocity and tangential stress are set to zero, or symmetry conditions may be imposed such that the calculation is effectively confined to the upper portion of a convection cell (as in the models of Nordlund and Steffen). Finally, the effects of ionization on opacity can be modelled by making κ a function of position, so that ς increases with increasing depth, as described by Hurlburt in these Proceedings.

3. Boussinesq magnetoconvection

Boussinesq convection in an imposed magnetic field has been reviewed at length by Proctor & Weiss (1982). Most detailed studies have been concerned with two-dimensional or axisymmetric configurations which reveal some fundamental properties of magnetoconvection. Flux segregation occurs for $\varsigma < 1$, with stagnant flux sheets in which the field is limited by dynamical processes. Astrophysical implications of this effect are discussed by Galloway & Weiss (1981).

The bifurcation problem for Boussinesq magnetoconvection provides the essential skeleton without which nonlinear behaviour cannot be understood. Convection may occur as steady overturning motion or as finite-amplitude oscillations, corresponding to trapped hydromagnetic waves. In Figure 1 the rms velocity U is plotted against the Rayleigh number R, which is a dimensionless measure of the superadiabatic temperature gradient. There are two solution branches. A branch of steady solutions emerges subcritically from the bifurcation point at $R = R^{(e)}$. This branch is initially unstable but gains stability at the turning point (through a saddle-node bifurcation); the upper portion corresponds to stable steady overturning convection. However, convection first set in as oscillations (represented by the mean value of U): the branch of oscillatory solutions emerges from an oscillatory (or Hopf) bifurcation at $R = R^{(o)}$ and terminates on the unstable portion of the steady branch (in a heteroclinic bifurcation). Compared with steady overturning motion, oscillatory convection is generally rather ineffectual at transporting heat.

Most studies have assumed fixed lateral boundaries on which the horizontal velocity is set to zero. If this constraint is relaxed by imposing periodic boundary conditions then two solution branches, corresponding to travelling waves and standing waves, appear at the oscillatory bifurcation. Dangelmayr & Knobloch (1986) have shown that in certain circumstances standing waves (corresponding to the oscillations described above) are preferred in strong magnetic fields but the competition with travelling waves has not yet been fully investigated.

Kinematic effects have been explored in three dimensions (Galloway & Proctor 1983). In a hexagonal cell with fluid rising at the centre, flux is swept aside at the top to produce strong fields at the corners. Nevertheless, inflow at the base of the cell maintains a central concentration which extends to the upper boundary, where it may contain more flux than that in the corners though the field is relatively weak. It is not known whether axial flux tubes persist in the dynamical regime, though magnetic buoyancy resists the outflow at the upper surface (Schmidt et al. 1985).

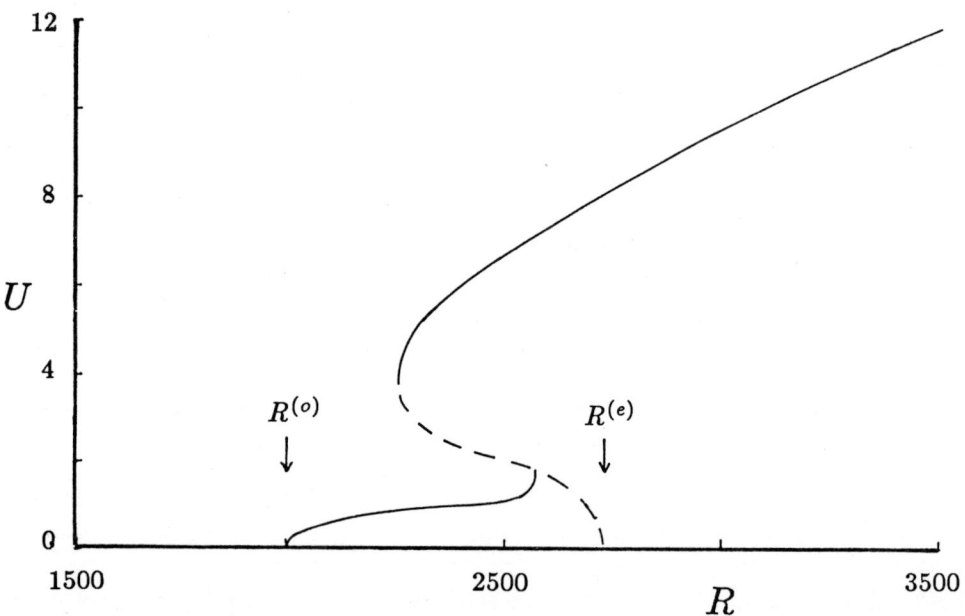

Figure 1. Bifurcation diagram for Boussinesq magnetoconvection. Root mean square velocity U as a function of Rayleigh number R for two-dimensional numerical solutions with $Q = 10\pi^2$, $\varsigma = \sigma = 0.2$ and $\lambda = 1.0$. Broken line indicates the unstable portion of the steady branch which is conjectured to connect it to the linear bifurcation point.

4. Steady Compressible Convection

In a Boussinesq fluid upward and downward motion are equivalent. Compressibility breaks this symmetry. Density stratification affects the velocity through the continuity equation, distinguishing diverging updrafts from converging downdrafts. This leads to a preference for cells with hot fluid rising near their axes at the onset of convection (Massaguer & Zahn 1980). Pressure fluctuations also change the density in compressible fluids. In nonlinear convection pressure maxima occur at stagnation points, which are on the axis and at the corners on the top and bottom boundaries of the cell. Hot fluid rising at the axis experiences a sharp rise in the pressure fluctuation; if this relative increase in pressure exceeds that in temperature, the density will attain a relative maximum, resulting in buoyancy braking which impedes the motion (Massaguer & Zahn 1980; Hurlburt et al. 1984). Conversely, increased pressure cooperates with reduced temperature to enhance the motion in a sinking plume. The downward velocity at the corners of a cell is therefore greater than the upward velocity at the centre. Numerical simulations of three-dimensional compressible convection show narrow, rapidly descending plumes which develop into miniature tornadoes (Graham 1977; Nordlund 1984).

It is not clear what determines the scale of granular convection. Linear theory predicts that convection cells should span the entire depth of the layer and most nonlinear results show a single layer of cells extending over many scale heights. The only numerical experiments that show multiple vertical scales are those of Chan et al. (1982), Sofia & Chan (1984) and Chan & Sofia (1986), which differ from other treatments in using a nonlinear sub-grid scale viscosity to represent the Reynolds stresses. At present we cannot say whether the disparity between granules and supergranules results from shear instabilities, from an unstable thermal boundary layer, or from some other property of a region where all scale heights are small.

The asymmetry between upward and downward motion is also affected by magnetic fields. Where flux is swept together the magnetic pressure rises and continuity of the total pressure leads to a reduction in the thermal pressure of the gas. As a result, regions of intense magnetic field are partially evacuated. In a stratified layer this effect is most pronounced at the top, where the flow converges on a downdraft. Figure 2 shows the velocity **u** and the magnetic field **B** for steady, two-dimensional convection in a layer of depth d across which the density ρ increases by an order of magnitude (Hurlburt & Toomre 1987). Both **u** and **B** are confined to the xz-plane and solutions are periodic in the horizontal x-direction with a wavelength of $2\lambda d$. The temperature T is fixed at the upper boundary and the heat flux is fixed at the bottom, while the vertical velocity and tangential stress vanishes at $z = 0, d$. The configuration is defined by four dimensionless parameters, the Rayleigh number $R = g\Delta\Theta d^3/\kappa\nu T$, the Chandrasekhar number $Q = B_o^2 d^2/\mu_o\rho\eta\nu$ and the Prandtl numbers $\varsigma = \eta/\kappa$ and $\sigma = \nu/\kappa$, all evaluated at the middle of the layer. Here $\Delta\Theta$ is the potential temperature difference across the layer, B_o is the mean vertical field, and ν is the viscous diffusivity. For the results illustrated in Figure 2 $R = 10^5$, $Q = 66$, $\sigma = 1$, $\varsigma = 1.5$ and $\lambda = 1.5$ so that $\beta = 1818$. Figure 2(c) shows the variation of the gas pressure P, the magnetic pressure $B^2/2\mu_o$ and the dynamic

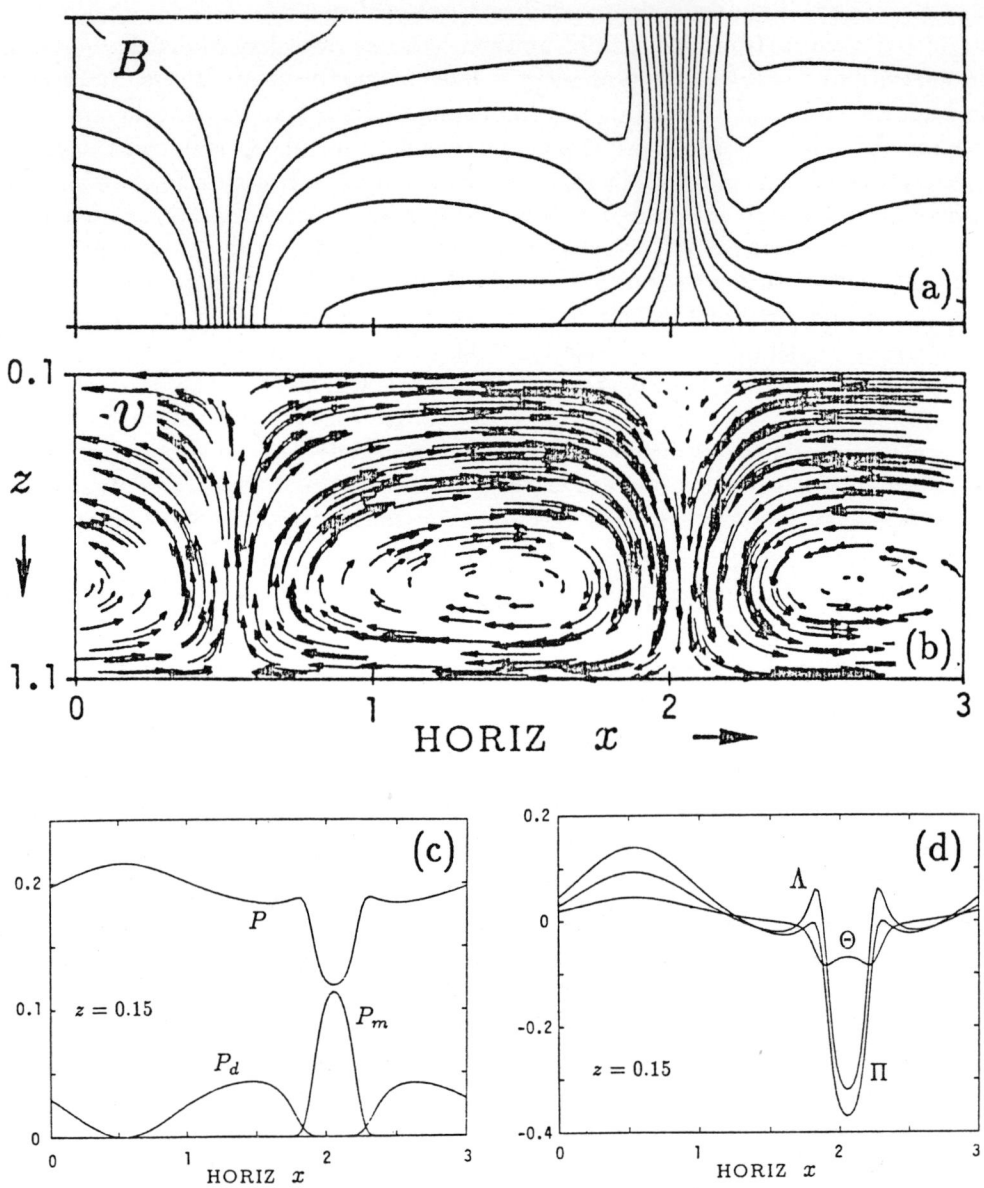

Figure 2. Steady compressible two-dimensional magnetoconvection with $R = 10^5$, $Q = 66$, $\varsigma = 1.5$, $\sigma = 1.0$ and $\lambda = 1.5$. (a) Magnetic field lines, (b) velocity streaklines, (c) gas pressure, magnetic pressure and dynamic pressure profiles, as functions of x for $z/d = 0.95$ and (d) relative fluctuations in P, ρ, and T at the same level.

pressure ρu^2 with the normalized horizontal coordinate x/d at a level 5% below the top of the layer. The most striking effect is the rise in magnetic pressure and the associated fall in thermal pressure. The relative changes in pressure, density and temperature are shown in Figure 2(d): here ρ follows P, while T shows a much smaller variation. The reduced density retards motion in the downward plume so that the upward and downward velocities are once more comparable in magnitude, as shown in Figure 2(b). More flux is concentrated in the sinking plume, where the strongest magnetic fields are found, as shown in Figure 2(a). Compressibility apparently favours flux concentration at the corners of a cell, where the velocity is directed downward, when the field is dynamically important.

5. Oscillatory convection in a strong magnetic field

At high β, fast and slow magnetoacoustic waves travel with speeds close to V_S and V_A respectively; since $V_S \gg V_A$ in the anelastic or Boussinesq approximations the fast waves can be filtered out and oscillatory convection corresponds to slow magnetoacoustic modes that are destabilised by the superadiabatic temperature gradient. When β is of order unity, V_S and V_A are comparable and sound waves cannot be ignored. The stability problem becomes more complicated and more difficult to explain. Cattaneo (1984) found that there is a regime where a polytropic atmosphere is unstable to oscillatory modes even when it is subadiabatically stratified. Moreover, the bifurcation may be subcritical, though preliminary computations suggest that oscillatory convection does not attain significant amplitudes in the nonlinear regime.

The transition from Boussinesq to compressible behaviour can be investigated by increasing B_o (i.e. decreasing β) in a layer with a modest variation of density. (Studies of convection in the absence of a magnetic field indicate that the Boussinesq approximation remains adequate provided that the density contrast $\bar{\rho}(d)/\bar{\rho}(0) < 5$, where $\bar{\rho}(z)$ is the horizontally averaged density.) We have studied two-dimensional oscillatory convection with fixed temperatures at the top and bottom boundaries for $128 > \beta > 6$ and find that results remain qualitatively unchanged for $\beta > 32$. As expected, the Boussinesq approximation is robust.

In Figures 3 and 4 we show nonlinear oscillatory solutions obtained for $\beta = 8$ and $\lambda = 1$. (In photospheric conditions this would correspond to an average field $B_o \approx 700$G which could only be found in an active region.) Figure 3 shows a periodic oscillation (corresponding to a standing wave) for a Rayleigh number $R = 2R^{(o)}$, where $R^{(o)}$ is the Rayleigh number at the oscillatory bifurcation for the same value of β. Velocity streaklines and magnetic lines of force are plotted at equal intervals of approximately one-fifth of the period of the oscillation. The dynamics are clearly dominated by the magnetic field, which is only slightly distorted by the motion. The velocity reverses during the oscillation, remaining almost symmetrical about a mirror plane at $x = d$. Although the distortion of the magnetic field by the convection is slight, it is sufficient to generate variations in the magnetic pressure which are comparable to the mean gas pressure. This generates large fluctuations in both the gas pressure and density. At the point of maximum field strength the

VELOCITY MAGNETIC FIELD

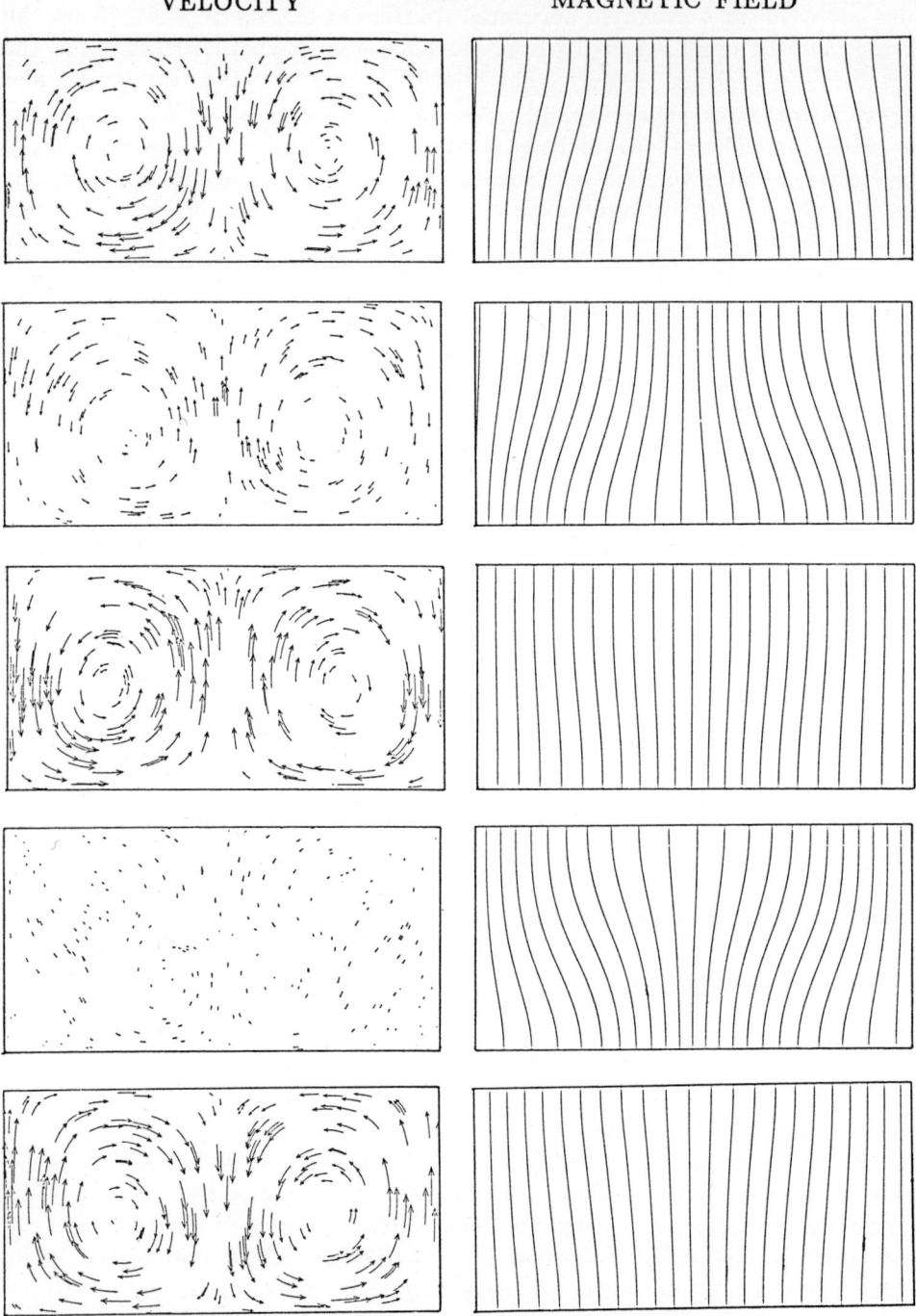

Figure 3. Oscillatory magnetoconvection: Standing wave pattern for $\beta = 8$, $R = 2R^{(o)}$. Velocity streaklines and magnetic field lines at equally spaced intervals during one complete cycle.

| VELOCITY | MAGNETIC FIELD |

Figure 4. Oscillatory magnetoconvection: Travelling wave pattern for $\beta = 8$, $R = 8R^{(o)}$. As for Figure 3.

density can drop to 5% of its undisturbed value. The resulting buoyancy forces give an extra kick to the flows in the upper portion of the cells as they begin to rise. Hence the upward velocities in this simulation have larger peak values than the downward velocities.

The results for $R = 8R^{(o)}$ in Figure 4 show more interesting spatial structure. At any instant there are four (rather than two) rolls in the box. Each roll is triangular (rather than square) in cross-section and alternate triangles have vertices at the upper and lower boundaries respectively. Rising and sinking plumes move obliquely, producing an almost continuous jet with motion from left to right across the box. The field is compressed at the narrow vertices of the rolls but weak at their broad bases, while the velocity is parallel to the field in the plumes but transverse to it at the upper and lower boundaries. As time proceeds the whole pattern moves to the right but the velocity does not drop to zero and reverse, as in Figure 3. This solution provides an example of a finite amplitude travelling wave; there is also a companion solution with the jet stream and the pattern moving to the left.

These numerical experiments show that standing wave and travelling wave solutions are both possible in a magnetic field. Further work in needed in order to establish the full bifurcation pattern and to explore the transitions from one form of solution to another as well as from oscillatory to steady convection.

6. Fine structure of photospheric magnetic fields

In the last two sections we have presented a report on work in progress at Boulder and at Cambridge. So far we have concentrated on two-dimensional magnetoconvection but a fully-compressible, three-dimensional code is on the way. In future calculations we hope to reduce the magnetic diffusivity ς so as to allow the formation of isolated sheets or tubes of flux which are partially evacuated owing to convective collapse. As before, our philosophy will be to use numerical experiments to establish the essential physical processes that are involved and then to form simplified models which may allow extrapolation to solar conditions.

Comparison between computational results and solar observations raises a number of questions. One concerns the choice of some idealized boundary condition at the base of the convecting region. Is it bettter to assume a free boundary at which $u_z = 0$ or to impose some symmetry condition that eliminates the counterflow, or would yet another condition be more appropriate? Numerical experiments might provide an answer to this question if they could reliably explain the different scales of motion in the convection zone. Meanwhile we must turn to observations. The fragmentation of "exploding granules" (Mehltretter 1978; Title et al. 1986) suggests that most granules are formed by a local instability (cf. Jones & Moore 1979) and this implies the presence of a counterflow below the photosphere. Magnetic fields may indicate the presence of inflow towards the centre of a convection cell and so the question could be answered by following the motions of small magnetic flux tubes (Schmidt et al. 1985).

Finally, there is the issue that most concerns this meeting. How far is the stucture of small-scale magnetic fields in the photosphere controlled by the pattern of granular convection and to what extent can small flux tubes be regarded as autonomous? Certainly the fields evolve in response to changes in the granulation pattern with linear magnetic features in the intergranular lanes and isolated flux tubes at the corners in the intergranular network. At these corners, however, granular convection only provides an initial configuration from which intense magnetic fields can develop. Models of slender flux tubes generally assume a perfectly conducting gas but any small-scale turbulence within the tube will tend to disperse the flux unless there is a systematic flow across the field lines. This turbulent diffusivity provides a link between magnetoconvection and the physics of slender flux tubes. Although these topics have so far been treated separately, we expect that future calculations will demonstrate that they are different aspects of the same fundamental problem.

Acknowledgments

We are grateful for comments and suggestions from our colleagues. This research has been supported by grants from the UK Science and Engineering Research Council.

References

Cattaneo, F. (1984). Oscillatory convection in sunspots. *The Hydrodynamics of the Sun*, ed. T.D. Guyenne, pp.47-50. Noordwijk:ESA SP-220.

Chan, K.L., Sofia, S. & Wolff, C.L. (1982). Turbulent compressible convection in a deep atmosphere I. Preliminary two-dimensional results. Astrophys. J. **263**, 935-943.

Chan, K.L. & Sofia, S. (1986). Turbulent compressible convection in a deep atmosphere III. Tests on the validity and limitation of the numerical approach. Astrophys. J. **307**, 222-241.

Dangelmayr, R & Knobloch, E. (1986). Interaction between standing and travelling waves and steady states in magnetoconvection. Phys. Lett., **117A**, 394-398.

Galloway, D.J. & Proctor, M.R.E. (1983). The kinematics of hexagonal magnetoconvection. Geophys. Astrophys. Fluid Dyn., **24**, 109-136.

Galloway, D.J. & Weiss, N.O. (1981). Convection and magnetic fields in stars. Astrophys. J. **243**, 945-953.

Graham, E. (1977). Compressible convection. in *Problems of Stellar Convection*, ed. E.A. Spiegel & J.-P. Zahn, pp. 151-155. Berlin:Springer.

Hurlburt, N.E. & Toomre, J. (1987). Magnetic fields interacting with nonlinear compressible convection. Astrophys. J., submitted.

Hurlburt, N.E. Toomre, J. & Massaguer, J.M. (1984). Two-dimensional compressible convection extending over multiple scale heights. Astrophys. J. **282**, 557-573.

Jones, C.A. & Moore, D.R. (1979). The stability of axisymmetric convection. Geophys. Astrophys. Fluid Dyn.. **11**, 245-270.

Massaguer, J.M. & Zahn, J.-P. (1980). Cellular convection in a stratified atmosphere. Astron. Astrophys., **87**, 315-327.

Mehltretter, J. P. (1978). Balloon-borne imagery of the solar granultion II. The lifetime of solar granulation. Astron. Astrophys., **62**, 311-316.

Muller, R. (1983). The dynamical behavior of facular points in the quiet photosphere. Solar Phys., **85**, 113-121.

Nordlund, Å, (1982). Numerical simulations of solar granulation I. Basic equations and methods. Astron. Astrophys., **107**, 1-10.

Nordlund, Å, (1984). Magnetoconvection: The interaction of convection and small-scale magnetic fields. in *The Hydrodynamics of the Sun*, ed. T.D. Guyenne, pp. 37-40. Noordwijk:ESA SP-220.

Nordlund, Å, (1985a). The 3-D structure of the magnetic field and its interaction with granulation. in *Theoretical Problems in High-Resolution Solar Physics*, ed. H.U. Schmidt, pp. 101-123. Munich: Max-Planck Institut für Astrophysik MPA-212.

Nordlund, Å, (1985b). Solar convection. Solar Phys., **100**, 209-235.

Parker, E.N. (1963). Kinematical hydromagnetic theory and its application to the low solar photosphere. Astrophys. J., **138**, 552-575.

Priest, E.R. (1982). *Solar Magnetohydrodynamics*, Dordrecht:Reidel.

Proctor, M.R.E. & Weiss, N.O. (1982). Magnetoconvection. Rep. Prog. Phys., **45**, 1317-1379.

Schmidt, H.U., Simon, G.W. & Weiss, N.O. (1985). Buoyant magnetic flux tubes II. Three-dimensional behaviour in granules and supergranules. Astron. Astrophys., **148**, 191-206.

Sofia, S. & Chan, K.L. (1984). Turbulent compressible convection in a deep atmosphere II. Two-dimensional results for main-sequence A5 and F0 type envelopes. Astrophys. J. **282**, 550-556.

Title, A.M. & Tarbell, T.D. (1986). On the relation between magnetic field structures and granulation. Preprint.

Title, A.M. Tarbell, T.D. Simon, G.W. and the SOUP team (1986). White-light movies of the solar photosphere from the SOUP instrument on Sacelab 2. Preprint.

Weiss, N.O. (1964). Convection in the presence of restraints. Phil. Trans. Roy. Soc. A, **256**, 99-147.

Weiss, N.O. (1985). Magnetoconvection. in *Solar System Magnetic Fields*, ed. E.R. Priest, pp. 156-171. Dordrecht:Reidel.

A 2D STUDY OF COMPRESSIBLE GRANULAR FLOW AND PREDICTED
SPECTROSCOPIC PROPERTIES

M. Steffen
Institut für Theoretische Physik u. Sternwarte,
Universität Kiel, Olshausenstr. 40, D-2300 Kiel, F. R. G.

1 INTRODUCTION

In order to obtain quantitative information on the magnitude of granular velocities and temperature fluctuations, inhomogeneous models have to be constructed to reproduce the observed spectroscopic properties of granulation, such as intensity contrast and line asymmetry. For purely empirical models the number of free parameters is so large that no unique models can result. Semi-empirical models (e. g. Nelson, 1978) need only few free parameters but have to rely upon more or less arbitrary assumptions. Clearly, fully self-consistent, theoretical models of granular convection are of great interest in this context. Using the anelastic approximation, Nordlund (1982, 1984) has obtained 3-dimensional, time-dependent models of this kind. After reducing the temperature fluctuations in the optically thin layers by a factor of 10 to compensate for the missing radiative exchange in the spectral lines, impressive agreement between observed and calculated line asymmetry is achieved (Dravins et al., 1981).
The essentially parameter-free, numerical simulations described below avoid the anelastic approximation and have greater spatial resolution than Nordlund's models, but are restricted to 2 dimensions in cylindrical co-ordinates. Predicted spectroscopic properties are in remarkable agreement with solar observations.
In order to investigate the immediate consequences of the reported variation of the granular scale with solar activity (Macris et al., 1984) on the spectrum, we compare models of different horizontal dimensions. The results are discussed with respect to observed long term variations of solar photospheric lines (Livingston & Holweger, 1982).

2 CALCULATIONS
2.1 Physics

The framework of the calculations is given by the time-dependent, non-linear equations of hydrodynamics prescribing the conservation of mass, momentum, and energy. They are treated fully compressibly and include turbulent viscosity as a free parameter. Axial symmetry is assumed. Magnetic fields are not considered. Additional thermodynamical relations explicitly account for temperature- and pressure-dependent ionization of hydrogen as an essential factor controlling the strength of convection. Finally, radiative transfer plays an important role for a realistic description of granular convection. In order to avoid the diffusion approximation, which is not acceptable for this kind of calculation, we solve the equation of radiative transfer along a large number of rays crossing the model with

various inclinations. We use the grey approximation in LTE with appropriate Rosseland mean opacities as a function of temperature and pressure (Kurucz, 1979).
The resulting equations are solved numerically by the method of bi-characteristics with a modified version of the code described by Stefanik et al. (1984).

2.2 The models

The models presented here extend vertically over approximately 6 scale heights, from 250 km below to 650 km above optical depth unity. The corresponding grid width varies from 20 km in the lower to 40 km in the upper part of the computational domain. The diameter of the convection cell is a free parameter. Up to now, diameters between 650 and 1750 km have been studied. 30(vertical)x36(horizontal) grid points, fixed in space, are used. Time step is typically less than 1 second.
The upper boundary is transmitting simple acoustic waves (Stefanik et al., 1984). At the location of the lower boundary nearly 100% of the total flux is transported by convection. For this reason an open lower boundary condition was devised which allows a free flow of gas out of and into the model, assuming constant gas pressure along the bottom. In this way the horizontal position of up- and downflows is determined by the flow itself and is variable in space and time.

3 RESULTS
3.1 Convection and oscillations

Starting with more or less arbitrary initial conditions we observe that the time evolution always proceeds towards a stationary state. The resulting characteristic flow pattern (fig.1), which seems to develop independent of initial conditions, exhibits a strong downdraft at the axis of symmetry, surrounded by a broader ring-like upflow with smaller velocity that again turns into a downflow in the outer ring. In this example, convective velocities extend at least 250 km above $\tau=1$, whereas according to local mixing-length theory convection stops very close to $\tau=1$ (Z=0).
The corresponding thermal structure (fig. 1) is characterized by large horizontal temperature fluctuations and by steep vertical temperature gradients which occur where the hot ascending gas reaches the surface and looses its excess energy within a short time by efficient radiative cooling. Density shows a marked inversion in these regions.
Interestingly, superposed on the convective flow and throughout the upper convectively stable parts of the atmosphere, we find oscillations with a well defined period of about 260 sec which are nearly undamped in time and may have some relation to the solar 5-min oscillations (for more details see Steffen, 1986).

3.2 Spectroscopic consequences of granular convection

In order to make possible a direct comparison of numerical simulations and solar observations we have calculated the spectrum resulting from our model convection cells corresponding to observations at disc center. In the continuum we see a bright ring with a dark center, surrounded by an outer dark lane. This picture suggests that the model may represent a typical phase in the evolution of an "exploding granule". In reasonable agreement with observation we obtain a rms intensity contrast of about 13% in the continuum at 5000 Å. Moreover, the computed wavelength dependence of granular contrast closely reproduces observations.

Spatially resolved synthetic spectra computed from the models show that line profiles originating from the bright granular regions show a blue-shifted core and a considerable asymmetry, the line wings being more blue-shifted than the core, while spectral lines originating from the dark intergranular regions are red-shifted and extremely asymmetric in the opposite direction (see also Steffen & Gigas, 1985). The horizontally averaged (unresolved) profile is only weakly asymmetric because opposite asymmetries of granular and intergranular contributions cancel to a large degree. This is demonstrated once again by plotting the corresponding line bisectors (fig. 2). The different slopes of the spatially resolved line bisectors reflect a different depth dependence of temperature and convective velocity. The velocity variation derived from the spatially resolved spectrum is roughly a factor 10 greater than the velocity fluctuation obtained from the bisector of the unresolved line profile. Although this picture is quite different from the "classical" explanation of the C-shape, the average bisector resulting from the numerical simulations closely reproduces observation (fig. 3). The sharp blueward turn in the immediate vicinity of the continuum,

Fig. 1: Stationary flow resulting from numerical simulation of granular convection. Arrows indicate direction and magnitude of the local velocity. Upward velocities are less than 2.5 km/s, while maximum downward velocity is about 7 km/s. Thermal structure is represented by lines of constant temperature in steps of 200 K. Maximum temperature difference between hot and cool regoins is more than 4000 K just 100 km below the visible surface. Vertical temperature gradient exceeds 40 K/km near the top of the hot ascending parts of the flow, causing a strong density inversion in these regions. Z=0 corresponds to optical depth unity. Only the lower part of the model is shown.

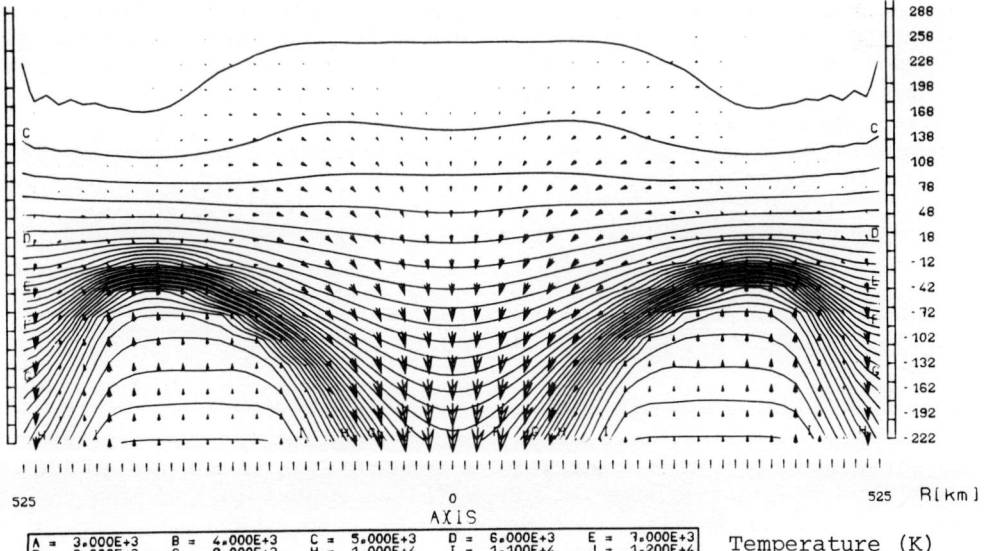

distinctly visible in calculation and observation, is **not** the result of a blend, but is typical for synthetic bisectors of stronger lines! From test calculations we conclude that this feature does not reflect a special depth dependence of the flow but seems to be related to the marked transition from the Doppler core to the damping wings in the Voigt profile of the line absorption coefficient. This conclusion is supported by the fact that bisectors calculated by Kostyk (1985) from a completely different model show the same behavior near the continuum. Another interesting prediction of the numerical simulation is that the maximum redshift of the line bisector near the continuum should be about twice as large (in km/s) near 1.6 µ than in the visual. This can probably be explained by the fact that due to the reduced continuous opacity the line profiles originate in deeper atmospheric layers where the downward flow is even stronger. Indeed the predicted bisectors can be observed in the infrared (Brandt, 1986). While this agreement gives us some confidence in the models it must be kept in mind that quite different models can produce quite similar average line bisectors. For a more crucial test of the models, observations of high spatial resolution are of great interest.

3.3 Investigation of granulation with different horizontal scale

To find an answer to the question how properties of convection depend on horizontal dimension, we have calculated 5 models of different cell size ($\varnothing$= 650...1750 km). The main result of this study is that, while topology is similar for all diameters, larger granules produce a larger effect on the line formation regions (overshoot) than smaller granules although their subsurface velocities are lower. The reason is that horizontal radiative exchange becomes increasingly inhibited for larger granules which consequently can ascend to greater heights before their excess heat is lost. Furthermore, from these calculations there is no indication that small granules ($\leq$0.8 arc sec) are in principle different from larger ones; they may well be convection cells.

The corresponding synthetic spectra imply the following spectroscopic differences between large and small granules: (i) Continuum contrast is enhanced for larger granules, because they can maintain greater horizontal temperature differences near the visible surface than the smaller ones. (ii) Line asymmetries (bisector amplitudes) are more pronounced for larger granules, since temperature- and velocity fluctuations in the line formation regions are increasing with horizontal scale of granulation. (iii) Calculated equivalent widths show a non-monotonic dependence on cell size. The strength of the high-excitation CI line λ 5380 Å, and that of average FeI lines is found to change in opposite directions. Moreover, it seems that the ground state MnI line λ 5394 Å, is one of the most sensitive indicators for changes in the thermal structure of the solar atmosphere.

If magnetic fields have no significant effect on the dynamics of the convective flow we may apply these results to observations reported from Pic-du-Midi (Macris et al., 1984) indicating that the size distribution of solar granulation is shifted to smaller scales as solar activity rises from minimum to maximum. From (i) we expect that during the same time interval **intensity contrast** should have decreased. Just the opposite seems to be observed by Alissandrakis et al. (1982). On the

Fig. 2: Bisectors of spatially resolved line profiles calculated from the model displayed in fig. 1. Numbers refer to the horizontal position of the line of sight (**1** = axis, **36** = outer ring). **A** denotes the bisector of the horizontally averaged line profile.

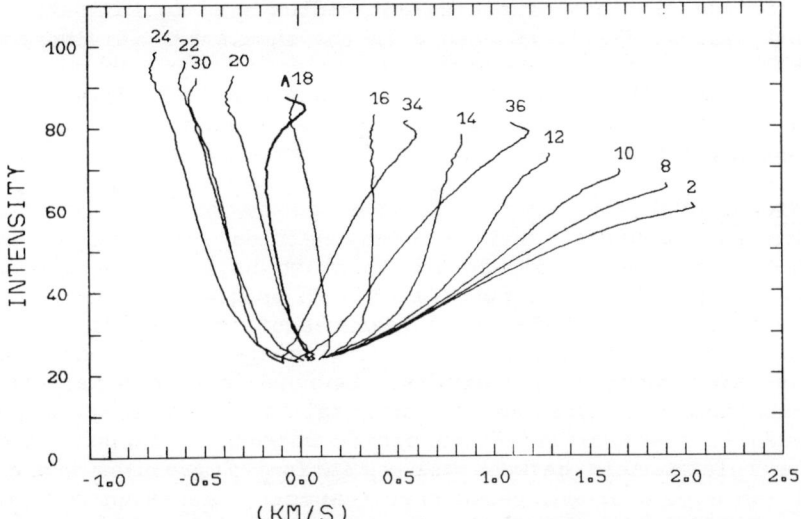

Fig. 3: Comparison of calculated and observed line bisector for Ti II, λ 5336.7 Å. To eliminate the effect of oscillations the calculated spectrum was averaged in time. The resulting bisector is given on an absolute velocity scale, but the observed bisector is arbitrarily diplaced. Model diameter: 1750 km. Observation from Barambon (1977).

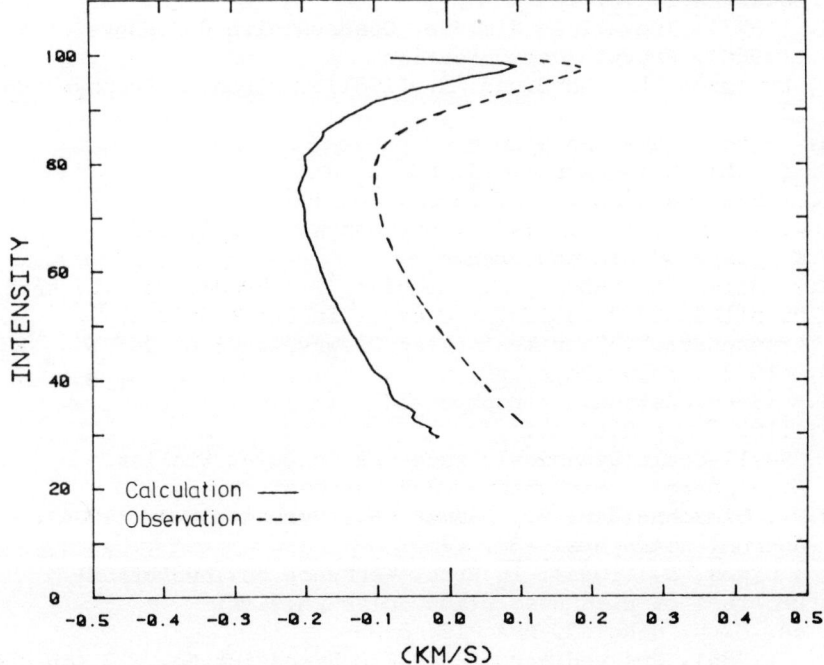

other hand, **line asymmetry** is predicted to increase from 1980 (maximum) to 1986 (minimum) according to (ii). Observations by Livingston (1986) show a slight tendency in this direction which may not be significant, however. In contrast to what is predicted from the models (iii), observed long term variations of **equivalent widths** seem to indicate that CI λ 5380 Å and typical FeI lines change in the same sense (Livingston & Holweger, 1982), but in re-analysed data covering a longer period of time (Holweger, 1986) this correlation is less certain. Finally, MnI λ 5394 Å indeed exhibits the strongest variations in equivalent width as expected (Holweger, 1986).

4 CONCLUSIONS

Granular contrast and line asymmetries predicted by the numerical simulations presented here are found to be in close agreement with observations (disc center). Spectra of high spatial resolution will provide an important check of the basic features of the models.

The computed spectroscopic consequences of a change in the horizontal scale of granulation cannot fully explain observed long term variations of the solar spectrum. In view of the uncertainties in observation and models it presently seems difficult to decide whether it is possible to understand the relationship between the variation of granular scale on the one, and variations of the photospheric spectral lines on the other hand, in terms of the dependence of convective properties on horizontal cell size as derived from the study of different model diameters, or if more complicated explanations, explicitly involving magnetic fields, are necessary.

5 REFERENCES

Alissandrakis, C.E., Macris, C.J., Zachariadis, G. (1982). Solar Phys. 76, 129
Barambon, C. (1977). Travail de diplome, Observatoire de Geneve
Brandt, P.N. (1986). Private communication
Dravins, D., Lindgren, L., Nordlund, Å. (1981). Astron. Astrophys. 96, 345
Holweger, H. (1986). This conference
Kostyk, R.I. (1985). Sov. Astron. 29, 65
Kurucz, R.L. (1979). Astrophys. J. Suppl. 40, 1
Livingston, W. & Holweger, H. (1982). Astrophys. J. 252, 375
Livingston, W. (1986). This conference
Macris, C.J., Muller, R., Rösch, J., Roudier, T. (1984). In S.L. Keil, "Small-Scale Dynamical Processes in Quiet Stellar Atmospheres", Sacramento Peak Observatory, p. 265
Nelson, G.D. (1978). Solar Phys. 60, 5
Nordlund, Å. (1982). Astron. Astrophys. 107, 1
Nordlund, Å. (1984). In S.L. Keil, "Small-Scale Dynamical Processes in Quiet Stellar Atmospheres", Sacramento Peak Observatory, p. 180
Stefanik, R.P., Ulmschneider, P., Hammer, R., Durrant, C.J. (1984). Astron. Astrophys. 134, 77
Steffen, M. & Gigas, D. (1985). In Proc. Workshop on Theoretical Problems of High Resolution Solar Physics, ed. H. U. Schmidt, MPA 212, p.95
Steffen, M. (1986). Proceedings of the IAU Symposium No. 123 (in press

ON THE CORRELATION OF THE C-SHAPE OF THE FeI LINE λ5576 Å WITH THE BRIGHTNESS OF Ca^+-K-FACULAE.

S. Immerschitt and E.H. Schröter
Kiepenheuer-Institut für Sonnenphysik, Freiburg, FRG

Abstract. We inquired into the basic properties of the line-profile of FeI λ5576 Å, particularly the c-shape, in a large active region. We find that systematic changes of the form of the c-shape, of the residual intensity, of the full-width-at-half-minimum (fwhm) and of the equivalent-width take place in active regions. These results constitute strong evidence that magnetic activity modifies the convective temperature and velocity pattern within the photosphere.

Introduction

It is known that the c-shape of any solar line, when observed with low spatial resolution, results from a correlation of temperature- and velocity-fluctuations ($\Delta T \otimes \delta v$). The first interpretations of solar line-shifts and asymmetries were published independently by Voigt (1956) and Schröter (1957). Beckers and Nelson (1978) presented much more advanced model calculations, based on stationary two-dimensional granulation models. The present aim of this study is, to determine the influence of solar magnetic fields on the structure of convective motions in the solar photosphere. As shown by Livingston (1981, 1982) magnetic fields appear to influence the form of the c-shape. Further investigations of Kaisig and Schröter (1983) and Cavallini, Ceppatelli, Righini (1982, 1986) confirm that magnetic fields may well modify the granular structure. For our investigations we used the diagnostics of the c-shape of the magnetically insensitive FeI line λ5576 Å. Our aim was to determine the changes of the c-shape in active and quiet solar regions and to conclude on changes of convective temperature- and velocity-fluctuations due to magnetic fields.

Observational material

The observational material consisted of photographic spectra of the region around the line FeI λ5576 Å from one of the largest active regions of the recent cycle, exposed at the observatory of Capri in June 1982. An area of 265 arcsec x 185 arcsec was covered by a 3-arcsec-equally-spaced sample of 86 spectra (see fig. 1). An iodine tube put on a part of the focal plane of the spectrograph gave us the reference-wavelengths.

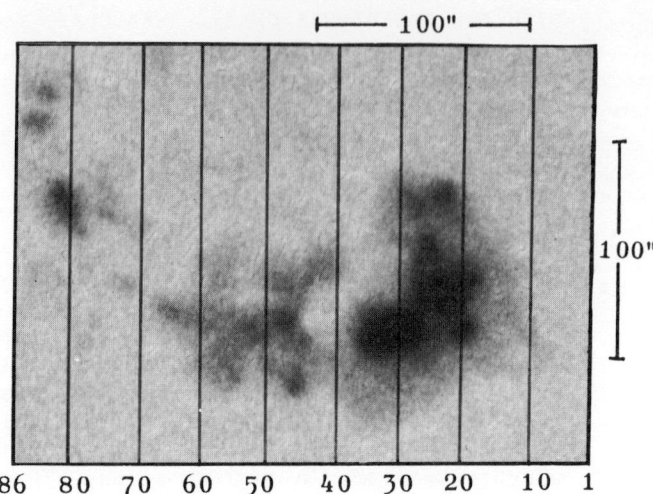

Fig. 1: The active region under study from a white-light-photo

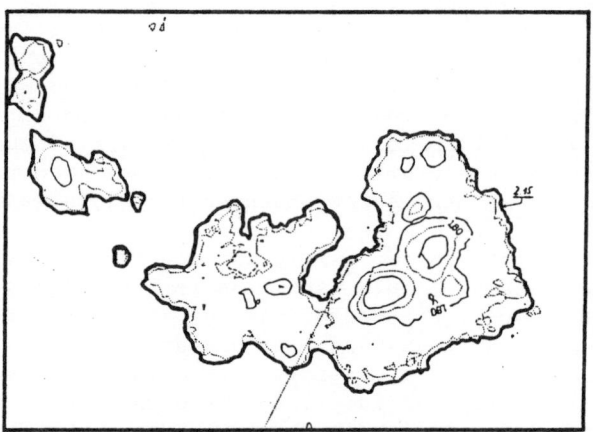

Fig. 2: The same active region with the spot-regions excluded from the study

Methods: As an indicator of magnetic activity we used a Ca^+-K-spectroheliogram from the same day, obtained at the Wendelstein-Observatory of the Universitätssternwarte Munich. This allowed us to correlate any point of the calcium-image with the line-profile measured at the corresponding point on the sun, particularly the c-shape, of the FeI line λ5576 Å. An additional white-light-image of the active

region, exposed on Capri, allowed us - again by a point-to-point-correlation - to exclude umbral and penumbral regions, which are known to exhibit other types of dynamical processes, from our investigations (see fig. 2). We inquired into the behaviour of the following characteristics of the line-profile with increasing Ca^+-K-brightness, which we understand as increasing magnetic activity:
- the c-shape (curvature, wavelength of the line-minimum)
- the residual intensity
- the full-width-at-half-minimum (fwhm)
- the equivalent-width (W_λ)

We measured about 5100 single line-profiles. These were 66% of profiles remaining after excluding those within spotregions.

Results:

C-shape: We find that the bisector-curvature decreases with increasing Ca^+-K-brightness, i.e. with increasing magnetic activity. The upper part of the bisector, with respect to the line-minimum shows a systematic redshift, whereby the line-minimum is raised in its residual intensity (see fig. 3 and 4). A small blueshift of the wave-

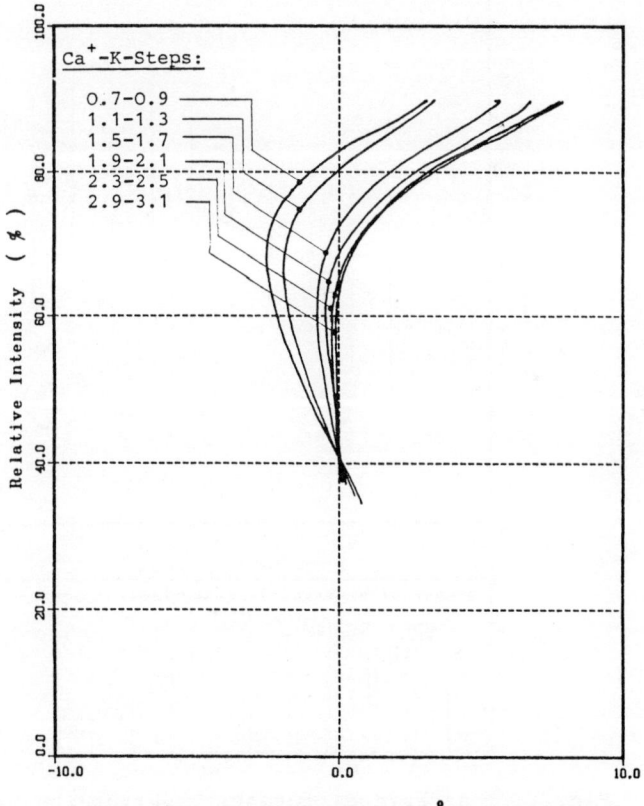

Fig. 3: The C-shape for different Ca^+-K-brightness

note: The wavelength of any individual bisector was set to zero at 40% of the continuum intensity. The brightness of the Ca^+-K-features is normalized in terms of the "mean intensity of quiet sun". This norm was achieved by measuring the average brightness of quiet regions in the center-of-sun.

length at minimum-intensity of the order of 1.3 mÅ seems to be implied (see fig. 5). Despite an overall redshift in the range $0.75 < I < 0.90$% the constant inclination of the upper part of the bisector is a striking characteristic: This part of the line-profile represents the dynamics of the lower layers of the solar photosphere and its constant behaviour will require further investigations. On the other hand the middle and the upper photospheric layers show a distinct dependence of the ($\Delta T \otimes \delta v$)-correlation on the brightness of Ca^+-K-faculae.

<u>Residual intensity:</u> The residual intensity of the line is raised with increasing Ca^+-K-brightness, i.e. with increasing magnetic

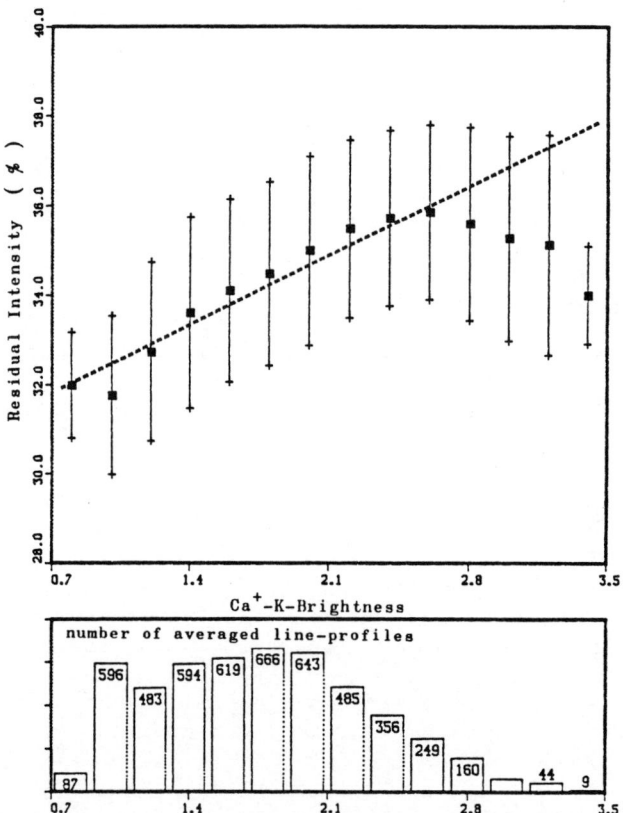

<u>Fig. 4:</u> The residual intensity versus the brightness of the Ca^+-K-faculae. The straight line was calculated by a linear least-square-fit through all datapoints, i.e. the mean residual intensity of any step was weightened by the number of datapoints that were averaged.

activity. We find an increased residual intensity of about 4% (see fig. 4).

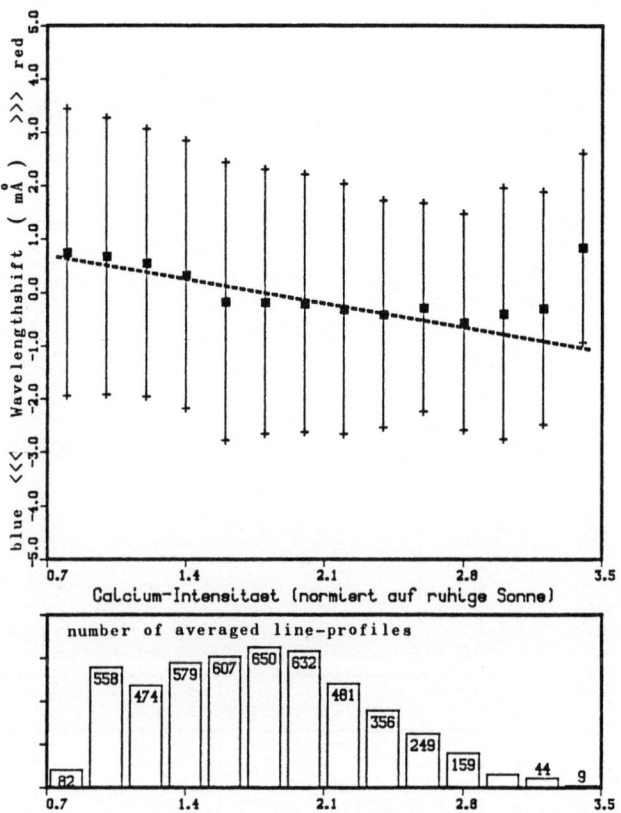

Fig. 5: The wavelength-drift of the line-minimum versus the brightness of Ca^+-K-faculea, with respect to a mean wavelength of all about 5100 single line-profiles.

Fwhm: A 3%-increase in fwhm is measured from quiet regions to the most active ones (see fig. 6).

Equivalent-width: As the equivalent-width is a basic property of the observed flux, we searched for the changes of its behaviour. However, blends in the wings of the line forced us to define a modified W^*_λ by integration only to 85% of continuum intensity (see fig. 7a). With the reservation that changes of intensity in the wings of the line over 85% intensity are not taken into account, we find a decrease in W^*_λ of about 7% as compared active to quiet regions (see fig. 7).

A striking fact of the line-profile properties is that for the brightest Ca-features there is a new behaviour. This can be seen distinctly in the case of the equivalent-width: It decreases with increasing facular-brightness up to an intensity of about 2.4, whereas raising Ca^+-K-brightness further leads to an increase of W^*_λ. An explanation for this behaviour may be found in the inexact definition of our equivalent-width (85%). But we find an analogous behaviour of the residual intensity and of the wavelength-drift of the line-minimum (see fig. 4 and 5). Additionally the c-shape seems to come too to a "saturation" at this Ca^+-K-brightness 2.4. This particular behaviour for the brightest faculae needs further treatment.

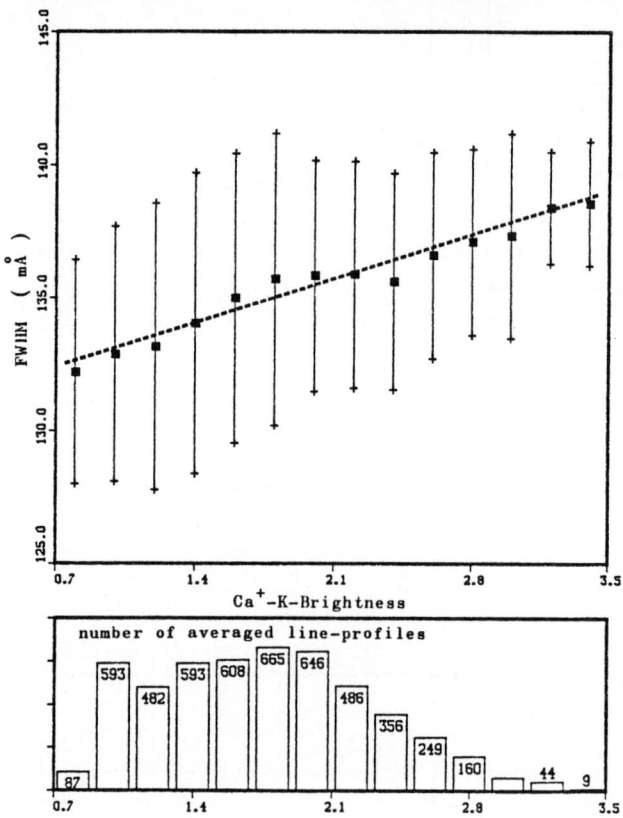

Fig. 6: The fwhm versus the brightness of the Ca^+-K-faculae

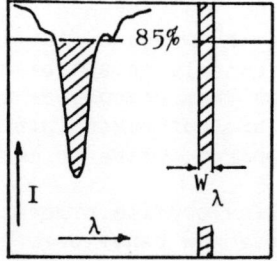

Fig. 7a: The definition of our modified equivalent-width W_λ^*

Recent studies of Brandt and Solanki (1986), who correlated among others the c-shape of the FeI line λ5576 Å with the magnetic filling-factor confirm our results, and in addition justify the use of Ca^+-K-brightness as an indicator of magnetic activity.

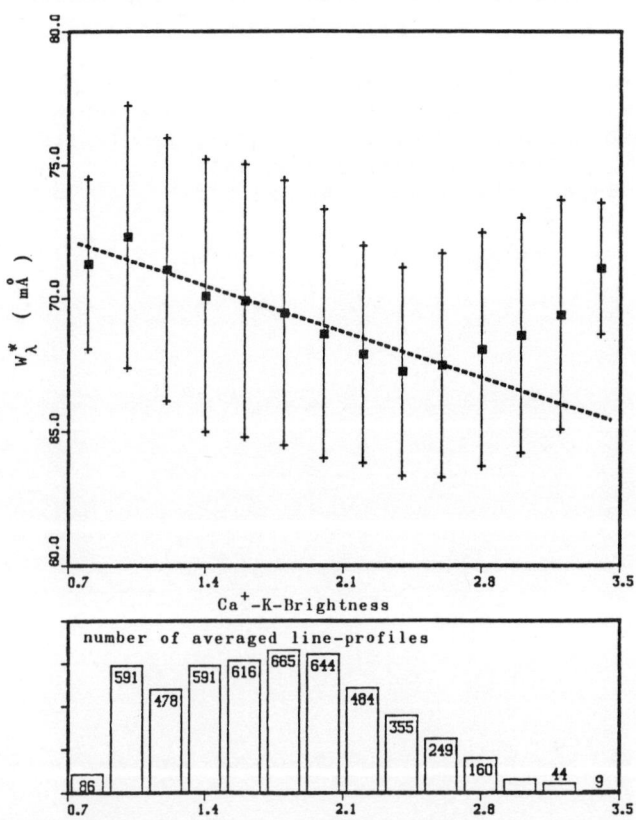

Fig. 7: W_λ^* versus the brightness of the Ca^+-K-faculae

Conclusions:
The behaviour of the fwhm and equivalent-width is hardly explained by either changing micro- or macroturbulence alone. On the other hand the assumption of a reduction of the total line-absorption in active regions leads to an explanation of the experimental results. This will require more calculations, from which it will have to be determined, which thermodynamical parameters of the solar atmosphere are to be changed in order to explain the behaviour of the other parameters, particularly the c-shape profile of the FeI line λ5576 Å observed in active regions.

References

Beckers, J.M., Nelson, G.D.: 1978 Solar Phys. 58, 243
Brandt, P.N., Solanki, S.: 1986 This conference
Cavallini, F., Ceppatelli, G., Righini, A.: 1982 Astron. Astrophys. 109, 233
Cavallini, F., Ceppatelli, G., Righini, A.: 1986 Astron. Astrophys. 158, 275
Kaisig, M., Schröter, E.H.: 1983 Astron. Astrophys. 117, 305
Livingston, W.C.: 1982 Nature 297, 208
Livingston, W.C.: 1983 IAU Symp. No. 102 "Solar and Magnetic Fields: Origins and Coronal Effects", J. Stenflo (ed.), 149
Schröter, E.H.: 1957 Z. Astrophys. 41, 141
Voigt, H.H.: 1956 Z. Astrophys. 40, 157

Mitteilungen aus dem Kiepenheuer-Institut Nr. 272

JOINT DISCUSSION ON TOPICS OF SESSIONS 1 AND 2, SUMMARIZED
BY THE CHAIRMEN

T. Roca Cortés
Instituto de Astrofisica
La Laguna

R. Rosner
Center for Astrophysics
Cambridge, Mass.

QUESTIONS SESSION 1. (Chairman: T. Roca Cortés)

Questions following presentation by H. Holweger and W. Livingston

S. Solanki: Is the total magnetic flux measured at Mt. Wilson corrected for inclination of the field lines?

H. Holweger: The answer is no.

P. Brandt: Could you comment on the possibility that the filigree, if unresolved, could lead to fictitiously large granular diameters?

H. Holweger: The Pic-du-Midi observers say that they are able to identify filigree or network bright points and granules separately on their photographs taken under very good seeing conditions, which were used to investigate the cycle dependence of the surface density of granules.

E. Priest: Both the network facular points and Ca^+K emission are a result of magnetic flux tubes, but the facular points are out of phase with the sunspot cycle while the Ca^+K emission is in phase. So, what is the solar cycle variation from magnetograms of the size and flux in network magnetic elements, both in the quiet sun and the sunspot zones?

H. Holweger: The full disk K_3 intensity receives contributions from the quiet sun network. Network bright points are anticorrelated with sunspot number and plage areas. The latter probably dominate around sunspot maximum while the former does it at minimum.

F.L. Deubner: With regard to the dependence of the granular scale on the distance from the sunspots position for which you have shown two rather conflicting sets of measurements: Can you specify the sizes of the sunspots chosen for the 1986 study? Could the null result of this study be attributed to a lack of sufficiently large spots?

H. Holweger: The observations of Collados et al. (1986) at Izaña were made in the years 1980, 1983 and 1984. Large spots were available.

QUESTIONS SESSION 2. (Chairman R. Rosner)

Questions following presentation by E.H. Schröter

H. Wöhl: What was the time difference between the Wendelstein Ca^+ images and your Capri spectra?

E. Schröter: Sorry, I forgot to mention that exposing the target of the spectra of Fe I $\lambda 5576$ took almost four hours (from morning to noon), whereas the Wendelstein Ca^+ K-spectroheliogram was obtained around 8 o'clock in the morning. It can be therefore not excluded that particularly the very bright Ca^+ K-emission points were the most active ones, with an increased brightness when we took the spectra. This could perphaps explain the change of the slope of the correlation curves for $I_{Ca^+K} > 2.5$.

S. Solanki: If we assume that Ca II brightness is correlated with the magnetic filling factor, then the decrease in rest intensity and the increase in equivalent width of $\lambda 5576$ above a certain level may be explained by the models of flux tube temperature which I have calculated. For the small sample of regions underlying these calculations it is found that the flux tubes in the high filling factor regions are cooler than those in low filling factor regions.

E. H. Schröter: At present, we are not in a position to exclude (or accept) this type of interpretation. As I already mentioned in my talk, we are now starting to look at what type of model calculations can best fit the observed dependence of the properties of the Fe I $\lambda 5576$ line on the brightness of the Ca^+ emission points.

A. Righini: These results are very similar to those previously obtained at the Arcetri Solar tower in active regions as far as the bisector is concerned. The straightening of the lower part of the bisector may be due to a decrease of the convection while the higher part is affected by the line profile arising from magnetic flux tubes.

E. H. Schröter: This could be one of many possible explanations. Another plausible assumption could be that intergranular lanes beneath bright Ca^+ emission points are even less transparent (cooler) than they are already in quiet regions and that the contribution of their deeper layers to the composite line profile overwhelms that of granules, reducing the blue-shift for $I \approx 0.4-0.8$, and yielding a larger redshift in the wings. I personally prefer, at the moment, the latter interpretation, but we shall see very soon what the model calculations teach us.

Questions following presentation by M. Steffen

N.O. Weiss: Two questions: first, what exactly are your lower boundary conditions? Second, have you considered the possible effect of non-axisymmetric instabilities on your solutions? For the analogous Boussinesq problem, Jones and Moore showed that the axisymmetric roll was unstable to non-axisymmetric perturbation (corresponding to Busse's wavy instabilities). This result suggests that a non-axisymmetric calculation in cylindrical geometry might provide a good description of the evolution of an exploding granule.

M. Steffen: In response to the first question, the lower boundary condition is a kind of an open boundary where gas is allowed to flow out along the bottom. As for the second question, we have not performed a corresponding stability analysis. But to me it seems quite probable that the ringlike structure will tend to break up into several pieces if the flow is no longer required to be axisymmetric.

H. Schmidt: Can you comment on center-to-limb variations of your results?

M. Steffen: The center-to-limb variations of the spectrum have not yet been investigated. Of course, corresponding calculations are possible in principle, but are expected to be more uncertain than for solar disk-center because then you have to make some assumption of how the flow continues horizontally outside the domain of computation.

F. Busse: Does the outer descending motion disappear in the granule model when the width is decreased?

M. Steffen: No. For all diameters considered so far the models show a downflow at the axis of symmetry, surrounded by a broader ringlike upflow which again turns into a downflow near the periphery of the cylindrical model.

F. Kneer: Does the overshoot depend on the turbulence parameter?

M. Steffen: Variation of the turbulent viscosity parameter over a factor of 15 has only marginal effects on the flow and resulting spectrum. Overshoot depends much more strongly on the horizontal scale of granulation.

Questions following presentation by F. Busse

H. Hurlburt: What governs the relationship for the frequency of the oscillations in your model? Is it perhaps the circulation velocity of the participating cells?

F. Busse: The frequency of the oscillations is related to the frequency of trapped, magneto-acoustic waves. However, the viscous and magnetic diffusivities significantly modify this frequency.

Questions following presentation by N.O. Weiss

E.R. Priest: On a linear theory both standing waves and propagating waves are possible, so why physically in your nonlinear computation are the propagation modes preferred? Also, what is their nature; are they gravity or magnetoaccoustic modes?

N.W. Weiss: The mathematical theory of the interaction between travelling and standing waves in a nonlinear regime has been developed by Knobloch ond others. One might conjecture that with a horizontal magnetic field, hydromagnetic waves would propagate horizontally with the field acting as a waveguide, and travelling waves are indeed preferred. With a vertical field one would expect waves to propagate vertically, giving a preference for standing waves. This holds for normal choices of parameters, but there are ranges where travelling waves are preferred. We plan to investigate this behaviour systematically in the Boussinesq regime.

U. Anzer: Would you expect these oscillations to occur in the Sun?

N.O. Weiss: Such oscillations will not be found where the

mean flux density is low, but we would expect them to occur in sunspot umbrae, where there is a strong magnetic field, or in pores. This will be discussed in more detail by Neal Hurlburt later in these Proceedings.

J. Zirker: What factors do you think are most important in generating small-scale convection from larger scales? (There is no evidence of such small scales in your calculations so far.)

N.O. Weiss: Almost all computations show cells that span the whole depth of the convecting layer in a compressible medium. The only exception is the work of Chan et al. (1984), where the small-scale motion seems to be produced by their recipe for dealing with sub-grid scale diffusion. Perhaps our calculations have viscosities that are too high, perhaps we need to include three-dimensional instabilities. Intuitively, one would expect that the boundary layers just below the photosphere, where everything varies rapidly, would become unstable, leading to motion on a granular or sub-granular scale. But something more than a small density scale height is needed to produce this effect.

Questions following presentation by N. Hurlburt

R. Rosner: Could you comment on the values of ϱ/κ and ν/κ used in your calculations, and on the sensitivity of the results to these parameters?

N. Hurlburt: Our magnetic and viscous Prandtl numbers were both 0.10 in these simulations. The behaviour should be similar for smaller values. For larger magnetic Prandtl numbers the magnetic field's coupling to the motions is reduced, so the amount of magnetic buoyancy present should also be reuced.

B. Roberts: How do periodicities and phase speeds of the oscillations compare with sunspot values?

N. Hurlburt: Model calculations typically yield periods that are short compared with the lifetime of umbral dots. However, by varying the parameters in a nonlinear regime, one can obtain periods that are arbitrarily long. So it is, in principle, possible to match the observed timescales.

General Discussion following sessions 1 and 2

R. Rosner: I wonder whether Dr. Busse could comment on the issue of modeling sub-grid dynamics?

F. Busse: I am not an expert on sub-grid scale modeling and other methods to take into account the effects of turbulence of unresolved scales. Even though there is no sound theoretical basis to support the eddy viscosity concept, it works quite well in many cases. The theory of isotropic turbulence suggests that all eddy diffusivities should be approximately equal in the limit of high Reynolds number with the consequence that the eddy Prandtl number, or the eddy magnetic Prandtl number be about unity. On the other hand , models of solar convection or of convection in the major planets suggest that more appropriate Prandtl numbers are those which deviate from unity in the direction of the molecular values. The diffusive processes on the molecular level, or by radiation in the case of the thermal diffusivity, still appear to be important enough, perhaps in boundary layers or internal diffusion interfaces between eddies, to influence the transport of heat, momentum, or magnetic flux when the turbulent transport dominates in the bulk of the fluid.

H. Spruit: A comment to N. Weiss' and N. Hurlburt's presentations concerning the choice of thermal boundary condition (b.c.) at the base of the calculation (flux constant vs. fixed temperature): the real Sun does not have a boundary at the position where the calculation puts one. If one is interested in modeling the real Sun, a boundary condition has to be devised that interferes as little as possible with the process one tries to calculate. Which boundary condition is appropriate cannot be decided without making some kind of model for the medium below the boundary. One needs to know in some approximate way its response to what happens inside the computational volume. If the model extends deep into the convection zone, a constant temperature condition would perhaps be more appropriate than constant flux, because the high effective ("turbulent") heat conductivity limits temperature fluctuations to small values. But it would be easy to make more sophisticated b.c.'s, for example, by using a "turbulent diffusion" model for the external medium. One could then calculate its response to the boundary values analytically, and stick those into the numerical calculation.

N. Hurlburt: Shouldn't constant flux be used because solar flux is constant? The solar flux must be constant only globally, and then only on time scales large compared with the thermal time scale (of the convection zone, i.e., 10^9 years, if the calculations refer to things happening in the convection zone). On shorter time scales (i.e., almost all of those that one is interested in), the very large heat capacity of the gas in the solar interior becomes apparent. It would very easily (i.e., with only small T-variations) supply the heat flux variations that a computational boundary would request.

C. Zwaan: As mentioned by Dr. Holweger, the number of X-ray bright points (XBP) varies in antiphase with the solar cycle. From this it does not follow, however, that (some type of) ephemeral regions vary out of phase with the cycle. Karen Harvey showed that the He I 10830 dark points (which are proxies for XBP's) do correspond to magnetic bipoles, but only one-third of these are ephemeral regions. The remaining two-thirds of the dark points correspond to "chance encounters" of previously unrelated magnetic poles. Harvey pointed out that the peak in the appearance of XBP's = dark HeI points ≃ "chance encounters" coincides with the maximum coverage by mixed-polarity network, which occurs around the time of minimum activity (as established by Giovanelli).

W. Mattig: I have some remarks regarding the papers given by M. Steffen and E.H. Schröter. First, my comment for Steffen: By analyzing the asymmetries of highly resolved spectra we (K. Fleig and W. Mattig) have found that the asymmetries are much larger than in spatially unresolved spectra. The asymmetry is about one order of magnitude larger, in full agreement with the calculations by Steffen. That means the normal C-shape is at first a mixture of strong C-shapes in the resolved spectra. The old picture for the C-shape has to be revised; this change in our understanding is a consequence of the strongly decreasing convective velocities with height in the atmosphere. Second, my comment for Schröter: From the analysis of highly resolved spectra, we have found (a) that the RMS velocities near sunspots in Ca^+ plage regions are smaller than the velocities in the non-Ca^+ regions, and (b) that in the NAR (non-active regions) near sunspots, the RMS velocities are smaller near the sunspots (1 sunspot radius) compared to the velocities far away (3 sunspot radii). This is a result from a power analysis in the spatial region <3.5 arcsec. To summarize all this,

$$V_{RMS}(Ca^+) < V_{RMS,NAR}(1 R_\odot) < V_{RMS,NAR}(3 R_\odot)$$

S. Solanki: W. Livingston and H. Holweger showed us viewgraphs displaying the magnetic flux averaged over the whole Sun (Harvey 1986; Bruning and LaBonte 1985). Did these authors take into account the projection effect of the (probably) vertical fields near the solar limb?

W. Livingston: No.

S. Solanki: In that case, we should be very careful in taking these diagrams at face value, since during sunspot ninimum the field is more dipolar in nature, and will give a smaller apparent flux than during sunspot maximum when strong active regions are often present near disk center.

STRUCTURE OF MAGNETIC FLUXTUBES AS DERIVED FROM OBSERVATIONS OF MODERATE SPATIAL RESOLUTION

S.K. Solanki

Institute of Astronomy, ETH-Zentrum, 8092-Zürich, Switzerland

1. INTRODUCTION

The small scale structure of the magnetic field outside sunspots has been extensively studied over the last two decades, but progress has often been slow due to the small size of these structures, usually referred to as magnetic elements or fluxtubes. The diameters of most of them are known to be smaller than the best presently available spatial resolution of approximately 0.3″ (Ramsey et al. 1977). Although this property presently limits the success of direct methods of study (i.e. methods which attempt to spatially resolve the structures), it has stimulated the development of a number of powerful indirect (i.e. spectral) techniques. Almost without exception these methods rely on observations of the Stokes parameters, in general Stokes V. Therefore, before continuing, the Stokes parameters shall be described briefly.

A system of four quantities is required to describe light in an arbitrary state of polarization. An example of such a system are the four Stokes parameters I, Q, U, and V. Stokes I is the total intensity and can be measured without any polarization analyzing equipment, while

$$Q = I_{\text{lin}}(\varphi = 0) - I_{\text{lin}}(\varphi = \pi/2),$$
$$U = I_{\text{lin}}(\varphi = \pi/4) - I_{\text{lin}}(\varphi = 3\pi/4),$$
$$V = I_{\text{circ}}(\text{right}) - I_{\text{circ}}(\text{left}).$$

In the above equations I_{lin} denotes the intensity of linearly polarized light with azimuth φ, defined by the direction of the analyzer, while I_{circ} is the intensity of circularly polarized light.

Fig. 1 shows I, V, and Q of the Fe I 5250.2 Å line observed in an active region at $\mu = 0.28$ ($\mu = \cos\theta$ is the cosine of the heliocentric angle). This is just a small section of a much larger spectrum containing thousands of lines, which was obtained with a Fourier transform spectrometer (FTS). V is proportional to the line of sight component of the magnetic field, while Q and U increase as the magnetic field component perpendicular to the line of sight increases. Of great importance, as far as their diagnostic capabilities are concerned, is the fact that whereas Stokes I has contributions from both inside and outside the fluxtubes (since they cover only a small percent of the observed area, this means that I is formed mainly in the non-magnetic atmosphere), Q, U, and V are formed exclusively inside the magnetic elements. Therefore, by observing in polarized light, we can isolate the contribution from the fluxtubes even with moderate spatial resolution.

2. COMPARISON OF DIFFERENT OBSERVATIONAL AND ANALYSIS PROCEDURES

Ideally, observations should combine high spatial, temporal, and spectral resolution, and cover a large spatial, temporal, and spectral range simultaneously in all four Stokes parameters. In reality, due to the limitations set by seeing and instrumentation and in order to reach a reasonable signal to noise ratio, compromises have to be made. Therefore, two main types of observational polarimeter data exist. Firstly, those with high spatial and temporal resolution (both are required simultaneously due to the everchanging seeing) and one or two spatial dimensions, but with only very limited spectral information (magnetograph type, or spectra in one, or at the most a few

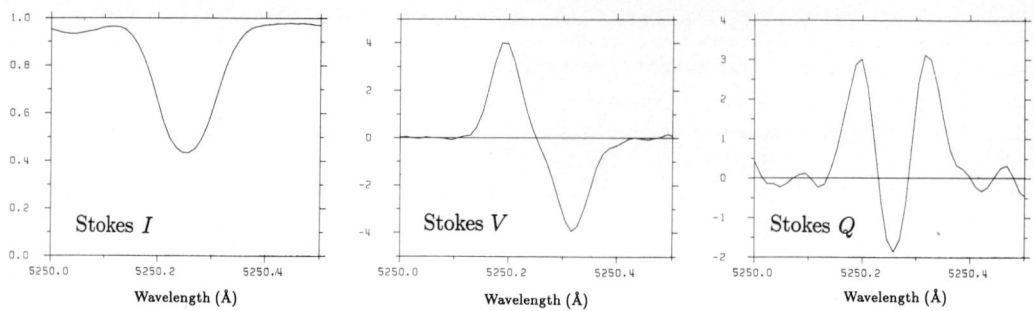

Fig. 1: Stokes I, V, and Q profiles of Fe I 5250.2 Å observed with an FTS in an active region plage at $\mu = 0.28$.

TABLE 1

	High spatial resolution in one spectral line	*Low spatial resolution FTS spectrum*
Internal horizontal variation of fluxtube parameters	NO (later ?)	PERHAPS[1]
Height variation of fluxtube properties	PERHAPS[2]	YES
Range of fluxtube properties	YES (between individual fluxtubes)	YES (between regions of varying α)[3]
Interaction of fluxtubes with their surroundings	YES (direct)	YES (bisectors)
Fluxtube diameters	YES ?	PERHAPS (indirect)[4]
Evolution of fluxtubes (lifetimes)	YES ?	NO
Waves, oscillations etc.	5-minutes: YES Rest: PERHAPS	YES (line widths)
Geometry	YES (model dependent)	Average: YES Individual: NO
User friendliness for quantitative interpretation	Modeller's nightmare, but doable[5]	Fodder for Ph.D. students

spectral lines). Secondly, those with high spectral resolution and a broad spectral range but with only moderate spatial and temporal resolution (as obtained for example with an FTS). Of course observations with properties intermediate to these two extremes exist as well.

Table 1 contrasts the two types of data to each other. Note that the table is based on present capabilities, and that future data or analysis procedures may be able to realise things now seemingly impossible. In the following a few remarks to the table are listed.

[1] The observation that the IR line Fe I 15648.5 Å has a Stokes V profile whose σ components are much broader than the complete I profile, may be due to a range of field strengths distributed horizontally across the fluxtube diameter (Stenflo *et al.* 1986b). However, at present this is not

the only possible explanation and much more work is required to decide this point.
[2] It may be possible to obtain information on the Height variation of fluxtube properties from observations at various distances from the limb, or by comparing IR with visible observations, etc.
[3] α is the filling factor.
[4] It may be possible to determine diameters of fluxtubes with low resolution data off disk centre when 2-D models are used, since the line profiles may depend strongly on the fluxtube diameter and angle of inclination (cf. Van Ballegooijen, 1985b).
[5] The work of Brants (1985a, b) is a good example of how such data can be quantitatively interpreted.

As is nicely illustrated by Table 1, we need both high and low resolution data in order to obtain a maximum of information on the fine scale structure of solar magnetic fields. We may also generalize from the table that high spatial resolution methods give information on the distribution, morphology, and evolution of magnetic features, while the low resolution spectra are more useful for determining their internal structure.

For the rest of this review we shall concentrate on data with moderate spatial resolution and on the internal structure of fluxtubes derived from such data. We shall also restrict ourselves to the solar photosphere. Some of the applications and limits of data with high spatial resolution are discussed by Title (1986) and Wiehr (1986). The raw data alone are not very useful, since the information, although present, is usually well concealed. The vital next step is therefore to consider the type of analysis to be carried out with such data. Once more there are basically two approaches. We shall call them the 'few line analysis' (using typically 1–10 spectral lines) and the 'many line analysis' (ideally involving hundreds of different spectral lines). The advantages and disadvantages of these two techniques are summarized in Table 2. From this table it is clear that the two analysis procedures complement each other, and should, if possible, both be used to make the maximum out of a given data set.

3. MAGNETIC FIELD

It is now observationally well established that the magnetic field in fluxtubes has a value of 1–2 kG at the level at which photospheric spectral lines are formed. First indications of a concentration of magnetic fields into smaller structures were found by Sheeley (1966, 1967) who observed fields ranging from 200 to 700 G in both active and quiet regions by direct high spatial resolution measurements of Stokes V. Using the same technique, Beckers and Schröter (1968a) found field strengths of the order of 600–1400 G in active regions. With the help of the line ratio of Fe I 5250.2 Å to Fe I 5232.9 Å, Howard and Stenflo (1972) and Frazier and Stenflo (1972) showed that over 90% of the magnetic flux was concentrated into strong fields, although they could not give a specific number for the field strength. Stenflo (1973), from the line ratio of Fe I 5250.2 Å to Fe I 5247.1 Å, discovered that the field strength of the strong field flux was of the order of 1–2 kG everywhere. For a horizontal field distribution in the shape of a Gaussian, the peak value turned out to be close to 2000 G. These observations were confirmed by Harvey and Hall (1975) (see also Harvey, 1977), who determined a field strength of 1200–1700 G directly from the splitting of the $g = 3$ Fe I 15648.5 Å line in the infrared (IR).

Tarbell and Title (1977) also derived fields ranging from 1000 G to 1800 G from Fe I 5250.2 Å, via a single line technique based on the Fourier transform of the line profile (see also Title and Tarbell, 1975 and Tarbell and Title, 1976). The line ratio technique of Stenflo (1973) was extended by Wiehr (1978) to include three spectral lines, and resulted in field strengths of 1500–2200 G for the high excitation lines Fe I 6302.5 Å, 6336.8 Å, and 6408.0 Å. From a statistical analysis of a large number of Fe I lines Solanki and Stenflo (1984) were able to isolate the influence of the Zeeman effect on their widths and depths. They found values of the field in the range 1400–1700 G. By determining the line ratio 5250/5247 for regions of different magnetic flux, Stenflo and Harvey (1985) discovered a weak dependence of the field strength on filling factor, with the field strengths being roughly 800–1000 G in the quiet network and 1100–1200 G in active region plages near disk centre. Finally Stenflo et al. (1986b) showed that the field strength decreases continuously with height from approximately 1400 G at the height at which Fe I 15648.5 Å is formed (somewhere near $\tau = 1$) to approx 1100 G at

TABLE 2
Few Line Analysis:

Advantages	Disadvantages
Coverage of many μ positions and α values. Short integration times $\Rightarrow$ also usable for observations with good spatial and temporal resolution.	The height variation of fluxtube properties is often not determinable.
No Fourier transform spectrometer is required. It is ideal for telescopes at Tenerife and La Palma.	Results are often model dependent. One line parameter may depend on many fluxtube parameters.
Empirical model calculations require little computer time. In particular 2-D models with many lines of sight can be calculated.	Results might depend on the lines chosen, due to blends, or because the chosen lines may be insensitive to the desired fluxtube parameters.

Many Line Analysis:

Advantages	Disadvantages
Detailed dependence on the spectral parameters (line strength, excitation potential, etc.) can be studied and the dependence of fluxtube properties on height determined.	Large data flow and long integration times. Few μ and α values can be sampled. Low spatial and temporal resolution.
Blends are no problem, since they only increase the scatter in the data.	A Fourier transform spectrometer (or an Echelle grating) are required.
With a regression analysis the influence of the different physical quantities can be partially separated. Leads to a certain model independence.	Empirical model calculations eat up immense amounts of computer time and force restrictive assumptions on the modeller.

the height of formation of Fe I 5250.2 Å (in the vicinity of $\tau = 10^{-2}$). However, they could not, in the framework of their simple analysis, give the true geometrical height scale needed for a proper comparison with theoretical models.

Fig. 2 shows the variation of fluxtube cross-section with height, as determined from the IR line, in somewhat unnatural units. $1/\sqrt{B}$ is proportional to fluxtube radius if we assume flux conservation with height and a horizontally constant field within the fluxtube, while $1 - \mu$ is a measure of height in the atmosphere, since lines are in general formed at greater height near the limb.

To obtain more quantitative information, we need more and better model calculations. Much of the work described above is based on calculations using the simple analytical Unno (1956) theory of line formation in a magnetic field, or even more primitive models. We therefore require numerical radiative transfer calculations, for example of the line ratio of 5250/5247 using the best presently available empirical fluxtube models, including velocity broadening of the lines and, wherever known, the inclination of the magnetic field. Finally, we refer the interested reader to the fine reviews by e.g. Harvey (1977) and Stenflo (1978), which contain more detailed discussions of the measurement of solar magnetic fields.

4. TEMPERATURE

A decade after the discovery of a decrease in the line depths of Fe I Stokes I profiles

Fig. 2: Height variation of fluxtube radius, as determined from the centre to limb variation of the field strength B resulting from the splitting of Fe I 15648.5 Å (cf. text for more details).

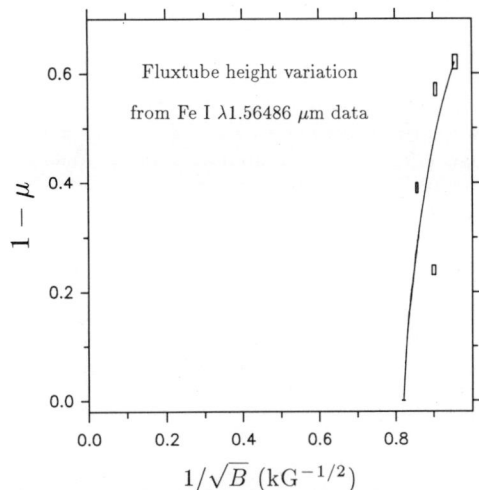

in faculae (so called 'line gaps') by McMath et al. (1956), Sheeley (1967) showed that the magnitude of this line weakening is correlated with the amount of magnetic flux. Chapman and Sheeley (1968) conclusively evinced that it can only be explained by a higher temperature in the magnetic elements. Harvey and Livingston (1969) used the ratio of the Stokes V profiles of Fe I 5250/5233 to determine the 'true' line weakening in fluxtubes, unaffected by the limited spatial resolution. This early work was mainly aimed at establishing the temperature at a single height in the atmosphere. Most of the later effort, as far as fluxtube temperature is concerned, has focussed on models of $T(\tau)$, i.e. the dependence of temperature on optical depth. In the following, a number of these so-called facular models are listed. They can be divided into two main families, the one-component models, which give the average temperature structure of faculae, and two-component models, which differentiate between the fluxtubes and their surroundings, supposing fluxtubes to cover a fraction α of the surface and the non-magnetic surroundings the rest, i.e., $1 - \alpha$. Since α is usually small, the two-component models come much closer to describing the internal temperature of fluxtubes than the single-component ones, and we shall therefore only review the former.

The two-component models can be further subdivided into three main classes according to the type of data they are based on. A first class of models is derived from the centre to limb variation (CLV) of continuum intensity. Examples of such models are those of Rogerson (1961), Chapman (1970), Wilson (1971), Muller (1975), and Hirayama (1978). A second class of models is based mainly on the Stokes I profiles of usually a few spectral lines. Sometimes these models have also additional constraints imposed by continuum data (e.g. the continuum contrast at disk centre). Examples are the models of Chapman (1977, 1979), Koutchmy and Stellmacher (1978), Stellmacher and Wiehr (1979), and Walton (1986). The model of Stellmacher and Wiehr (1979) is actually a three-component model, since it differentiates between granule centres and intergranular lanes in the non-magnetic atmosphere. Walton's models include the effects of fluxtube expansion and the possible presence of more than one fluxtube along the line of sight when observing away from disk centre. Finally, there are the models of Stenflo (1975) and Solanki (1984, 1986), which are based on an analysis of Stokes V profiles.

In Fig. 3 we plot $T(\tau)$ of some of the models listed above. As a glance at the figure shows, there appears to be no consensus among the empirical modellers regarding the fluxtube temperature. The differences in $T(\tau)$ are mainly due to the differences in data that the various models are based on. The latest additions to this series of models are shown in Fig. 4 (Solanki, 1986). These models are based on the Stokes V profiles of a large number of Fe I and II lines. They are also compatible with the disk centre continuum contrast results of Muller and Keil (1983) for the network, and Frazier and Stenflo (1978) for active plages.

As visible from the figure, different models have been derived for the fluxtubes in active region plages

Fig. 3: Temperature T vs. optical depth at 5000 Å, τ_{5000}, of the quiet sun model HSRA (Gingerich et al. 1971) and of a number of fluxtube models. HSRA: solid line, Stenflo (1975): -----, Hirayama (1978) model Z: ·····, Chapman (1979): — —, Stellmacher and Wiehr (1979): — · —, Solanki (1984) model 6K: — ·· —, model 6P: — ··· —.

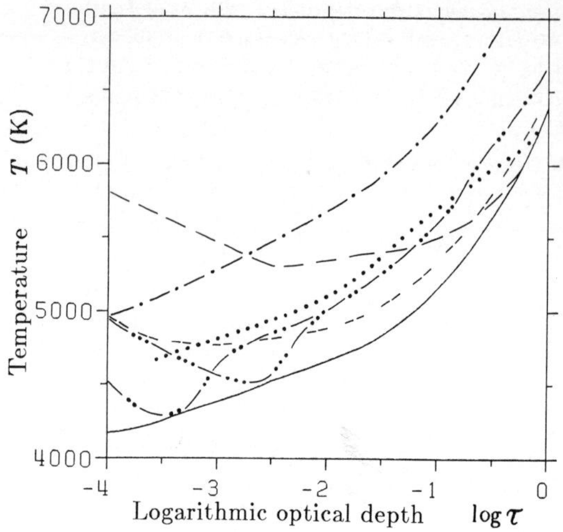

Fig. 4: T vs. τ_{5000} of the HSRA and a model each for network and plage fluxtubes, as derived from Stokes I and V data obtained with an FTS.

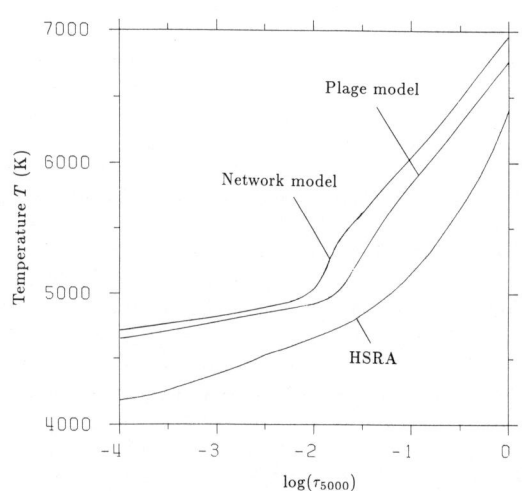

and in the quiet network, with the network model being hotter in its lower layers. This difference in temperature structure is suggested by our FTS data, which, however, only covers a few regions on the sun. A much larger sample of regions with different α is therefore required to test whether this difference in temperature is a universal property or not.

Each model depends critically on the data it is based on. As far as determining the internal structure of fluxtubes is concerned, both continuum contrast and Stokes I data have the disadvantage that the filling factor is a completely free parameter, and the empirically determined temperature structure depends critically on the value assumed for it, as has been shown by Walton (1986). In this respect the models based on Stokes V have a distinct advantage. However, in every model calculation assumptions, of often quite a drastic nature, have to be made. All empirical models so far assume LTE, most are only one-dimensional as well. So far only the models of Solanki (1986) take the quite sizable velocity broadening of the spectral lines inside the fluxtubes into account. An additional limitation is that the data the models are based on often cover only a limited height range, so that the temperature over some portion of τ only reflects the taste of the modeller. We can therefore safely conclude that the real temperature structure in fluxtubes may differ considerably from the temperature of the present generation of models in some layers.

Fig. 5: Downflow velocity, determined from the wavelength shift of Stokes V relative to I, as a function of height in the non-magnetic atmosphere (from Giovanelli and Slaughter, 1978). The upper curve represents the original data, the lower curve is obtained after subtracting the Stokes I blueshift due to granulation.

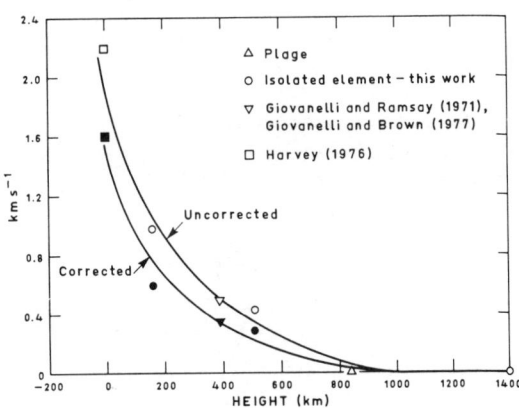

5. DOWNFLOWS

A correlation between magnetic fields and redshifts in Stokes I has been observed by many investigators (mostly with a Babcock type magnetograph used to measure the shift in the line wings) and has generally been interpreted as representing a downflow in the magnetic elements. A few random examples of such investigations are: Beckers and Schröter (1968a, b), Frazier (1970), Howard (1971), Simon and Zirker (1974), Skumanich et al. (1975), Tarbell and Title (1977), and Frazier and Stenflo (1978). The last mentioned authors suggest, via an indirect argument, that the downflows may be concentrated outside the magnetic elements themselves. Recent observations of full Stokes I line profiles in network and active regions by Miller et al. (1984) and Cavallini et al. (1984, 1986a, b), find no evidence for true downflows greater than 0.1 km sec^{-1}.

There have also been a number of measurements of the shifts of Stokes V, with varying results. Such investigations can be roughly divided into three camps. Those which show downflows, those which rule out downflows, and those which are inconclusive.

The observations of Giovanelli and Ramsey (1971) with the line centre magnetogram technique (LCM, which relies on the fact that for an antisymmetric V profile the signal from a single-slit magnetograph disappears when the slit is centred at the zero-crossing wavelength of the line), of Harvey (1977) of the zero-crossing wavelength of the IR line at 15648.5 Å, Giovanelli and Brown (1977) and Giovanelli and Slaughter (1978), both with the LCM technique, and most recently of Wiehr (1985) of the full line profile of Fe I 8468.4 Å show zero-crossing shifts of photospheric lines ranging from approximately 0.5 km sec^{-1} to over 2 km sec^{-1} with respect to Stokes I.

Fig. 5 shows the height dependence of the downflow velocity as derived from different lines by Giovanelli and Slaughter (1978). Of interest is the lower curve, which is obtained after correcting for the blueshift of Stokes I induced by the correlation of brightness and velocity in the granulation (Dravins et al. 1981, 1986). It is very difficult to explain the increase of the velocity with depth theoretically. From mass-conservation one would expect a decrease in velocity with depth due to the exponential increase in density, even for fluxtubes which flare out rapidly with height. As has been pointed out by Hasan and Schüssler (1985) and Schüssler (1986a), inflow of matter near the temperature minimum cannot explain these observations.

On the other side of the fence we have Stenflo and Harvey (1985), who have observed the V profiles of 5250 and 5247 for regions with varying α, Brants (1985b), who has analysed high spatial resolution spectra of Fe I 6302.5 Å, Solanki (1986) who has determined the shifts of hundreds of Fe I and II Stokes V profiles, and Stenflo et al. (1986a), who have studied the CLV of four lines near 5250 Å. These authors find no evidence for downflows greater than 0.2 km sec^{-1} inside magnetic fluxtubes from the difference between V and I wavelengths, after correcting for the Stokes I blueshift. In addition, Solanki (1986) has determined the Stokes V wavelength shifts relative to a laboratory wavelength scale. In Fig. 6 $v_V = c(\lambda_V - \lambda_{\text{lab}})/\lambda_{\text{lab}}$, the absolute shift of Stokes V (corrected for the different components of the motion between the observer and the observed solar region), is plotted

Fig. 6: Downflow velocity, v_V, as determined from the absolute wavelength shift of Stokes V vs. line strength, S_I, defined as the area of the lower half of the line. Plotted are the smoothed mean curves of the Fe I lines observed in two active region plages and in two network elements. The arrows denote the v_V values of the Mg I b lines at 5172.7 Å and 5183.6 Å in two of the regions.

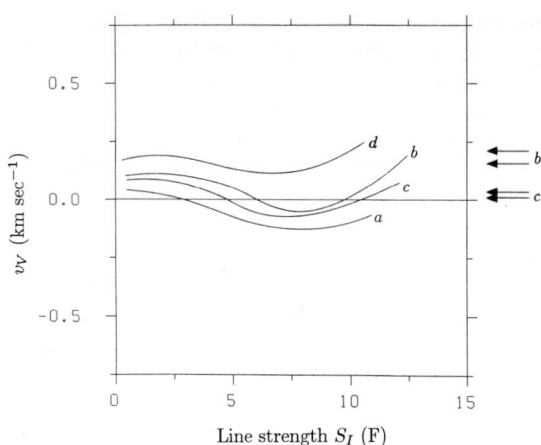

against the line strength, S_I, defined as the area of the lower half of the spectral line. Each of the curves in the figure represents the smoothed mean of the v_V values of the 150–200 Fe I lines in the spectrum of a particular region (marked by the letters a, b, c, and d respectively). The absolute accuracy of v_V is approximately 0.25 km sec^{-1}. Also indicated by arrows to the right of the figure are the v_V values of the Mg Ib lines at 5172.7 Å and 5183.6 Å for two regions. These observations are easily consistent with the sometimes quite large flows observed in the transition zone above photospheric active regions (10–20 km sec^{-1}, e.g. Dere et al., 1981; Feldman et al., 1982), even in the presence of magnetic canopies, whose presence has been reported by Giovanelli (1980), Giovanelli and Jones (1982), and Jones and Giovanelli (1983). See also the review by Jones (1985).

Finally, in the third camp, Scholier and Wiehr (1985) find that the Fe I 6301.5 and 6302.5 Å V profiles show sizeable redshifts in three regions and a blueshift in only one region. However, according to Pahlke and Wiehr (1986) the analysis of an additional seven regions results in an average shift of almost exactly zero relative to Stokes I for both lines. The other measurement which is inconclusive regarding the presence of downflows is the CLV of the zero-crossing of the IR-line at 15648.5 Å, which is redshifted by approximately 1 km sec^{-1} with regard to Stokes I (Stenflo et al. 1986b). Due to the absence of measurements of the absolute Stokes I wavelength it has not been possible to determine whether this redshift is due to a downflow in the fluxtube or if it is simply an artifact caused by a large convective blueshift of Stokes I. That the second explanation may be correct is suggested by the fact that the convective blueshift of Stokes I increases rapidly with increasing depth of formation of the line (Balthasar, 1985), and the IR line is formed at a level where the blueshift may approach 1 km sec^{-1}. However, measurements of the absolute shifts of either Stokes I or V in the IR region around 1.6μ are of critical importance to settle this question.

How can the differences between all these observations be reconciled? The interpretation proposed by Miller et al. (1984) for the Stokes I profiles is that no real downflows are present but rather the strength of the convection is reduced by the magnetic field. This leads to a decrease in the curvature of the line profile, so that measurements in the line wings (e.g. with a Babcock type arrangement) will lead to an apparant downflow. This interpretation is supported by the analysis of Cavallini et al. (1986a,b), who are able to reproduce the shapes and wavelengths, measured in an active region, of the line bisectors of three lines around 6300 Å by assuming no downflows in both the fluxtubes and their surroundings. They only assume an inhibition of convection in the surroundings. They are not able to reproduce their observations if they assume a downflow inside the magnetic elements.

The discrepancies between the observed Stokes V zero-crossing wavelength shifts may possibly be due to differences in the *spectral* resolution of the various measurements. A limited spectral resolution, when combined with the asymmetry of Stokes V (blue amplitude and area larger than red amplitude

Fig. 7: Stokes V profile of Fe I 5250.2 Å for different amounts of spectral smearing. The highest and narrowest profile is taken from the original FTS data. The other profiles have been convoluted with increasingly broader Gaussians representing instrumental smearing. The broadest Gaussian has $v = 150$ mÅ, where v is the e-folding width ('Doppler' width).

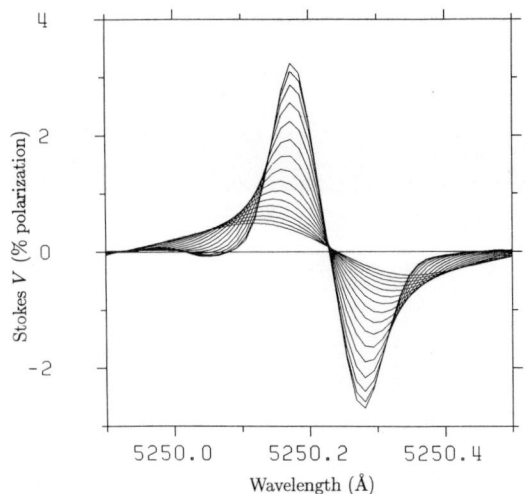

and area, cf. Sect. 7) gives rise to a fictitious redshift, as has been shown by Solanki and Stenflo (1986).
A simulation of how spectral smearing affects Stokes V is illustrated in Fig. 7, where the Stokes V profile of Fe I 5250.2 Å, observed in an active region plage with very high spectral resolution (the highest and narrowest profile in the figure), is plotted along with the same profile smeared with a Gaussian apparatus function having a 'Doppler width' $v = 10, 20, 30, \ldots, 150$ mÅ. Some of the effects of the smearing are clearly visible from the figure. The amplitude and area of each wing decreases, but also the zero-crossing wavelength shifts towards the red. Solanki and Stenflo (1986) were able to reproduce the magnitudes of the Stokes V shifts observed by various authors with the help of this effect, if they used the appropriate instrumental parameters.

6. NON-STATIONARY VELOCITIES

Although there are strong theoretical grounds for expecting non-stationary velocities to be present in fluxtubes, only a limited observational base for their presence exists at present. Tanenbaum et al. (1971) saw an oscillation of five minute period in the magnetic flux from an active region, when observing in the wings of Fe I 5250.2 Å with a Babcock type magnetograph. However, oscillations in velocity, temperature, or magnetic field may all contribute to the observed effect, so that the diagnostic potential of such an observation alone is quite limited. Using time series measurements with the LCM technique (cf. Sect. 5), Giovanelli et al. (1978) showed that 5-minute oscillations in the velocity are present in Stokes V with an amplitude of approximately 0.25 km sec^{-1} in photospheric lines. Wiehr (1985) also observed oscillations of a similar period and amplitude, in addition to some time dependent shifts and changes in the amplitude of Stokes V whose origin is not clear.

The widths of the I profiles inside the fluxtubes can also give information on the amplitudes of non-stationary motions. (A spatial resolution independent approximation of the I profile inside fluxtubes can be obtained by integrating Stokes V. See Solanki and Stenflo, 1984, 1985 for more details). Their analysis has yielded that unresolved motions with large amplitudes are present in fluxtubes (Solanki, 1986). However, in this manner, no direct information is obtainable on whether such motions are due to, e.g., oscillations, waves, or a mixture of up and downflows in different fluxtubes. Nor can anything be said on the frequencies or phases of such motions.

Fig. 8 shows the approximate velocity amplitudes derived by assuming macroturbulence to be the main broadening mechanism. It should, however, be noted that this is simply a convenient way of determining the velocity amplitude and does not imply the presence of truly turbulent motions

Fig. 8: ξ^V_{mac}, the macroturbulence velocity inside fluxtubes, as derived from the integrated V profile vs. S_I. Microturbulence $\xi_{\text{mic}} = 0$ is assumed. The curves represent both network and plage data. The ξ^V_{mac} values derived from Fe I lines with $\chi_e < 3$ eV are denoted by the solid curve, the ξ^V_{mac} values derived from Fe I, $\chi_e \geq 3$ eV lines by the dashed curve, and ξ^V_{mac} from Fe II by the dot-dashed curve. The Fe II curve is dotted between $S_I = 5$ and 9 F to indicate that it is interpolated in that region.

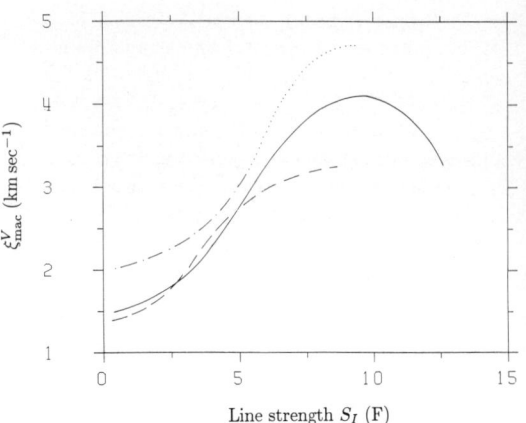

Fig. 9: $\xi^V_{\text{mac}} - \xi^I_{\text{mac}}$ vs. S_I, i.e. the macroturbulence excess in fluxtubes (from Stokes V data), as compared to the quiet sun (from Stokes I data). Symbols are the same as in Fig. 8.

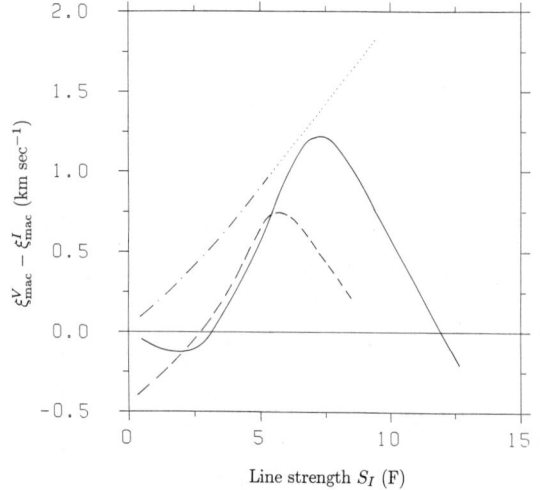

inside fluxtubes. ξ^V_{mac} is the 'Doppler' width of the Gaussian macroturbulence distribution, S_I is the line strength. The solid curve represents Fe I lines with excitation potential $\chi_e < 3$ eV, the dashed curve Fe I lines with $\chi_e \geq 3$ eV and the dot-dashed curve Fe II lines (the dotted part of this curve indicates that no suitable Fe II lines are available in the given S_I range). Maximum ξ^V_{mac} values of 4–5 km sec^{-1} are derived. Note that the rms velocity is a factor of $\sqrt{2}$ smaller than ξ^V_{mac}.

In Fig. 9, the difference, $\xi^V_{\text{mac}} - \xi^I_{\text{mac}}$, between the velocity amplitudes derived from Stokes V (fluxtubes) and Stokes I (observed in the quiet sun), is plotted vs. S_I. It is clear that the velocity amplitudes are in general larger for the fluxtubes than for the quiet sun.

The obvious question is, why do the line widths give such large velocity amplitudes while direct observations of time dependent Stokes V shifts give at best only equivocal evidence for any motions besides low amplitude five-minute oscillations. One possible answer may be that so far such time dependent measurements have not been able to isolate individual fluxtubes. Consequently only the global oscillations in which all the fluxtubes oscillate in phase are observed. Non-stationary motions excited locally in any one fluxtube (e.g. via the mechanism proposed by Hasan, 1984, 1985; or Venkatakrishnan, 1986), will most probably not be in phase with such motions in another fluxtube.

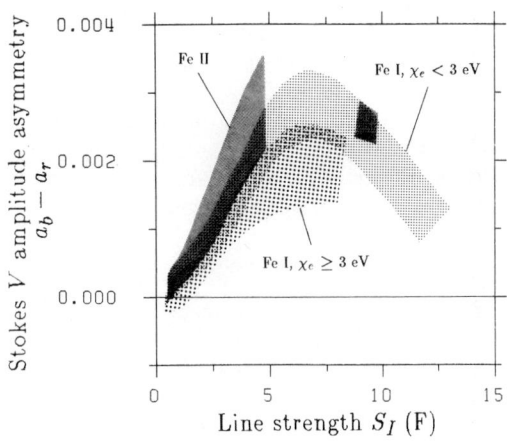

Fig. 10: $a_b - a_r$, the absolute amplitude asymmetry of Stokes V vs. S_I for an enhanced network region. a_b and a_r are the amplitudes of the blue and red wings of Stokes V in units of the adjacent continuum. The differently shaded portions of the figure represent Fe I and II lines of different excitation potential, as marked.

7. STOKES V ASYMMETRY

The Stokes V profile in both active region plages and the quiet network is now known to be asymmetric, with the amplitude and area of the blue wing being larger than the amplitude and area of the red wing. This asymmetry has been observed and analyzed by Stenflo et al. (1984), Solanki and Stenflo (1984, 1985), Stenflo and Harvey (1985), Wiehr (1985), Scholier and Wiehr (1985) and Stenflo et al. (1986a, 1986b). An example of such observations is shown in Fig. 10, where the amplitude asymmetry, $a_b - a_r$, of Fe I and II lines is plotted vs. S_I for an enhanced network region. The lines are grouped according to excitation potential and marked in the figure.

A number of mechanisms have been proposed to produce such asymmetries. Most of these proposals were originally intended to explain the broad-band circular polarization in sunspots observed by Illing et al. (1974a, b, 1975) and by Kemp and Henson (1983). However, they can also be applied to the observations outside sunspots. The first proposal was made by Illing et al. (1975), who suggested a combination of velocity and magnetic field gradient along the line of sight as the cause. Auer and Heasley (1978) noticed that if the angle between the line of sight and the magnetic field is not zero, then a velocity gradient alone is sufficient. However, a sizeable asymmetry is observed near disk centre, and since fluxtubes are thought to be almost vertical due to buoyancy (Schüssler, 1986a), one would expect this mechanism to play only a limited role, at least near disk centre. Pahlke and Solanki (1986) have shown with the help of model calculations that stationary downflows which reproduce the observed Stokes V asymmetry invariably also result in redshifts greater than approximately 1 km sec^{-1}, contradicting the newer observations of zero-crossing shift in fluxtubes. A steady flow therefore appears to be ruled out as the cause of such asymmetries. A somewhat modified version of the original proposal of Illing et al. (1975) was made by Solanki and Stenflo (1984), who suggested that instead of a steady flow an oscillation in velocity (with a height gradient) and a correlation with an oscillation in temperature and/or magnetic field is responsible. This proposal has the advantage that it need not necessarily give rise to large zero-crossing shifts like the steady flow gradient. On the other hand, it has not been worked out in detail yet.

A totally different approach was taken by Kemp et al. (1984) and Landi Degl'Innocenti (1985), who proposed that due to the anisotropy of the radiation field in a fluxtube, the different Zeeman sublevels are not equally populated, leading to a Stokes V asymmetry. This process requires that the lines be formed in NLTE and its description is quite involved, being strongly dependent on the details of the atomic transition.

Some indirect evidence exists that Stokes V asymmetry is closely related to the velocity structure inside fluxtubes (Solanki, 1985, 1986). This can be easily seen by comparing Fig. 8 with Fig. 10.

The marked similarity between the two figures suggests some relationship, and indeed the correlation coefficient between ξ_{mac}^V and $a_b - a_r$ is found to be of the order of 0.85. It therefore appears likely that velocity gradients are in some way responsible for Stokes V asymmetry, although it is by no means certain.

Let us summarize our present knowledge of velocities in fluxtubes. We expect no sizeable downflows, and in general only small amplitude 5-minute oscillations. However, motions with large amplitudes and probably a vertical velocity gradient are present in the fluxtubes or their immediate surroundings. The exact nature of such motions is unknown. Candidates are: stationary up- and downflows in different fluxtubes, so that on the average little vertical mass transfer takes place. Oscillations or waves in fluxtubes are another possibility, whereby these motions are not in phase in different fluxtubes. Finally, we should not forget that due to the geometry of the fluxtubes (expansion with height), it is possible that motions in their immediate surroundings may also affect Stokes V.

8. FUTURE CONSIDERATIONS

After this brief and incomplete review of the investigations carried out to data, there now follows an attempt at a very subjective preview of some of the work which needs to be done or is in the process of being carried out at the moment. Four main avenues come to mind.

Firstly, more data, both of high and of moderate spatial resolution are required in order to resolve a number of problems. Some examples of open questions which can be addressed with moderate spatial resolution data are listed in the following. i) Very little is known of the dependence of most of the fluxtube properties on filling factor α, the position on the solar disk, the age of an active region, the distance from a sunspot, etc.. A large sample of observations will be required to be able to determine these dependences with any measure of certainty. ii) The properties of both the deeper and the higher layers of fluxtubes are known only in their rudiments. This problem requires polarimetric data in chromospheric lines and in the IR near 1.6μ. iii) The data available to date are usually limited to Stokes I and V, but complete profiles of Stokes Q and U will also be required in future if we want to determine the geometry of the field and resolve the question of a possible inclination of the fluxtubes.

Additional data alone may not be sufficient to answer many of the presently open questions. Better diagnostic techniques are required as well. This is the second avenue which must be followed in future investigations. Although we already possess an impressive array of methods for obtaining information from polarized spectra, much of it is still hidden from us. In particular Stokes Q and U are almost virgin territory as far as fluxtube diagnostics are concerned (some exceptions are to be found in Hagyard, 1985). Besides being absolutely necessary for determining the direction of the field, Stokes Q and U are also capable of serving as diagnostics for the field strength, temperature, and perhaps even velocity inside the fluxtubes. In addition to new techniques, long established methods like the line ratio technique of Stenflo (1973) need to be studied further. In order to improve the diagnostics we also need to know the heights of formation of the spectral lines in fluxtubes. This may be done either with the method of Van Ballegooijen (1985a), or of Wittmann (1973, 1974). Such work is in progress.

So far all NLTE effects have been neglected in empirical models of fluxtubes, but in the long run they will have to be included. A few investigators have attempted to assess the influence of departures from LTE on line profiles in magnetic fluxtubes. Rees (1969) and Domke and Staude (1973) studied the general influence of NLTE on the Stokes profiles in a magnetic field. Later Stenholm and Stenflo (1977, 1978) and Owocki and Auer (1980) concentrated on the effects of 2-D radiative transfer in non-plane parallel geometry, as typifies fluxtubes. They found that 2-D effects can give rise to differences from 1-D profiles. However, Owocki and Auer (1980) showed that 1.5-D radiative transfer (i.e. radiative transfer along many parallel lines of sight, the profiles from all off which are summed to give the resultant) is often a sufficiently good approximation. Finally, Solanki and Steenbock (1987) use empirical models of the fluxtube temperature and a realistic iron model atom to study departures from LTE and their influence on the empirically determined temperature and velocity.

Finally, future *empirical* models must combine a 2-D MHD model of a fluxtube (without energy equation, e.g. thin tube model, the expansion model of Pneuman et al., 1986, or the exact solution of Steiner et al., 1986) with radiative transfer along many lines of sight (1.5-D). Specially away from disk centre, the effects of limited fluxtube diameter should play a major role. 1.5-D calculations of Stokes I, in conjunction with fluxtube models of varying sophistication, have been carried out by e.g. Chapman (1970), Caccin and Severino (1979), Owocki and Auer (1980), Chapman and Gingell (1984), Deinzer et al. (1983, 1984b), and Walton (1986). Only Van Ballegooijen (1985a) has studied the influence of the fluxtube geometry (including expansion, and finite diameter) on all four Stokes profiles. An additional investigation, which compares the Stokes profiles resulting in plane parallel, slab and cylindrical geometry with each other, is in preparation.

However, with increasing sophistication of empirical models, it will become increasingly time consuming to test a sufficiently broad range of parameters. The interpretation of the results will also become increasingly involved. Therefore, radiative transfer calculations in plane parallel atmospheres will still be important when a first idea of the diagnostic contents of the data are required, to separate influences of different physical parameters on the line profiles, or simply when the profiles of a large number of lines have to be calculated. We therefore have a similar situation with regard to empirical models as exists for theoretical models (as has been discussed by Schüssler, 1986a). Simple calculations serve to map the terrain, show which effects are particularly important, and advance the basic understanding of the processes involved. They thus simplify the task of constructing comprehensive and realistic models, which must reproduce a maximum amount of data.

An alternative, and in the future possibly quite promising approach, is to calculate the line profiles resulting from the most comprehensive, self-consistent theoretical models, including an energy equation (e.g. the models of Spruit, 1976; Deinzer et al. 1984a, b; Knölker et al. 1985, 1986, Nordlund, 1985), and compare with the observations directly. This approach becomes increasingly feasible as the theoretical models continue to improve. See the review by Schüssler (1986b) for comparisons between such models and the observations.

We are now at an exciting stage in the investigation of fluxtubes. Although some progress has been made in recent years, many promising possibilities beckon the enterprising observer, and we can look forward to new discoveries and a better understanding of these structures in the near future.

Acknowledgements : I wish to thank the Kiepenheuer Institute, and in particular Prof. E.H. Schröter for inviting me to give this talk. Many people have directly or indirectly contributed to the contents, foremost among them Prof. J.O. Stenflo, who has also critically read the manuscript. Their help is gratefully acknowledged. A part of the work presented here has been supported by the Swiss National Science Foundation grants Nos. 2.814-0.83 and 2.666-0.85.

REFERENCES

Auer, L.H., Heasley, J.N.: 1978, *Astron. Astrophys.* **64**, 67
Balthasar, H.: 1985, *Solar Phys.* **99**, 31
Beckers, J.M., Schröter, E.H.: 1968a, *Solar Phys.* **4**, 142
Beckers, J.M., Schröter, E.H.: 1968b, *Solar Phys.* **4**, 165
Brants, J.J.: 1985a, *Solar Phys.* **95**, 15
Brants, J.J.: 1985b, *Solar Phys.* **98**, 197
Caccin, B., Severino, G.: 1979, *Astrophys. J.* **232**, 297
Cavallini, F., Ceppatelli, G., Righini, A.: 1985, *Astron. Astrophys.* **143**, 116
Cavallini, F., Ceppatelli, G., Righini, A.: 1986a, *Astron. Astrophys.* in press
Cavallini, F., Ceppatelli, G., Righini, A.: 1986b, These proceedings
Chapman, G.A.: 1970, *Solar Phys.* **14**, 315
Chapman, G.A.: 1977, *Astrophys. J. Suppl. Ser.* **33**, 35
Chapman, G.A.: 1979, *Astrophys. J.* **232**, 923
Chapman, G.A., Gingell, T.A.: 1984, *Solar Phys.* **91**, 243
Chapman, G.A., Sheeley, Jr., N.R.: 1968, *Solar Phys.* **5**, 442
Deinzer, W., Hensler, G., Schmitt, D., Schüssler, M., Weisshaar, E.: 1983, in *Solar and Stellar Magnetic Fields: Origins and Coronal Effects*, J.O. Stenflo (Ed.), IAU Symp. **102**,

p. 67
Deinzer, W., Hensler, G., Schüssler, M., Weisshaar, E.: 1984a, *Astron. Astrophys.* **139**, 426
Deinzer, W., Hensler, G., Schüssler, M., Weisshaar, E.: 1984b, *Astron. Astrophys.* **139**, 435
Domke, H., Staude, J.: 1973, *Solar Phys.* **31**, 291
Dravins, D., Larsson, B., Nordlund, Å.: 1986, *Astron. Astrophys.* **158**, 83
Dravins, D., Lindegren, L., Nordlund, Å.: 1981, *Astron. Astrophys.* **96**, 345
Frazier, E.N.: 1970, *Solar Phys.* **14**, 89
Frazier, E.N., Stenflo, J.O.: 1972, *Solar Phys.* **27**, 330
Frazier, E.N., Stenflo, J.O.: 1978, *Astron. Astrophys.* **70**, 789
Gingerich, O., Noyes, R.W., Kalkofen, W., Cuny, Y.: 1971, *Solar Phys.* **18**, 347
Giovanelli, R.G.: 1980, *Solar Phys.* **68**, 49
Giovanelli, R.G., Brown, N.: 1977, *Solar Phys.* **52**, 27
Giovanelli, R.G., Jones, H.P.: 1982, *Solar Phys.* **79**, 267
Giovanelli, R.G., Livingston, W.C., Harvey, J.W.: 1978, *Solar Phys.* **59**, 49
Giovanelli, R.G., Ramsay, J.V.: 1971, in *Solar Magnetic Fields*, R. Howard (Ed.), IAU *Symp.* **43**, p. 293
Giovanelli, R.G., Slaughter, C.: 1978, *Solar Phys.* **57**, 255
Hagyard M.J.: 1985, (Ed.), *Measurements of Solar Vector Magnetic Fields*, NASA Conf. Publ. 2374, p. 322
Harvey, J.W: 1977, in *Highlights of Astronomy*, E.A. Müller (Ed.), Vol. **4**, Part **II**, p. 223
Harvey, J.W., Hall, D.: 1975, *Bull. Amer. Astron. Soc.* **7**, 459
Harvey, J.W., Livingston, W.: 1969, *Solar Phys.* **10**, 283
Hasan, S.S.: 1984, *Astrophys. J.* **285**, 851
Hasan, S.S.: 1985, *Astron. Astrophys.* **143**, 39
Hasan, S.S., Schüssler, M.: 1985, *Astron. Astrophys.* **151**, 69
Hirayama, T.: 1978, *Publ. Astron. Soc. Japan* **30**, 337
Howard, R.: 1971, *Solar Phys.* **16**, 21
Howard, R., Stenflo, J.O.: 1972, *Solar Phys.* **22**, 402
Illing, R.M.E., Landman, D.A., Mickey, D.L.: 1974a, *Astron. Astrophys.* **35**, 327
Illing, R.M.E., Landman, D.A., Mickey, D.L.: 1974b, *Astron. Astrophys.* **37**, 97
Illing, R.M.E., Landman, D.A., Mickey, D.L.: 1975, *Astron. Astrophys.* **41**, 183
Jones, H.P.: 1985, in *Chromospheric Diagnostics and Modelling*, B.W. Lites (Ed.), National Solar Obs., Sacramento Peak, NM, p. 175
Jones, H.P., Giovanelli, R.G.: 1983, *Solar Phys.* **87**, 37
Kemp, J.C., Henson, G.D.: 1983, *Astrophys. J.* **266**, L69
Kemp, J.C., Macek, J.H., Nehring, F.W.: 1984, *Astrophys. J.* **278**, 863
Koutchmy, S., Stellmacher, G.: 1978, *Astron. Astrophys.* **67**, 93
Knölker, M., Schüssler, M., Weisshaar E.: 1985, in *Theoretical Problems in High Resolution Solar Physics*, H.U. Schmidt (Ed.), MPA, Munich, p. 195
Knölker, M., Schüssler, M., Weisshaar E.: 1986, These proceedings
Landi Degl'Innocenti E.: 1985, in *Theoretical Problems in High Resolution Solar Physics*, H.U. Schmidt (Ed.), MPA, Munich, p. 162
Leroy, J.-L.: 1962, *Ann. Astrophys.* **25**, 127
McMath, R.R., Mohler, O.C., Pierce, A.K., Goldberg, L.: 1956, *Astrophys. J.* **124**, 1
Miller, P., Foukal, P., Keil, S.: 1984, *Solar Phys.* **92**, 33
Muller, R.: 1975, *Solar Phys.* **45**, 105
Muller, R., Keil, S.L.: 1983, *Solar Phys.* **87**, 243
Nordlund, Å.: 1985, in *Theoretical Problems in High Resolution Solar Physics*, H.U. Schmidt (Ed.), MPA, Munich, p. 101
Owocki, S.P., Auer, L.H.: 1980, *Astrophys. J.* **241**, 448
Pahlke, K.D., Solanki, S.K.: 1986, *Mitt. Astron. Gesellschaft* **65**, 162
Pneuman, G.W., Solanki, S.K., Stenflo, J.O.: 1986, *Astron. Astrophys.* **154**, 231
Ramsey, H.E., Schoolman, S.A., Title, A.M.: 1977, *Astrophys. J.* **215**, L41

Rees, D.E.: 1969, *Solar Phys.* **10**, 268
Rogerson, J.B.: 1961, *Astrophys. J.* **134**, 331
Scholier, W., Wiehr, E.: 1985, *Solar Phys.* **99**, 349
Schüssler, M.: 1986a, in *Proc. Workshop on Small Magnetic Flux Concentrations in the Solar Photosphere*, Göttingen, Oct. 1–3, 1985, in press
Schüssler, M.: 1986b, These proceedings
Sheeley, Jr., N.R.: 1966, *Astrophys. J.* **144**, 723
Sheeley, Jr., N.R.: 1967, *Solar Phys.* **1**, 171
Simon, G.W., Zirker, J.B.: 1974, *Solar Phys.* **35**, 331
Skumanich, A., Smythe, C., Frazier, E.N.: 1975, *Astrophys. J.* **200**, 747
Solanki, S.K.: 1984, in *The Hydromagnetics of the Sun*, T.D. Guyenne and J.J. Hunt (Eds.), ESA SP-220, p. 63
Solanki, S.K.: 1985, in *Theoretical Problems in High Resolution Solar Physics*, H.U. Schmidt (Ed.), MPA, Munich, p. 172
Solanki, S.K.: 1986, *Astron. Astrophys.* **168**, 311
Solanki, S.K., Steenbock, W.: 1987, *Astron. Astrophys.* to be submitted
Solanki, S.K., Stenflo, J.O.: 1984, *Astron. Astrophys.* **140**, 185
Solanki, S.K., Stenflo, J.O.: 1985, *Astron. Astrophys.* **148**, 123
Solanki, S.K., Stenflo, J.O.: 1986, *Astron. Astrophys.* in press
Spruit, H.C.: 1976, *Solar Phys.* **50**, 269
Steiner, O., Pneuman, G.W., Stenflo, J.O.: 1986, *Astron. Astrophys.* in press
Stellmacher, G., Wiehr, E.: 1979, *Astron. Astrophys.* **75**, 263
Stenflo, J.O.: 1973, *Solar Phys.* **32**, 41
Stenflo, J.O.: 1975, *Solar Phys.* **42**, 79
Stenflo, J.O.: 1978, *Rep. Prog. Phys.* **41**, 865
Stenflo, J.O., Harvey, J.W.: 1985, *Solar Phys.*, **95**, 99
Stenflo, J.O., Harvey, J.W., Brault, J.W., Solanki, S.K.: 1984, *Astron. Astrophys.* **131**, 33
Stenflo, J.O., Solanki, S.K., Harvey, J.W.: 1986a, *Astron. Astrophys.* in press
Stenflo, J.O., Solanki, S.K., Harvey, J.W.: 1986b, *Astron. Astrophys.* in press
Stenholm, L.G., Stenflo, J.O.: 1977, *Astron. Astrophys.* **58**, 273
Stenholm, L.G., Stenflo, J.O.: 1978, *Astron. Astrophys.* **67**, 33
Tanenbaum, A.S., Wilcox, J.M., Howard, R.: 1971, in *Solar Magnetic Fields*, R. Howard (Ed.), IAU Symp. **43**, p. 348
Tarbell, T.D., Title, A.M.: 1976, *Solar Phys.* **47**, 563
Tarbell, T.D., Title, A.M.: 1977, *Solar Phys.* **52**, 13
Title, A.M.: 1986, These proceedings
Title, A.M., Tarbell, T.D.: 1975, *Solar Phys.* **41**, 255
Unno, W.: 1956, *Publ. Astron. Soc. Japan* **8**, 108
Van Ballegooijen, A.A.: 1985a in *Measurements of Solar Vector Magnetic Fields*, M.J. Hagyard (Ed.), NASA Conf. Publ. 2374, p. 322
Van Ballegooijen, A.A.: 1985b, in *Theoretical Problems in High Resolution Solar Physics*, H.U. Schmidt (Ed.), MPA, Munich, p. 167
Venkatakrishnan, P.: 1986, *Solar Phys.* **104**, 347
Walton, S.R.: 1986, *Astrophys. J.* in press
Wiehr, E.: 1978, *Astron. Astrophys.* **69**, 279
Wiehr, E.: 1985, *Astron. Astrophys.* **149**, 217
Wiehr, E.: 1986, These proceedings
Wilson, P.R.: 1971, *Solar Phys.* **21**, 101
Wittmann, A.D.: 1973, *Ph.D. Thesis*, University of Göttingen
Wittmann, A.D.: 1974, *Solar Phys.* **35**, 11

FTS MEASUREMENTS OF SOLAR LINE ASYMMETRIES IN QUIET AND ACTIVE REGIONS

P. N. Brandt
Kiepenheuer-Institut, Schöneckstr. 6, D-7800 Freiburg, F.R.G.

S. K. Solanki
Institut für Astronomie, ETH-Zentrum, CH-8092 Zürich, Switzerland

1. INTRODUCTION

Spectrograms of the solar photosphere almost never show a higher spatial resolution than 0.5 arcsec. Therefore, the properties of solar finestructures, like the granulation, facular points or umbral dots cannot be derived directly with sufficient reliability. Instead, as was demonstrated e.g. by Stenflo et al. (1984), spatially averaged Fourier transform spectra (FTS) with their well known high spectral resolution, their highly symmetric apparatus profile, low scattered light and high S/N ratio can be used as a complementing tool for the diagnosis of atmospheric parameters. The possibility to use many lines measured strictly simultaneously and covering a wide range of excitation potentials, heights of formation etc. represents another advantage of the FTS and at the same time reduces possible errors due to line blending.

The main motivation for the present investigation of plage versus quiet sun line profiles came from the exciting results by Livingston & Holweger (1982) and by Livingston (1983) on the possible cycle dependence of line equivalent widths and asymmetries, as measured in integrated sunlight. It seemed unclear to what extent these effects could be caused by the varying contribution of plages during the solar cycle.

The results shown and discussed in the following represent only the first step in the evaluation of the vast amount of material obtained - and are therefore to be regarded as preliminary.

2. OBSERVATIONS

In the period June 1 to 13, 1984 a series of spectra was observed at the McMath main solar telescope feeding the Fourier transform spectrometer designed and described by Brault (1978). Aided by an improvised slit jaw viewing arrangement (using an Hα Daystar filter) and the daily magnetograms, the entrance slit of the spectrometer, 5 by 25 arcsec^2 wide, was "scanned" through various plage regions, avoiding sunspots and pores, and regions of no perceptible activity. For the present investigation a set of 25 spectra taken at $\cos \theta > 0.9$, including 5 spectra recorded at disk center, was selected.

The spectra cover a usable spectral range from λ5050 to λ6650 Å and show a maximum resolving power of 180 000. Most spectra were integrated for 13.7^m, a few of them twice as long. In this way S/N ratios beween 2000 and 3000 were achieved.

3. DATA EVALUATION

To estimate the magnetic filling factors α the method first described by Stenflo & Lindegren (1977) was applied. In principle it represents a multi-dimensional regression of a line parameter sensitive to Zeeman splitting (e.g. line width or depth) vs. line strength, excitation potential and effective Landé factor g_{eff}; using a large number of lines of the same ion (here: 180 FeI lines), the generally small effects due to the magnetic field can be isolated and thereby some information obtained on α.

A choice of appropriate values for the line weakening, δ, and the magnetic field strength, B, in fluxtubes yields values of α ranging from 0.001 to 0.22 for the 25 spectra analysed (cf. Figure 1), with a **relative** accuracy of approx. ± 0.02. The **absolute** accuracy of the α values depends on the choice of δ and B. Schüssler (1986, this conference) pointed out, that also the continuum intensity of the fluxtube emission has to be considered. This fact and a comparison of the present results with those obtained from Stokes I and V measurements by Stenflo & Harvey (1985) leads us to surmise that our α values may be overestimated by a factor of 1.5 to 2, i.e. that the strongest plage region observed in this set had a filling factor α between 0.11 and 0.16. For the rest of the paper we shall use the originally determined α values, but ask the reader to keep this probable correction factor in mind.

For the present analysis 32 weak to medium-strong lines (mostly FeI) were selected. They comprise 3 FeI lines around λ6300 Å studied extensively by Cavallini et al. (1985,1986), a set of 11 lines in the range λ6265 to λ6232 Å used by Gray (1982) in a study of convective velocities in stars, and 17 FeI lines used by Livingston (1983) and Kaisig & Schröter (1983).

4. RESULTS

4.1 Changes of FWHM, equivalent width and line depth with α

An analysis of the relevant line parameters as a function of the filling factor α increasing from 0 to 0.22 yields the following results:
i) Most lines show an **increase** of the full width at half of the maximum line depth (FWHM) of 2 to 8% (after subtraction of the Zeeman splitting). A broadening of ≃ 5% is found for the three g=0 lines λ 5576, 5434, 5123 Å, one of which is shown in Figure 1.
ii) A **decrease** of the equivalent width, W, is found, ranging from 5 to 10% for weak lines and from 0 to 3% for strong lines, as is shown in Figure 2.
iii) The line depth **decreases** by between 5 and 17%, where the latter value applies to the FeI line λ5250.2 Å.

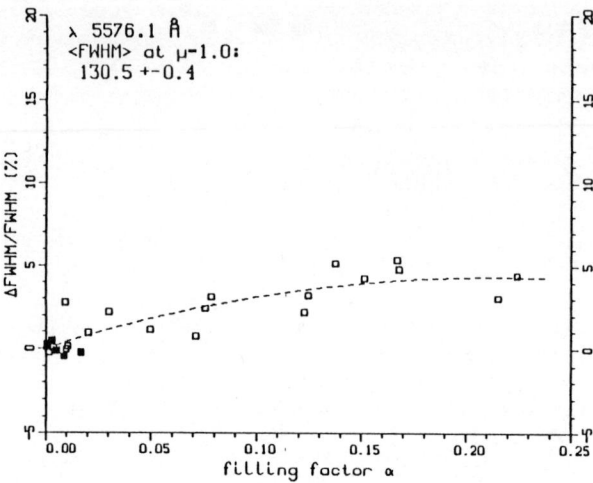

Figure 1: Variation of line full width at half maximum (FWHM) with magnetic filling factor α. The variations are referred to the average $\langle FWHM \rangle$ of the 5 spectra taken at disk center ($\alpha \approx 0.01$, solid squares).

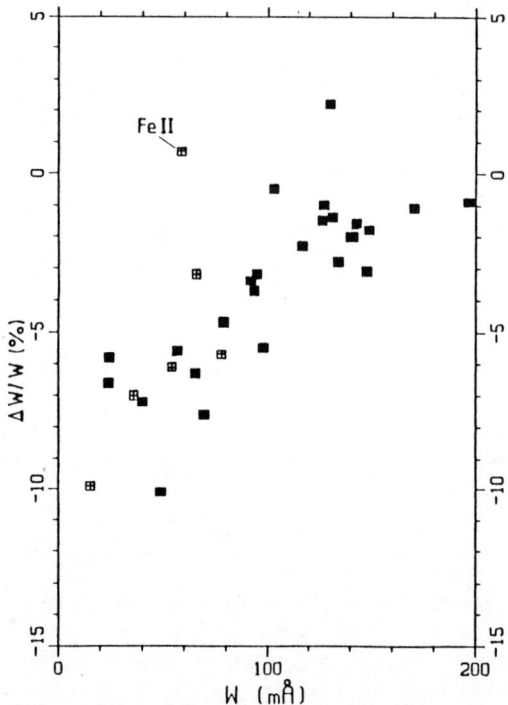

Figure 2: Difference between equivalent widths for filling factor $\alpha \approx 0.22$ and $\alpha \approx 0.0$ as function of the equivalent width. Solid squares refer to FeI lines.

4.2 Changes of the line bisector

Basically two aspects are of interest, when discussing bisector variations: their absolute wavelength position and their shape.

A preliminary analysis of our data indicates that the lowest bisector points coincide within ± 1 mÅ for spectra of small, intermediate and high filling factors.

For the investigation of the **shape** of the bisectors the following definition, characterizing the lower and the upper part, was introduced:
i) the wavelength difference between the average of the lowest three bisector points and the point at relative intensity 0.7 (i.e. $\lambda_{min} - \lambda_{0.7}$) was calculated and denoted ∇;
ii) the wavelength difference between the bisector points at relative intensities 0.7 and 0.9 (i.e. $\lambda_{0.7} - \lambda_{0.9}$) was calculated and denoted Δ.

A typical result is presented in Figure 3, referring to the FeI line $\lambda5250.7$ Å. The rather smooth and consistent decrease of the bisector "curvature" in the lower part ("∇") from 5 to 2 mÅ is clearly shown (dashed line), as well as an increase of the bisector wavelength difference in the line wing ("Δ", dotted line) from -7 to -9.5 mÅ. In Figure 4 the change of the C-shape with α is shown as a function of the equivalent width for all lines investigated. Similar to the definition given above the open symbol "∇" denotes $\lambda_{min} - \lambda_{0.7}$ for $\alpha \approx 0$ and the bar ending in the filled symbol "▼" its variation towards $\alpha \approx 0.22$, whereas the symbol "Δ" represents the value $\lambda_{0.7} - \lambda_{0.9}$ for $\alpha \approx 0$ and the corresponding bar ending at "V" its variation towards $\alpha \approx 0.22$. All lines of W>100 mÅ show the well known "straightening" of the lower part of the bisector, as already seen by Livingston (1982), Kaisig & Schröter (1983), Brandt & Schröter (1984), Cavallini et al. (1985) and others. However, as a novel feature a conspicuous "steepening" of the bisector in its upper part is found: more than one dozen of the line bisectors exhibit an increase of the red-shift in the line wing by 2 to 4 mÅ.

5. DISCUSSION AND CONCLUSION

The results presented here are in agreement with those obtained by Immerschitt & Schröter (1986, this conference), who studied the behaviour of the FeI line $\lambda5576.1$ Å in plage regions of different Ca^+-K-strength.

Two components may be responsible for the modification of the averaged line profiles in active regions: a possibly modified structure of the convection pattern around the fluxtubes or the contribution of the fluxtubes themselves - or both. In an investigation of velocity fields using high spatial resolution spectrograms, Mattig & Nesis (1976) claim to have found higher r.m.s. velocities in small scale structures in active regions. The increase of line width we observe is consistent with the results by Mattig & Nesis (1976). On the other hand, Cavallini et al. (1986) can explain the results of their interferometer spectra of low spatial re-

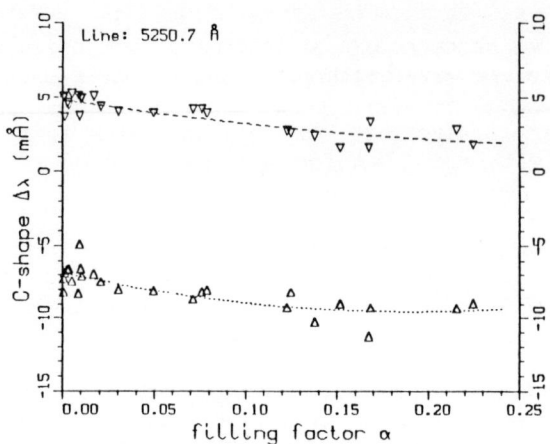

Figure 3: Differences between bisector wavelengths at different intensity levels as function of the filling factor α. Upper curve: $\lambda_{min}-\lambda_{0.7}$; lower curve: $\lambda_{0.7}-\lambda_{0.9}$. For definition cf. text.

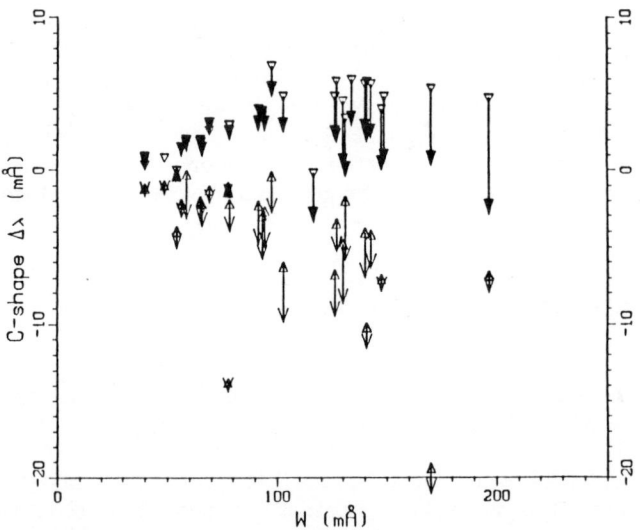

Figure 4: Change of average bisector C-shape from α≈0.0 to α≈0.22 as function of equivalent width.
Symbols: "▽" = $\lambda_{min}-\lambda_{0.7}$ averaged for spectra of α≈0.0; "▼" = same for α≈0.22; "△" = $\lambda_{0.7}-\lambda_{0.9}$ averaged for spectra of α≈0.0; "V" = same for α≈0.22. For details cf. text.

solution only by conjecturing "fluxtubes with zero downflow and partially inhibited convection", thus partially contradicting the findings by Mattig & Nesis (1976).

In searching for an explanation of the increased bisector red-shift found in the line wings one should bear in mind that this part of the bisector very probably stems from the contribution of the intergranular lanes. Therefore, the effect found here may hint at increased downward flows in intergranular lanes in active regions - the fluxtubes being excluded as candidates for downflows by Solanki (1986). Such downflows in field free regions near fluxtubes had been postulated earlier by Frazier & Stenflo (1978) and in the theoretical models of Deinzer et al. (1984).

Acknowledgements. We are grateful to J. Brault, B. Graves, R. Hubbard and G. Ladd of the N.S.O. (Tucson) for assistance in the operation of the McMath telescope and the FTS system, and for carrying out the data transformation.

REFERENCES

Brandt, P.N., Schröter, E.H.: 1984, in Proc. Workshop Small-Scale Dynamical Processes in Quiet Stellar Atmospheres, ed. S. Keil, NSO Conf., 371
Brault, J.W.: 1978, in Proc. JOSO Workshop: Future Solar Optical Observations - Needs and Constraints, G. Godoli, G. Noci, A. Righini, eds., Osserv. Mem. Oss. Astrofis. Arcetri, No. **106**, 33
Cavallini, F., Ceppatelli, G., Righini, A.: 1985, Astron. Astrophys. **143**, 116
Cavallini, F., Ceppatelli, G., Righini, A.: 1986, Astron. Astrophys., in press
Deinzer, W., Hensler, G., Schüssler, M., Weisshaar, E.: 1984, Astron. Astrophys. **139**, 435
Frazier, E.N., Stenflo, J.O.: 1978, Astron. Astrophys. **70**, 789
Gray, D.F.: 1982, Astrophys. J. **255**, 200
Immerschitt, S., Schröter, E.H.: 1986, this conference
Kaisig, M., Schröter, E.H.: 1983, Astron. Astrophys. **117**, 305
Kurucz, R.L., Furenlid, I., Brault, J., Testerman, L.: 1984, Solar Flux Atlas from 296 to 1300 nm, National Solar Observatory, Sunspot, NM
Livingston, W.C.: 1982, Nature **297**, 208
Livingston, W.C.: 1983, in Solar and Stellar Magnetic Fields: Origins and Coronal Effects, J.O. Stenflo, ed., IAU Symp. **102**, 149
Livingston, W.C., Holweger, H.: 1982, Astrophys. J. **252**, 375
Mattig, W., Nesis, A.: 1976, Sol. Phys. **50**, 255
Schüssler, M.: 1986, this conference
Solanki, S.: 1986, Astron. Astrophys., **168**, 311
Stenflo, J.O., Lindegren, L.: 1977, Astron. Astrophys. **59**, 367
Stenflo, J.O., Harvey, J.W.: 1985, Solar Phys. **95**, 99
Stenflo, J.O., Harvey, J.W., Brault, J.W., Solanki, S.: 1984, Astron. Astrophys. **131**, 333
Nr. 273 - Mitteilungen aus dem Kiepenheuer-Institut

NUMERICAL SIMULATIONS OF UMBRAL STOKES-V PROFILES CONSIDERING INFLUENCES OF UNRESOLVED DOTS

K.-D. Pahlke
Universitäts-Sternwarte, D-3400 Göttingen

Introduction

For sunspot finestructures which can not be fully resolved in the spectra, a study of their magnetic fields and their temperature stratifications might be done by 'indirect methods' using oppositely temperature sensitive lines which allow to discriminate the hotter and the cooler component. For this purpose, a line with a low excitation potential (as e.g. Ti 6149.74 Å) being largely strengthened in the interdot background is compared to a line with a high excitation potential being largely weakened (as e.g. Fe+ 6149.25 Å).

Table 1: Equivalent widths of the lines Fe+ 6149.25 Å and Ti 6149.74 Å using different temperature stratifications

		Fe+ 6149.3Å	Ti 6149.7Å
	χ_{low}	3.89 eV	2.16 eV
	g_{eff}	1.333	1.000
model			
photosphere	$W_\lambda (T_{eff} = 5780\ K)$	40.00 mÅ	1.75 mÅ
hotter dots	$W_\lambda (T_{eff} = 5200\ K)$	26.96 mÅ	4.47 mÅ
cooler	$W_\lambda (T_{eff} = 4300\ K)$	5.34 mÅ	18.09 mÅ
umbral background	$W_\lambda (T_{eff} = 3800\ K)$	0.71 mÅ	38.14 mÅ

Table 1 shows, that Ti 6149.74 Å from unresolved spectra of umbral finestructures obtains line-contributions from both, the brighter and the darker component: particularly if the temperature of the bright component is lower than that of the photosphere. In contast, the Fe+ line obtains no significant contribution from the interdot background.

Calculations

For the lines Fe+ and Ti Stokes-V profiles were calculated in units of the neighbouring continuum using the computer program by Wittmann (1974) in order to determine the contribution of bright and dark umbral regions, BR and DR, to observed V-profiles (2 component model). The measured apparent V-profiles, V_{app}, are thus simulated as an additive superposition of the intrinsic V-profiles, V_{BR} and V_{DR}, weighted with the corresponding continuum intensities, I(BR) and I(DR), and with the filling factors F_{BR} and F_{DR} (Stellmacher and Wiehr, 1980/81):

$$(1) \quad V_{app} = \frac{V_{BR} \cdot I(BR) \cdot F_{BR} + V_{DR} \cdot I(DR) \cdot F_{DR}}{I(BR) \cdot F_{BR} + I(DR) \cdot F_{DR}}; \quad F_{BR} + F_{DR} = 1$$

In order to calculate the intrinsic profiles for the interdot background, V_{DR}, the umbral model M4 with T_{eff} = 3800 K and I(DR)=0.1 I(phot) is used (Kollatschny, Stellmacher, Wiehr and Falipou, 1980). For the bright spot component, BR, the intrinsic V-profiles are calculated with different models representing stratifications with:
a) T_{eff} = 5780 K and I(BR) = $1.0 \times$ I(phot)
b) T_{eff} = 5200 K and I(BR) = $0.55 \times$ I(phot)
c) T_{eff} = 4300 K and I(BR) = $0.27 \times$ I(phot)
   (model c fits the darkest dots in the observation by Grossmann-Doerth, Schmidt and Schröter, 1986)

Results and discussion

Among the numereous calculations those are selected which fit the simultaneous measurement of both lines in an umbral region with clearly visible dots using a ϕ=1" measuring hole:
$\Delta\lambda_V$(Ti)=48.0 mÅ, $\Delta\lambda_V$(Fe+)=56.0 mÅ, V_{max}(TI)/V_{max}(Fe+)=2.61
For the observational method and the instrumentation see Scholier and Wiehr (1985).

A) Fig. 1 shows the superposition of the intrinsic V-profiles, V_{BR} and V_{DR}, of Fe+ and Ti, assuming the photospheric model for the dots. The intrinsic V-profile of Fe+ (Fig. 1a) nearly vanishes in the dark umbral background and is strong in the bright regions. (This behaviour is contrary to that of Ti; Fig. 1c). The nearly vanishing intrinsic V-profile of Fe+ from Fig.1a is additionally reduced when multiplying with the low continuum intensity for the interdot background (Fig.1b). The superposed V-profile thus shows nearly the same shape as the intrinsic profile from the hot component. The apparent splitting of the Fe+ line is therefore equal to the intrinsic splitting $\Delta\lambda_V^{BR}$.

For the Ti line, the intrinsic V-profile from the interdot background is also largely reduced by multiplication with the low continuum intensity; however, the intrinsic profile from the hot component remains unchanged. It is evident that, in contrast to Fe+, for Ti the resulting apparent splitting is inbetween the intrinsic values, $\Delta\lambda_V^{BR}$ and $\Delta\lambda_V^{DR}$.

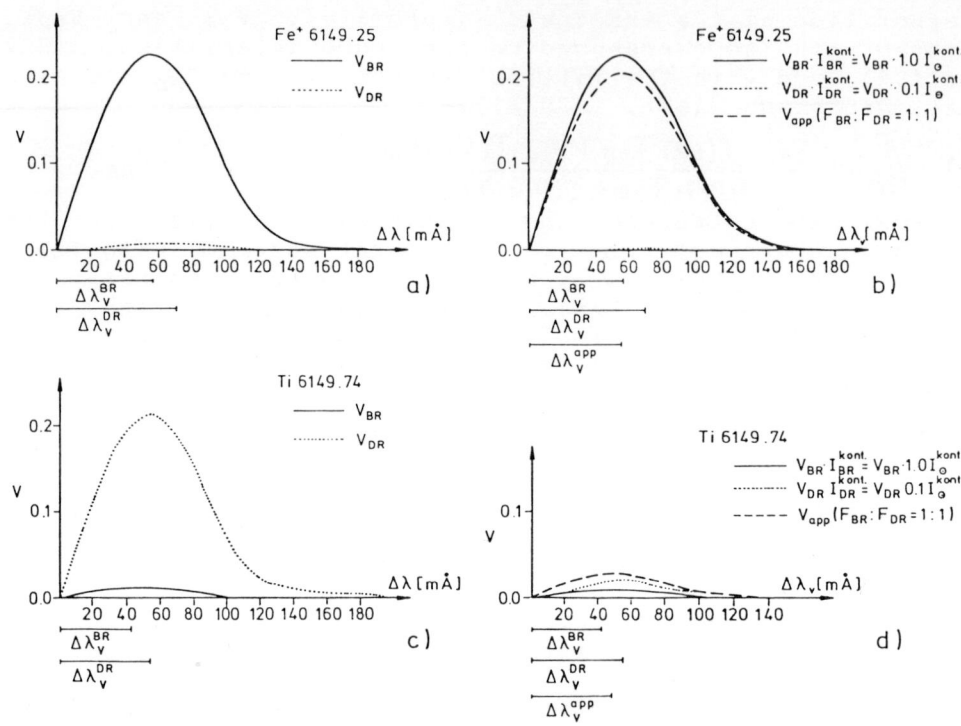

Fig.1: Intrinsic V-Profiles, V_{BR} (T_{eff} =5780 K model) and V_{DR} (T_{eff} =3800 K model), and their additive superposition V_{app} according to formula (1).

The behaviour of these 2 lines changes, when assuming lower than photospheric intensities for the hot component (Fig. 2). Here, the superposition yields for the Fe+ line a mixed V-profile with the same apparent splitting as in the case of photospheric dot intensities (Fig. 2b). However this is not the case for the Ti line: calculating the superposed V-profiles with the same magnetic field strengths as for the photospheric model, one obtains a lower splitting. Consequently, for a given (measured) splitting a higher magnetic field strength must be assumed in the dark interdot background. In turn, the magnetic field strength for the dots is given by the splitting of the Fe+ line (see above), which depends on the hot- but not on the cool-component model (Fig. 2d).

The former discussion shows, that, one is forced to assumed larger differences of the true magnetic field strengths in the two umbral components if the dots are cooler than the photosphere. This demonstrates the large dependence of the interpretation of unresolved Stokes-V splittings on the assumed atmospheres.

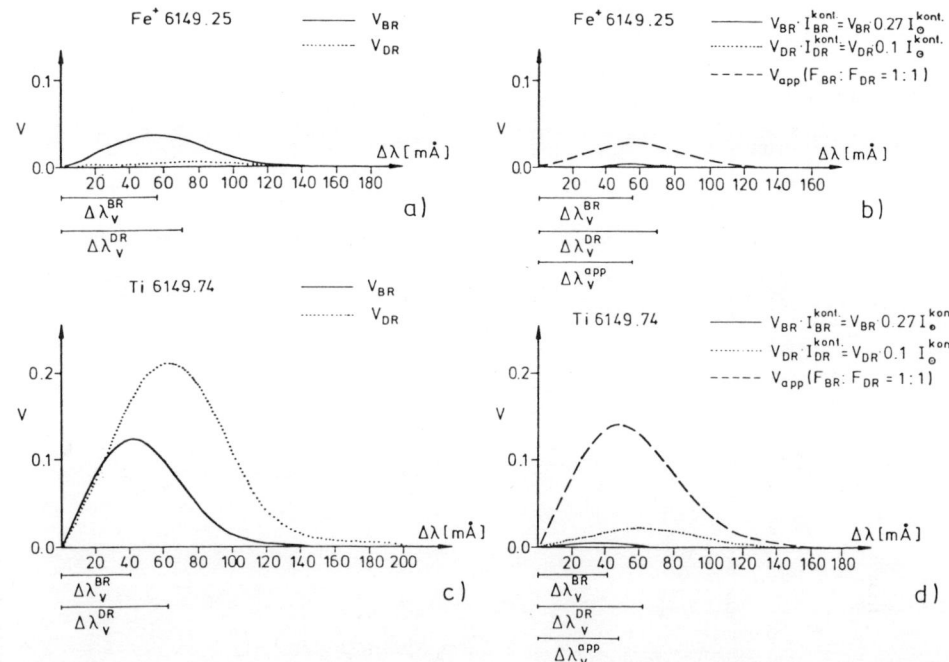

Fig.2: Same as Fig. 1, however the T_{eff} = 4300 K stratification is used to calculate V_{BR}. The magnetic field strengths H_{BR} and H_{DR} are chosen to obtain the same $\Delta\lambda_V^{app}$ as in Fig. 2 representing the measured values (see text).

The intrinsic magnetic field strengths obtained for the bright and dark umbral component are:
  dot model a) H_{BR} = 2350 Gs , H_{DR} = 3000 Gs
  dot model b) H_{BR} = 2350 Gs , H_{DR} = 3200 Gs
  dot model c) H_{BR} = 2350 Gs , H_{DR} = 3500 Gs

The results illustrated in Fig.s 1 and 2 are calculated with F_{BR} : F_{DR} = 1 : 1; the calculations with other filling factors yield similliar results.

B) Fig. 3 shows the ratio of the Stokes-V amplitudes of the Ti- and the Fe+ lines calculated with different filling factors and with the three models for the hot component (see above). When using model c it is not possible to explain the measured ratio. The other two curves fit the measurement for 3% (model a) and, resp., for 9% filling (model b). If the 1"x 1" measuring hole is smeared by seeing corresponding to 3"x 3", the number n of dots in the measuring hole can estimated:
1) model a (3% filling):
   i) area of 1 dot= $0.06"^2$ (180 km x 180 km) ==> n=4
   ii) area of 1 dot= $0.16"^2$ (290 km x 290 km) ==> n=2

```
    iii) area of 1 dot= 1"²        (725 km × 725 km)    ==>  n=1/4
2) model b (9% filling):
    i)   area of 1 dot= 0.06"²                          ==>  n=13
    ii)  area of 1 dot= 0.16"²                          ==>  n=5
    iii) area of 1 dot= 1"²                             ==>  n=3/4
```

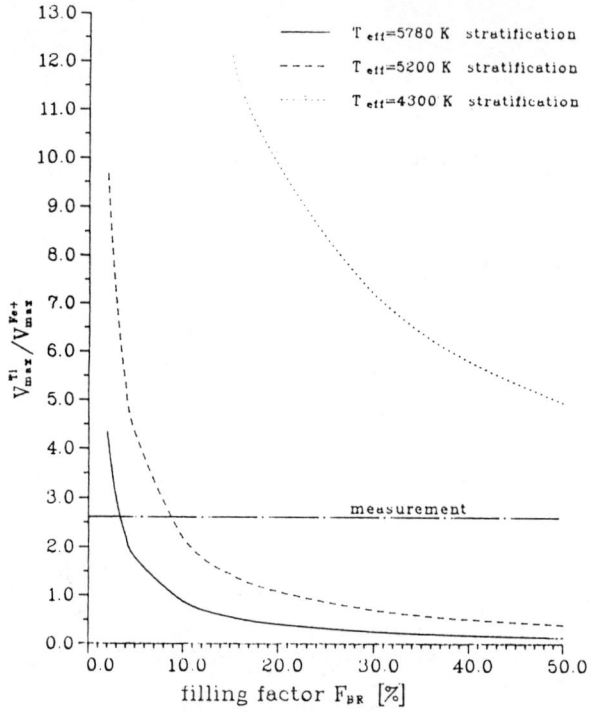

Fig. 3: Line ratio $V_{max}^{Tl}/V_{max}^{Fe+}$ versus the filling factor F_{BR} [%]; the models, used for the dots, are indicated.

Acknowledgement

I would like to thank Dr. E. Wiehr for the fruitful discussion and for providing me with the observational data. The numerical calculations have been performed using the Sperry 1100/83 computer of the G.W.D.m.b.H. in Göttingen.

REFERENCES

Grossmann-Doerth,U.,Schmidt,W.,Schröter,E.H.:
    1986, Astron.Astrophys. 156 , 347
Kollatschny,W.,Stellmacher,G.,Wiehr,E.,Falipou,M.A.:
    1980, Astron.Astrophys. 86 , 245
Scholiers, W., Wiehr, E.: 1985, Solar Phys. 99 , 349
Stellmacher,G.,Wiehr,E.: 1980,Astron.Astrophys. 82 ,157
Stellmacher,G.,Wiehr,E.: 1981,Astron.Astrophys. 95 ,229
Wittmann, A.: 1974, Solar Phys. 35 , 11

NEEDS AND LIMITS OF MAGNETIC AND VELOCITY FIELD MEASUREMENTS WITH SUB-ARCSEC RESOLUTION

E.Wiehr
Universitäts-Sternwarte
D-3400 Göttingen

Velocity fields on the solar surface may be divided into two groups: those existing permanently as granular, supergranular or oscillatory fields, and those connected with magnetic fields e.g. in sunspots, faculae, prominences. A particular characteristic of such features in active regions is their pronounced finestructure with typical dimensions of a few hundred kilometers. Models based on spatially averaged observations are therefore of rather limited relevance. Instead, a deep understanding of solar activity requires the detailed investigation of finestructures and in particular of their velocity and magnetic fields.

This clearly demonstrates the NEEDS for sub-arcsec resolution measurements. On the other hand, LIMITS of such observations are much more difficult to discuss since modern techniques (such as adaptive optics, speckle methods, space observations) allow to avoid most of the disturbing influences from the earth's atmosphere. As the aim of this workshop is not the discussion of future technological possibilities, this talk shall be given under the following "boundary conditions":

a) exclusive consideration of velocity and magnetic field in finestructures of sunspots, faculae and prominences;
b) observations using the evacuated 45cm Gregory-Coudé telescope with its f=10 m Czerny type spectrograph (Brückner et al., 1967; Wiehr et al.,1980; Schröter et al.1985); Wiehr(1987).

Magnetic and velocity fields in finestructures of active regions may be investigated by means of Zeeman and, resp., Doppler filtergrams. The interpretation of Zeeman polarization in terms of accurate magnetic field parameters is impossible without detailed knowledge of a model atmosphere which, for the finestructures, requires subarc-sec spectroscopy. Also the detailed investigation of velocity fields incl. its depth gradient is best done with the knowledge of whole line profiles.

Sub-arcsec spectroscopy is largely limited by the spatial resolution achieved in the spectra. This is due to 1)essentially larger exposure times required for spectra as compared to white light or filtergram pictures yielding larger influences of seeing; and 2) marked reduction of the resolution by the spectrograph's MTF. Fig.1 shows the empirical MTF of the f=10m spectrograph at the Gregory telescope using artificial light. The spectrograph had been perfectly adjusted for zero astigmatism (c.f.Wiehr,1987), the focus had been carefully determined by the Hartmann method. The figure shows that the spatial frequency pattern (100% contrast) is dropped by the spectrograph to 50% for 7.2 linepairs/mm = 0.55" and to a limiting resolution of 10% for 11.1 lp/mm = 0.36" or 250 km on the sun.

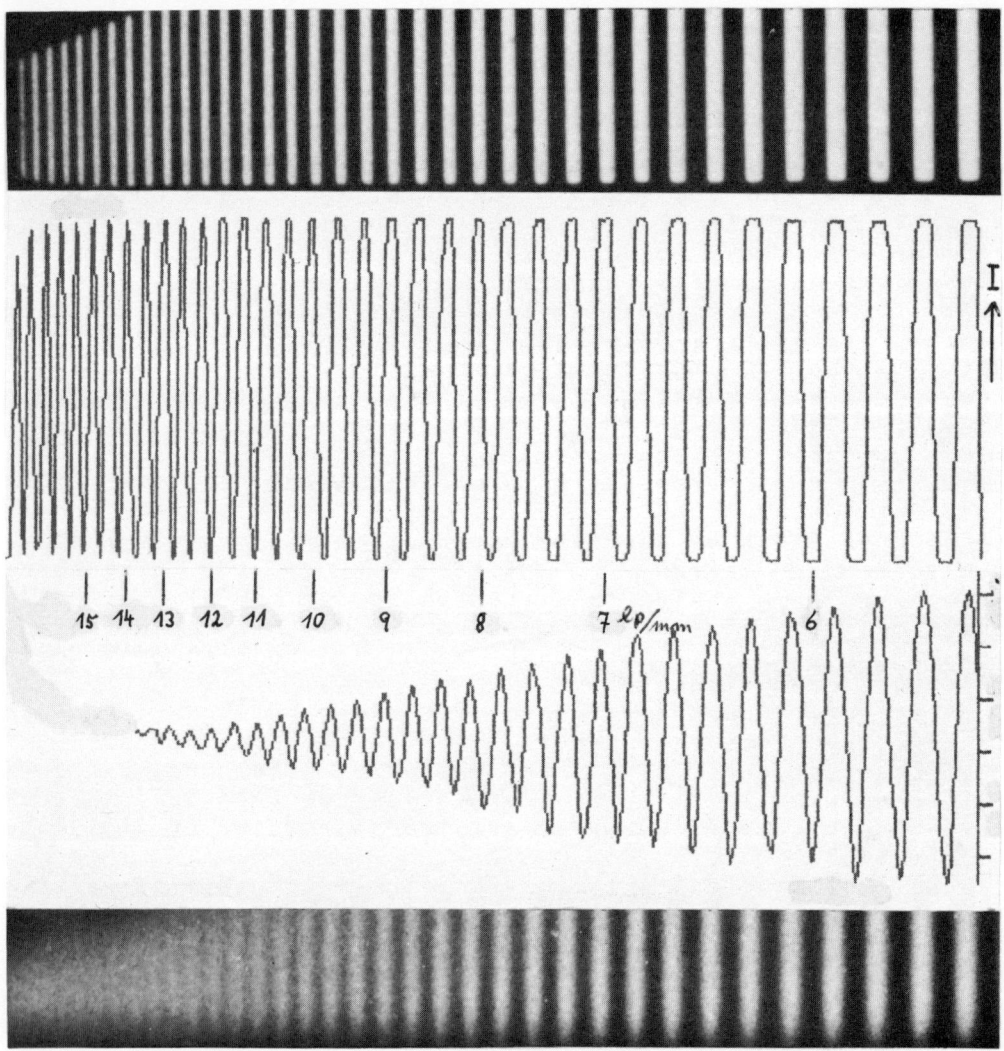

Fig.1: Monotonically increasing spatial frequency pattern as seen through a spectrograph with f=10m and h=12.8cm limiting height of the grating; for comparison the superposed original pattern.

This MTF of the spectrograph has to be multiplied by the one introduced by seeing integrated over the exposure time required for finestructure spectroscopy (a few sec). For Stokes polarimetry further influences are to be expected from the polarization optics.

If the combined influence of all optical components of the particular instrument (here the 45cm Gregory) does not allow to resolve the active region finestructures smaller than 0.4 arcsec, it may be suggestive to apply "indirect-methods" yielding information about active region finestructures without fully resolving them.

1.) SUNSPOT FINESTRUCTURE

For sunspots, an indirect method would be the investigation of differently excited lines in spatially smeared spectra: high-excitation lines as a probe for the hotter and low-excitation lines for the cooler finestructure component. Numerical calculations however, show that such oppositely temperature sensitive lines do not allow a clear distinction since the hotter and cooler sunspot components have not sufficiently large temperature differences:

Concerning the umbra, Grossmann-Doerth et al.(1986) find strong evidence that dots are cooler than the undisturbed photosphere. Wiehr and Stellmacher (1984) show that inter-dot temperatures may vary from one spot to the other, since different umbral intensities measured in clean continuum windows, cannot be explained by a different population with dots.

For the penumbra, Muller(1973) finds that the bright structures are almost of photospheric intensity, dark structures showing roughly half of that brightness. Grossmann-Doerth et al.(1981) and Wiehr et al.(1984) show that these brightness values might be largely violated by individual penumbral structures; the denotation 'bright' or 'dark' being only locally defined, i.e. dark structures do exist which are brighter than non-neighbouring bright structures.

These observed intensity differences in umbral and penumbral finestructures have the following consequences for an application of the "indirect method": On the one side, high excitation lines (as Fe+) disappear largely in the cool umbral inter-dot atmosphere and are sufficiently enhanced in dots - even at the lower dot temperatures suggested by Grossmann-Doerth et al.(1986). On the other side, low excitation lines (as Zr and Ti) are strong in inter-dot regions but do not sufficiently vanish in dots; - moreover if these are cooler than the photosphere.

The situation is even worse for penumbrae as compared to umbrae since the dark penumbral component is significantly hotter than dark umbral inter-dot background. Hence, in dark penumbral structures do neither the Fe+ lines vanish nor the Zr or Ti lines sufficiently strengthen.

Besides these aspects of line excitation, the intrinsic profiles of sunspot finestructures have to be weighted by their corresponding continuum intensity - and area - contributions. The intrinsic continuum intensities, however, diminish largely the contributions of the cooler sunspot component as compared to those from the hotter one. This affects mainly the low-excitation lines.

Hence, the "indirect method" only allows to deduce information about the bright umbral and penumbral finestructures (using high-excitation lines) - suitable information about the dark components being hardly obtained from low - excitation lines. This is not only valid for the determination of empirical models of finestructures but also for their velocity and magnetic fields. The "indirect method" can thus not solve the problems alone. On the other hand, the application of "direct methods" by fully resolving single sunspot finestructures in the spectrum was not successful until now; - even spatially best resolved spectra show conglomerates (c.f. Wiehr et al., 1987) rather than individual sunspot finestructures as known from white light pictures.

A possible solution of this dilemma is a combination of both methods. In this case, the area contribution from both finestructure components will deviate from the 50 : 50 value for totally unresolved sunspots; for spectra of "conglomerates" of finestructures these factors may even reach a 95 : 5 value (Pahlke,1987).

2.) FACULA FINESTRUCTURE

The situation appears to be less difficult for plages and network regions where isolated fluxtubes ("filigree") are imbedded in non-magnetic surroundings. As a consequence, observations of Zeeman profiles yield exclusive information about the atmosphere inside the fluxtube; (this being in contrast to individual structures of the sunspot magnetic field which probably are not imbedded in field-free surroundings).

Hence, spatially-unresolved Zeeman profiles allow to deduce the magnetic field strength (as well as the line-of-sight velocity). This "indirect method" for a measurement of isolated magnetic fluxtubes uses the saturation of the magnetograph signal when observing simultaneously either different lines with one exit slit (e.g.Stenflo,1973), or one line with different exit slits (Wiehr,1978). The so obtained "true" magnetic field strengths cover a limited range between 1000 and maximally 2000 Gauß equally for all non-spot magnetic fields. As a consequence, one can assume that the hierarchy of active region features, from the faintest Ca+K brightenings via enhanced network through plages up to small pores, reflects mostly an increase of the magnetic area rather than an increase of the field strength (Wiehr,1978).

These "indirect" observations, however, do not allow to decide wether that variable magnetic area might reflect an increasing number of single fluxtubes with an almost unique intrinsic field strength. In that case one would expect a lower limit of magnetic flux (one single fluxtube) as well as a "quantization" of flux. Both have been observed (Wiehr,1979; Livingston and Harvey,1969) but are so far neither established nor clearly disproved. If, instead, such individual fluxtubes were of different physical nature, the "indirect method" would only yield some "average information". This would then also hold for atmospheric models deduced from unresolved structures (e.g. Solanki,1986).

Hence, even the (more promising) non-spot magnetic fields on the solar disc require high spatial resolution. Among numerous of such "direct observations" (e.g. Harvey and Hall,1975; Tarbell and Title,1977; Koutchmy and Stellmacher,1978) none refers doubtless to one single isolated fluxtube (filigree point) but at best to enhanced network. This is due to the required minimum Zeeman signal, forcing the observer to select stronger fields.

As compared to such Zeeman spectroscopy,the much more sensitive Zeeman-polarimetry yields a rms noise of 10 Gauß for an entrance pinhole of 4 arcsec diameter when using the 45 cm Gregory telescope under optimal apparative conditions. Zeeman-polarimetry of single fluxtubes, however, would require entrance pinholes smaller than 0.5 arcsec yielding a noise level above 100 Gauß at integration times of several seconds. This additionally implies an enormous influence of atmospheric image degradation. Due to the smaller sensitivity, the situation is by far worse for transverse Zeeman polarimetry.

As far as Zeeman-spectroscopy is concerned,this has the advantage of yielding spatial information along the direction of the slit. As compared to photographic emulsions, matrix arrays are more sensitive but their finite distance of rows implies some problems when subtracting the left- from the right-handed spectrum for one and the same point on the solar disc within a small fraction of an arcsec.

The knowledge of the velocity-field inside single fluxtubes is of high interest because the mechanism producing high magnetic field strengths in small solar areas is supposed to be a convective collapse (Hasan, 1984). In this case one expects temporal variations of Doppler shifts inside the fluxtube. Attempts to deduce such Doppler shifts from spatially high-resolved intensity profiles ("direct method") meet large problems with interfering influences from velocity fields in the neighbourhood of the fluxtube. In contrast, measurements of the Zeeman profile (e.g. displacements of the Stokes-V zero crossing point as measured by Scholiers and Wiehr(1985) and by Wiehr(1985) allow to determine velocities inside a fluxtube even at low spatial resolution ("indirect method") . These results contradict findings with the Fourier transform spectrometer (Stenflo and Harvey, 1985). Recent observations with the Gregory Telecope at Tenerife are shown in Fig.2.

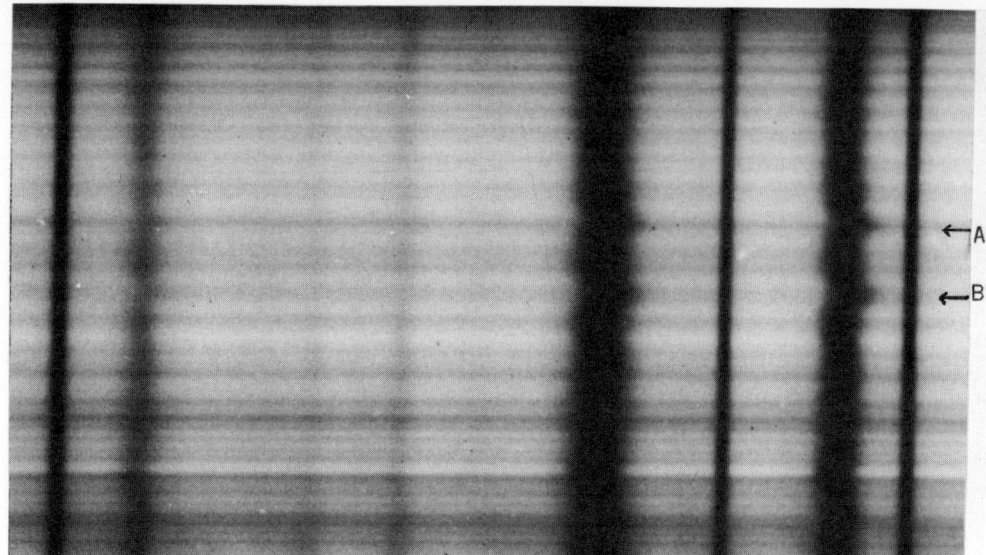

Fig.2: Left-handed Zeeman spectrum (via quarter-wave plate plus polarizer) covering two isolated Ca+K bright points on Sept.25, 1986; Vacuum Gregory - Coudé telescope at Tenerife; spatial range = 96 arcsec; spectral range 6299 - 6304 Å; Kodak-SO-115 film; 3.2 sec exposure.

Since the evaluation of these different methods with their contradicting results is highly informative for this discussion of NEEDS and LIMITS of high resolution velocity (and also magnetic) field measurements, first scans of the Stokes-V time se ries over 16 min. are given in Fig.3 - although not presented in the oral talk. These results Fig.3 clearly establish the Stokes-V redshifts mentioned above. The discrepancy with the FTS data might be explained by their spatial (10 * 10 arcsec) and temporal (30 sec and more) averaging.

This impressively demonstrates a problem of NEEDS and LIMITS of high spatial resolution velocity and magnetic field measurements: On the one side, isolated quiet region fluxtubes are much too small for direct spectroscopic observations. On the other side, the promising indirect measurements of Stokes-V profiles yield realistic informations only in case of sufficient spatial and temporal resolution. Hence, similar as for sunspot structures, also the tiny nonspot magnetic structures are best observed by the combined method of indirect observations at highest possible spatial resolution.

- -

Fig.3 (following page): Time series of Stokes-V from both structures A and B in Fig.2 showing significant redshifts up to 1 km/s relative to the quiet I-profile, as well as time variations of shift and V-asymmetry; spatial variations being indicated from the two neighbouring regions, A and B, for a given time.

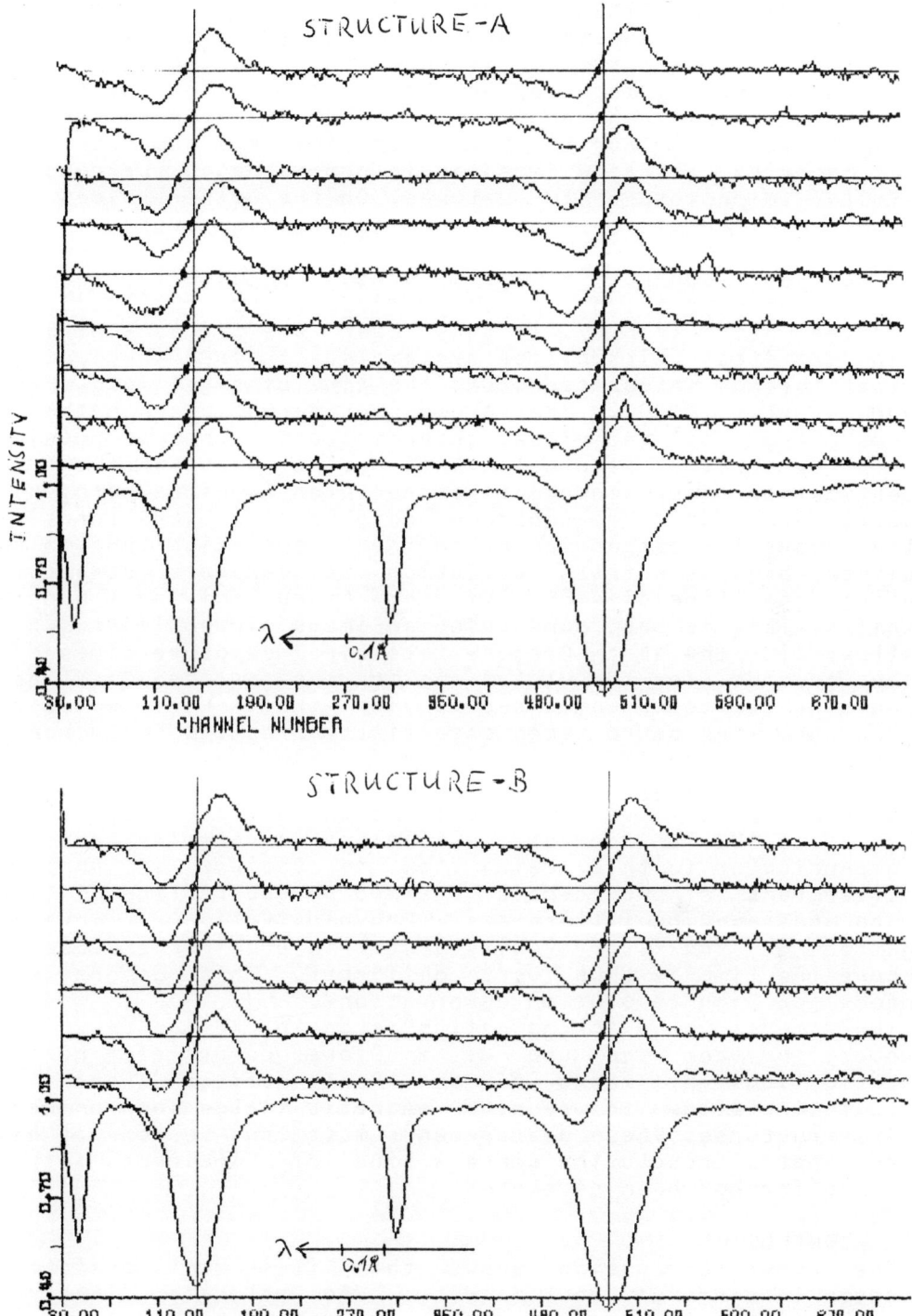

3.) PROMINENCE FINESTRUCTURE

Investigating finestructures of prominences above the limb ("threads"), the situation is somewhat inbetween that of spots and that of faculae: On the one side, the tiny threads project against an "empty background" (the corona does not emit those lines investigated in prominences) and can thus be regarded as being imbedded in non-magnetic surroundings similar to photospheric fluxtubes. On the other side, the prominence sheet is thick enough to contain several threads along the line-of-sight, thus yielding similar problems as unresolved sunspot finestructures. Individual threads in the line-of-sight can only be studied if they show different Doppler shifts (Fig.4). In this case one finds intrinsic line profiles which differ systematically from those of several threads integrated along the line of sight (e.g. Engvold et al., 1980). The so deduced thermal and non-thermal broadenings of individual threads can thus not be obtained with low spectral resolution.This demonstrates, that the investigation of prominence finestructures requires both, high spatial resolution to separate neighbouring structures in the direction perpendicular to the line-of-sight and, in addition, high spectral resolution to separate structures along the line-of-sight if they show individual Doppler shifts. This is best done using an image intensifier which allows (for the 45 cm Gregory telescope) exposure times of a few seconds for the violet and for the infrared Ca+ lines, and of a few tenths of a second for the H-Beta emission. Such short exposure times essentially drop the influence of seeing (c.f. Fig.4).

The _velocity-field_ of quiescent prominences is known to be oriented almost vertical to the solar surface (see Engvold et al., 1985). However, a significant velocity component perpendicular to this main flow is visible in Doppler-filtergrams of prominences obtained by subtraction of blue-wing and red-wing filtergrams (Engvold,1981). An investigation of the full velocity vector in prominences observed above the limb is thus very difficult. This demonstrates needs for high resolution observations.

The same is valid for _magnetic-fields_ in prominences. However, neither the use of a Stokes polarimeter nor investigations by means of the Hanle effect (Leroy et al.,1984) allow to resolve magnetic fields in prominence finestructures. These measurements must thus be done at highest spatial resolution since a kind of "indirect method" has so far not been developed.

CONCLUSION

The above discussion shows that it is well possible to deduce information about magnetic and velocity finestructures far below one arcsec even with a relatively small instrument (as e.g. the 45cm Gregory at Tenerife). If the telescope in connection with its spectrograph does not allow full spatial resolution of _single_ active region finestructures, and if "indirect methods" cannot solve the problem

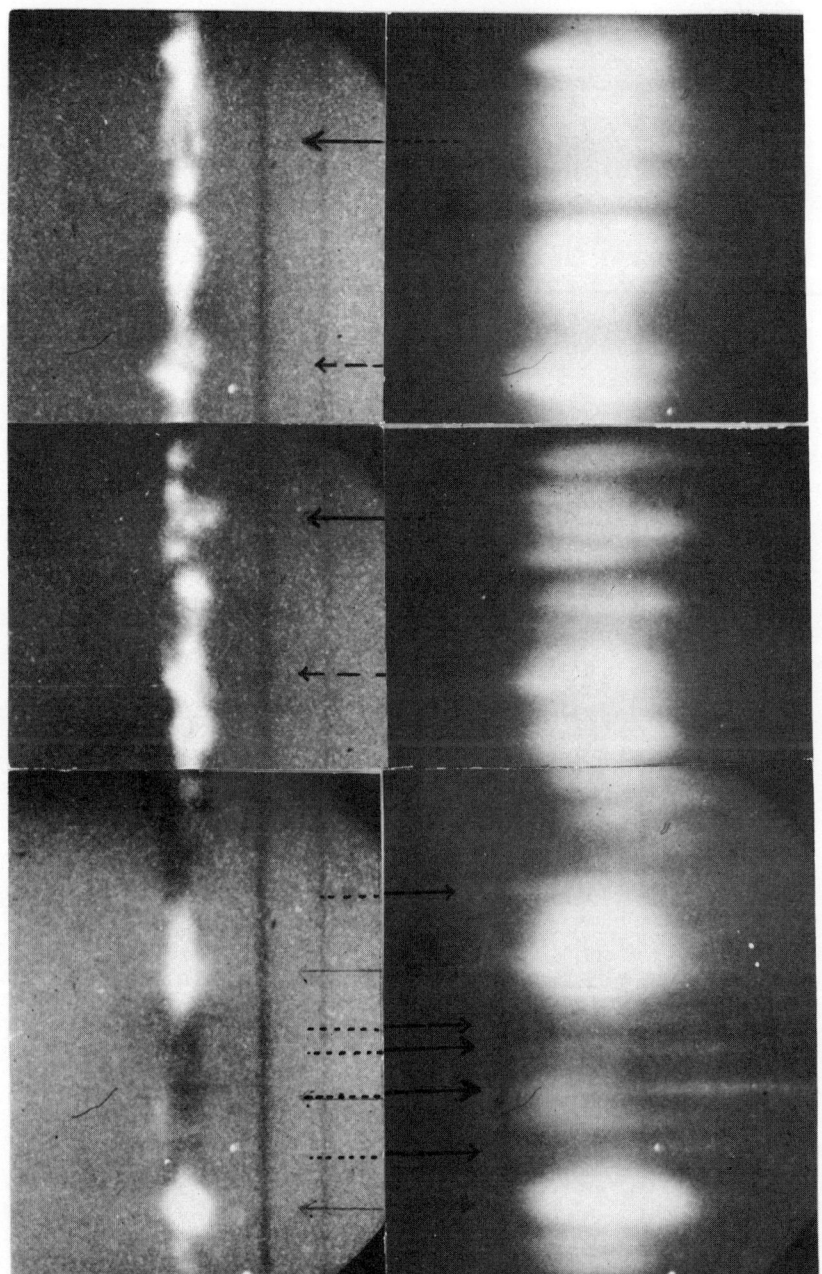

Fig.4: Spatial variation of H-Beta and Ca+8542 emissions in a quiescent prominence for three neighbouring slit positions using a 2-stage image - intensifier and exposure times of 0.8, resp., 2.2 sec: a) faint, Dopplershifted H-ß streaks not seen in Ca+IR; b and c) multiple emissions in Ca+IR from finestructures aligned along the line-of-sight exclusively temperature broadened and thus not resolved in H-ß.

alone, the combination of both methods offers a promising solution: indirect methods used at best spatial resolution in order to keep the filling factor as small as possible.

This has successfully been done e.g. by Stellmacher and Wiehr (1978) who showed that a filigree model is well able to explain spectroscopic facula observations obtained at moderate (1.5 arcsec) resolution if one carefully considers the rather complicated filling of few filigree within inter - granular spaces. The fact that recent models deduced from integrated Stokes-V profiles observed with modern FTS techniques yield roughly the same model atmosphere (Solanki, 1986) impressively demonstrates the suitability of the combination of direct and indirect methods for unresolved solar finestructures.

REFERENCES

Brückner,G.,Schröter,E.H.,Voigt,H.H.: 1967,Sol.Phys.**1**, 487
Engvold,O.: 1981, Sol.Phys.**70**, 315
Engvold,O.,Tandberg-Hanssen,E.,Reichmann,E.: 1985, Sol.Phys.**52**, 369
Engvold,O.,Wiehr,E.,Wittmann,A.: 1980,Astr.Astroph.**85**, 326
Grossmann-Doerth,U.,Schmidt,W.: 1981,Astr.Astrophys.**95**, 366
Grossmann-Doerth,U.,Schmidt,W.,Schröter,E.H.: 1986, Astr.Astrophys.**156**, 347
Harvey,J.W.,Hall,D.: 1975,Bull.Astr.Soc.Am.**7**, 459
Hasan,S.S.: 1984, Ap.J.**285**, 851
Koutchmy,S., Stellmacher,G.: 1978, Astr.Astrophys.**67**, 93
Leroy,J.L., Bommier,V., Sahel-Brechot,S.: 1984, Astr.Astrophys.**131**, 33
Livingston,W., Harvey,J.: 1969, Sol.Phys.**10**, 249
Muller,R.: 1973, Sol.Phys.**32**, 409
Pahlke,K.D.: 1987, oral contribution at this workshop
Schröter,E.H.,Soltau,D.,Wiehr,E.: 1985, Vistas in Astron.**28**, 519
Scholiers,W., Wiehr,E.: 1985, Sol.Phys.**99**, 349
Stellmacher,G., Wiehr,E.: 1979, Astr.Astrophys.**75**, 263
Stenflo,O.: 1973, Sol.Phys.**32**, 41
Solanki,S.K.: 1986, Astr.Astrophys.**168**, 311
Tarbell,D.T.,Title,A.M.: 1977,Sol.Phys.**52**, 13
Wiehr,E.: 1978, Astr.Astrophys.**69**, 279
Wiehr,E.: 1979, Astr.Astrophys.**73**, L19
Wiehr,E.: 1987, instrument-description in this draft
Wiehr,E.,Knölker,M.,Grosser,H.,Stellmacher,G.q987, poster at this workshop
Wiehr,E.,Koch,A.,Knölker,M.,Küveler,G.,Stellmacher,G.: 1984,Astr.Astrophys.**140**, 352
Wiehr,E., Stellmacher,G.: 1984, "High Resol.in Sol.Phys.", 8th IAU reg. meeting, Toulouse, p.254
Wiehr,E., Wittmann,A., Wöhl,H.: 1980, Sol.Phys.**68**, 207

PROPERTIES OF A CONCENTRATED MAGNETIC FIELD REGION

S. Koutchmy
Institut d'Astrophysique-CNRS, 98bis Bd Arago, F-75014 PARIS
now at: A.F.G.L./N.S.O. Sacramento Peak Obs. Sunspot NM 88349

G. Stellmacher
Institut d'Astrophysique-CNRS, 98bis Bd Arago, F-75014 PARIS

INTRODUCTION

Theoretic models of photospheric concentrated flux regions were worked out by different authors which appear to reproduce the most significant observations, see e.g. Spruit (1976) and Deinzer et al. (1984). At the present magneto-static models seem favoured accounting for the fact that no important flows are observed within the magnetic structures.

1. OBSERVATIONS

Further results concerning the fine structure of concentrated magnetic field regions or network elements, observed near disk center well outside of active regions, are presented. A very comprehensive set of observations, obtained at the Sac. Peak Vacuum Solar Telescope in 1975 and 1976, was used :

A. White light high speed photographic pictures in windows at λ 4680±20Å and λ 6000±30Å attaining a resolution up to 0.16 arcsec, the telescope's resolution limit; in order to measure the true contrast of filigrees, image restoration in the "long exposure time" approximation was performed, using the measured telescope MTF (Koutchmy, 1977).

B. Universal bi-refringent filter (UBF) pictures in continuum, NaD, Mgb$_1$ + 0.4, and Hα with a passband $\Delta\lambda \simeq 0.2$Å; achieved resolution up to 0.7 arcsec.

C. Spectrophotometric observations of different lines obtained simultaneously at the echelle spectrograph (0.125Å/mm, slit-width = 0.75x160 (arcsec)^2, exposure time 1.5s, S/N $\geqslant 10^3$) comprising : i) the magnetically sensitive lines Fe 6301.5 (g$_L$ = 1.67) and Fe 6302.5 (triplet, g$_L$ = 2.5) observed with a circular polarisation analyser (λ/4 + Wollaston) for the determination of B$_\parallel$(x); ii) the magnetically insensitive line Fe 5576 (g$_L$ = 0) for the determination of v$_\parallel$(x) and T(τ); iii) Hα spectra to monitor fibril activity and "moustache-like" features (not discussed here). Simultaneous slit-jaw pictures were obtained in white light and Ca$^+$ K ($\Delta\lambda$ = 0.3Å) in order to allow exact identification of the slit over the solar fine structures.

An example illustrating the achieved high resolution in the white light pictures is given in fig.1 showing well visible structured filigrees (rope-like crinkles) inside the intergranular space.

Fig.1 Rope-like filigrees reproduced from selected white light pictures. The telescope's resolution limit is attained on these "freezed" pictures; exposure time Δt = 0.1s. Note the higher contrast in the blue filigree picture.

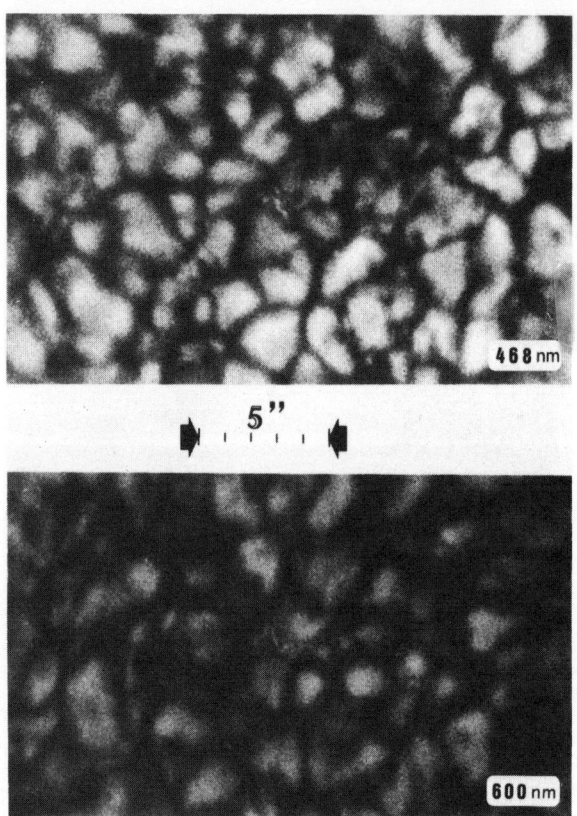

In the following we do not use the hypothesis of magnetic fluxes restricted to this highly structured "subarcsec" filigree structures; we rather assume, at least at the photospheric level, that the magnetic region spreads out well beyond these structures.

2. ANALYSIS

Two modes of analysis of selected best resolved spectra (up to 0.75 arcsec) are choosen : 2-dim scans for mapping a small region, see Dara-Papamargariti and Koutchmy (1983), and time sequences over the same region, covering ≥ 5 min for the study of velocity and magnetic field changes.

Applying the center of line method on both magnetically sensitive lines, apparent field values $B_\parallel^{app}$ are deduced. For $B_\parallel^{app}$ < 600 gauss the values obtained from the magnetically less sensitive line Fe 6301.5 are systematically higher when compared to the results from the more split line Fe 6302.5.

For $B_\parallel^{app}$ ⩾ 600 gauss both lines then yield the same results. A scatter-plot diagram for various magnetic regions is given in fig.2. A more quantitative determination based on the thermal filigree model, fig.3, together with the properly adapted filling factors, Koutchmy and Stellmacher (1978), allows to transform the apparent field strengths into true field strengths $B_\parallel^{true}$, entered in fig.2. The filling factor f_i is defined here by the ratio of light coming from the magnetic region over that from the nonmagnetic region including scattered light.

We note that in this model the magnetic field is not only confined to the filigree structure but extends over the surrounding intergranular space.

Examples of more recent synthetic profils calculated with this model for different filling factors are shown together with the corresponding Stokes V-profiles in fig.4; here a downdraft of ~ 0.5 km s^{-1} was assumed for the filigree structure and the surrounding intergranular space. One sees that for a smaller filling factor the V-profil is greatly reduced, approximately in proportion of the assumed filling factor ratio.

The velocity variations in time and space in the vicinity of two magnetic elements, obtained from the analysis of the Fe 5576 line, are displayed in the map of fig.5, taken from the extended analysis by Dara-Papamargariti et al. (1986). The diagram shows periodic motions of the order of 250s. No pronounced upward or downward motions are associated with the magnetic regions : it seems impossible to distinguish between the magnetic regions and their surroundings on the basis of the velocity field alone, both magnetic regions are rather regions with relatively low velocity amplitudes which do not coincide with the field peaks.

3. DISCUSSION

In view of the achieved high signal to noise ratio (~ 10^3) in the spectra a preliminary deconvolution of a spectrum over a magnetic region, fig.6, visible in the spectrum as a "tilted" gap, was attempted.

A simple restoration following the direction x along the slit was performed using the isophote maps of fig.6; the results are briefly resumed in fig.7. The already visible asymmetry in the observed profil which is not the same in the left and righthanded polarized spectrum of Fe 6302.5 reveals as a polarity of opposite sign in the restored scan. Accordingly a high value of the horizontal fieldgradient $\delta B_\parallel / \delta x$ up to 3 or 5 gauss km^{-1} is deduced. No noticeable continuum

Fig. 2 Scatter-plot diagram of observed field values $B_\parallel^{app}$ in concentrated field regions for the two magnetic sensitive lines Fe 6301.5 (g_L = 1.67) and Fe 6302.5 (g_L = 2.5). Calculated "true" field values $B_\parallel^{true}$ are indicated.

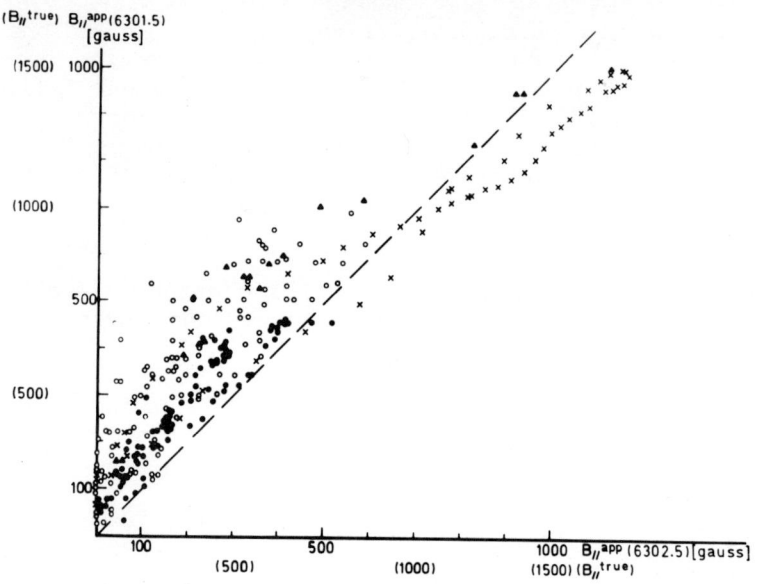

Fig. 3 Assumed thermal structure of a filigree (model f; Θ_{eff} = 0.74).

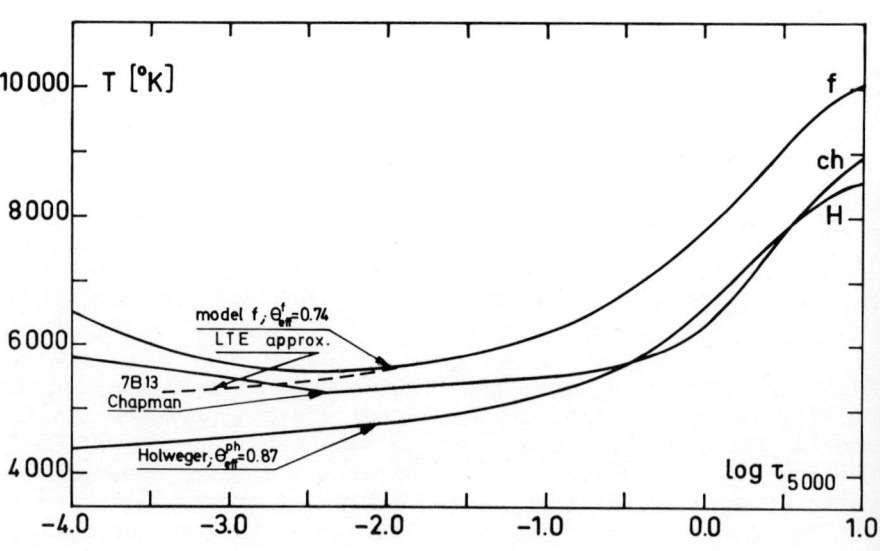

Fig. 4 Synthetic left and right handed polarized line profils and Stokes-V profils of Fe 6302.5 calculated for a magnetic region with filling factors $f_i = 0.75$ and $f_i = 0.15$. A downdraft of 0.5 km s^{-1} is assumed for the filigree and the surrounding intergranular region.

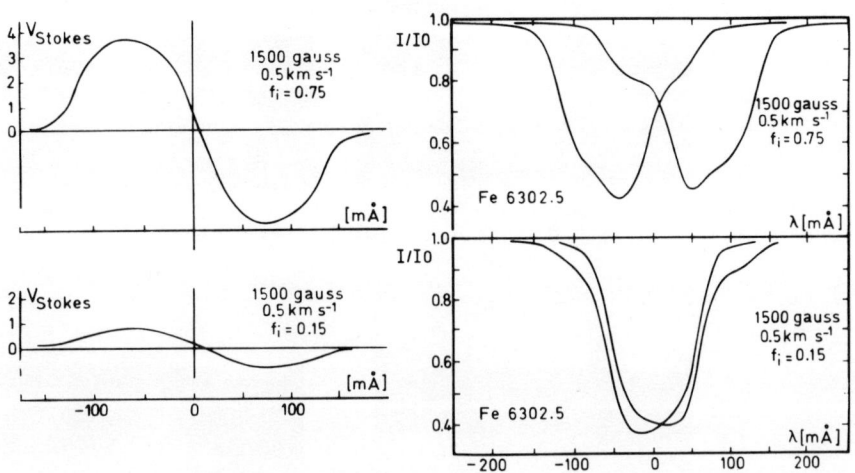

Fig. 5 Map showing the time variation of vertical velocitie in the vicinity of two magnetic regions A and B, deduced from the line Fe 5576 ($g_L = 0$) by Dara-Papamargariti et al. (1986).

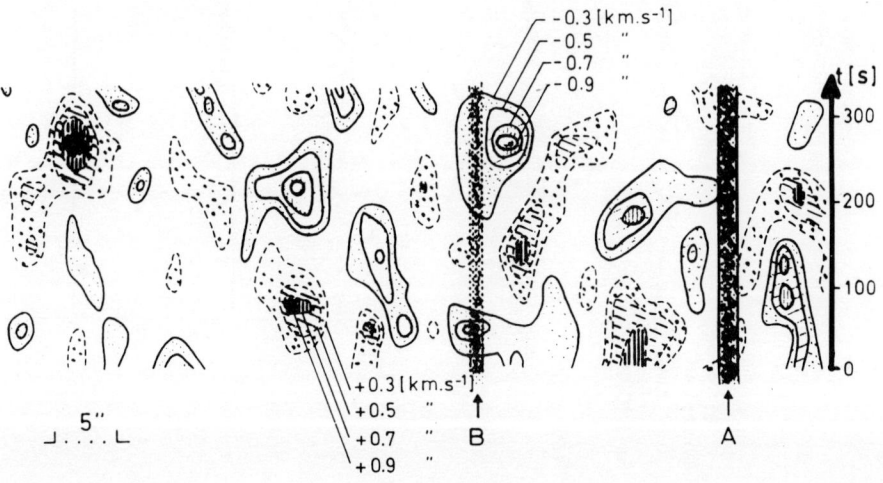

Fig. 6 Left and right handed polarized spectra (Fe 6302.5) over a magnetic region (bottom). An isophote display of these spectra is given at the top. Note the asymmetry (tilt) in the original spectra.

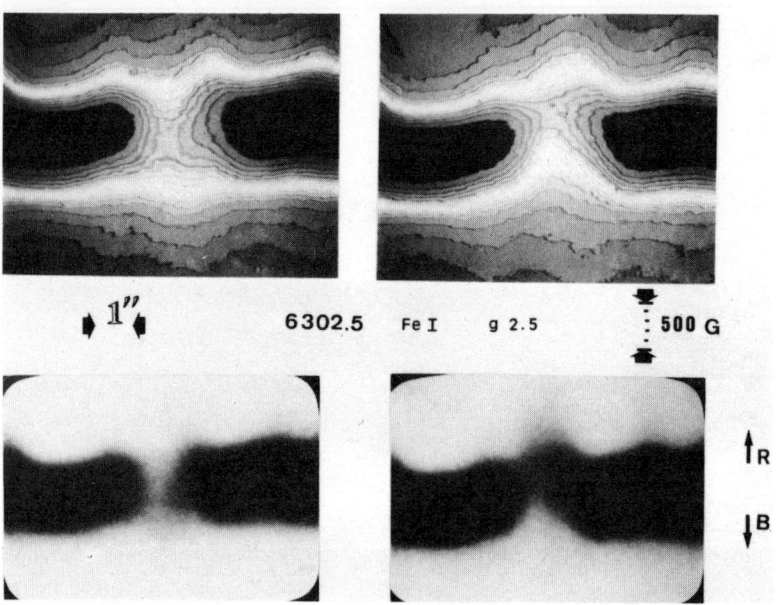

Fig. 7 Results of the temptative restoration of a spectrum of Fe 6302.5 observed in left and right handed polarization, see text for details.

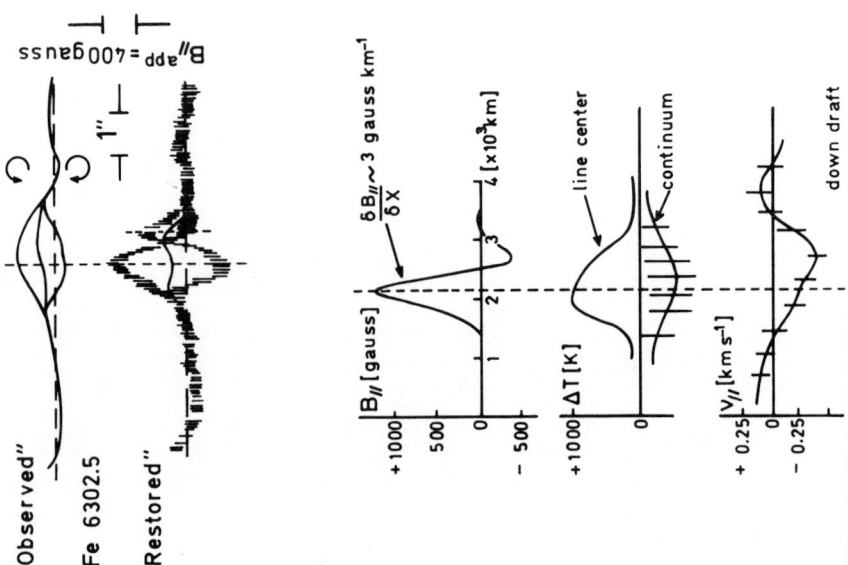

enhancement appears at this resolution level, the continuum appears rather depressed due to the location of the filigree in the intergranular space; the thermal enhancement of the filigree manifests only in the line center (gap). Again it is noted that no conspicuous velocity enhancement in the magnetic region is observed besides the small down-draft (~ 0.5 km s^{-1}) associated with the intergranular space surrounding the filigree.

The complex structures observed in the white light pictures over a concentrated magnetic field region, fig.1, as well as the presence of high horizontal fieldgradients including the occurence of both polarities, strongly suggest that these regions are filled with a magnetic rope rather than with a "point-like" flux-tube.

The results could thus favour the concept of segmented magnetic fluxes or ropes, see Childress and Soward (1985). More theoretical work on the formation of intense magnetic fluxes in quiet Sun regions are needed together with observable predictions.

REFERENCES

Childress, S., Soward, A.M. : 1985, J.R. Butchler et al (eds.), Chaos in Astrophysics, 223, Reidel Publishing Company.
Dara-Papamargariti, H., Koutchmy, S. : 1983, Astron. Astrophys. **125**, 280
Dara-Papamargariti, H., Alissandrakis, C., Koutchmy, S. : 1986, Solar Phys., submitted.
Deinzer, W., Hensler, G., Schüssler, M., Weishaar, E. : 1984, Astron. Astrophys. **139**, 426
Koutchmy, S. : 1977, Astron. Astrophys. **61**, 397
Koutchmy, S., Stellmacher, G. : 1978, Astron. Astrophys. **67**, 93
Spruit, H.C. : 1976, Solar Phys. **50**, 269

DRIFT VELOCITIES IN FLUX TUBES INFERRED BY SPATIALLY AVERAGED LINE BISECTORS

F. Cavallini, G. Ceppatelli
Osservatorio Astrofisico di Arcetri
Largo Enrico Fermi 5, 50125 Firenze, Italy

A. Righini
Istituto di Astronomia dell'Università di Firenze
Largo Enrico Fermi 5, 50125 Firenze, Italy

Abstract. The bisectors of the spatially averaged line profiles observed in active regions may be a useful tool to investigate the drift velocities within the flux tubes. We show that a simple heuristic model may account for the major effects observed in some Fe I line bisectors in active regions. The model satisfactorily reproduces the behaviour of the line shifts and asymmetry with the increasing magnetic field. We assume that an active region consists of static magnetic flux tubes embedded in an atmosphere where the convection is partially inhibited. However we cannot exclude a slight downflow up to 250 m/s in the flux tubes. We have also considered the case of non static flux tubes showing downflows as large as 1000 - 2000 m/s with disappointing results.

1 INTRODUCTION

Recent polarimetric observations, both in the enhanced network and in plages, have shown that the magnetic field has a discrete structure formed by tiny flux tubes (facular points). These magnetic elements show a field of about 1 - 2 kG, a size not larger than 200 km and they are 500 - 1000 K hotter than the surrounding photosphere (e.g. Beckers and Schröter, 1968; Frazier and Stenflo, 1972; Stenflo, 1973; Tarbell and Title, 1977; Wiehr, 1978; Koutchmy and Stellmacher, 1978; Solanki and Stenflo, 1984; Wiehr, 1985; Stenflo and Harvey, 1985). The value of the drift velocity inside the flux tubes is a very controversial matter. Stenflo et al. (1984) and Stenflo and Harvey (1985) find a practically null Doppler shift in flux tubes from the Stokes V zero crossing point together with a large Stokes V profile asymmetry. However, previous results by Giovanelli and Slaughter (1978) indicate a down draft of about 0.5 km/sec. This observation is confirmed by Wiehr (1985) and by Scholiers and Wiehr (1985) who find a downdraft not greater than 2 km/s inside the flux tubes with respect to the non magnetic surroundings, and a correlation between the amplitude of the blue-red asymmetry and the red-shift of the zero crossing of the Stokes V profiles. Recently Solanki (1986), discussing this problem, reached the conclusion that, within the accuracy of approximatively 250 m/s of his wavelength scale, non net flows are present in the photospheric layers of flux tubes at the disk center, in both active region plages and the quiet network.

An indirect mean for obtaining physical information on the flux tubes is the analysis of the line bisectors of spatially averaged line profiles in active regions. The observations show that in the network and in the active regions the line shift and the asymmetry are modified. In particular the characteristic "C-shape", shown in the quiet regions by medium-strong lines, is less pronounced and, in most cases, the whole line becomes shallower and shifts towards the red (e.g. Livingston, 1982; Kaising and Schröter, 1983; Koch, 1984; Miller et al., 1984; Cavallini et al., 1985a).

In this paper we discuss to what extent these changes of the line bisectors are compatible with systematic downflows in the flux tubes.

2 MEASUREMENTS

This study has been carried out on the three Fe I lines at 6297.8 Å, 6301.5 Å, and 6302.5 Å. The profiles of these lines have been repeatedly measured in quiet atmosphere at the disk center and in active regions with the spectro-interferometer installed at the Donati Solar Tower in Arcetri. Some of these measurements have been obtained by compensating the drift of the observed regions due to the solar rotation. In all cases a pointing accuracy of ±2 arcsec rms was secured by the automatic guiding system. Further details about the instrumental parameters and data reduction may be found in Cavallini et al. (1985a). We have no direct measurements of the magnetic field in the observed regions, but we may assume that, the larger the red-shift, the larger the magnetic field. This hypothesis is confirmed by the bisectors observed when, due to the solar rotation, the active region drifts (Cavallini et al., 1984, 1985a). We find that, for all the three investigated lines, there is a relationship between shifts and shapes. We have therefore collected the bisectors of each line in three groups characterized by a similar shape and shift, taking care that the bisectors of the three lines obtained in the same observations should fall in the same group. For each group we have superimposed the line bisectors in order to minimize the discrepancies in shape and we have assigned to each group the mean of the shifts relative to the individual bisectors. The resulting error on the shifts is about ±1.5 mÅ. The results of the measurements are summarized in Fig. 1, where the leftmost bisectors are the averaged ones obtained from all the measurements performed at the center of the disk in 1983.

3 CALCULATION OF LINE BISECTORS

We consider the atmosphere in the active region as a multicolumn structure. One column is always composed of normal convective photospheric plasma emitting the asymmetric line profile observed in quiet regions at the disk center (Cavallini et al., 1985b). The second column is the flux tube atmosphere. The line profiles arising from this "magnetic column" have been obtained by Solanki and Stenflo (1985) integrating the Stokes V profiles obtained in quiet network approximately at the disk center. The third column represents a non convective solar plasma emitting a Voigt profile passing through the core and the two points at half depth of the line observed in the quiet regions at the center of the disk.

The spatially averaged line profile is obtained by adding the line profiles emerging from each considered column. The result depends on the flux tube magnetic field, on the filling factors of the magnetic area and of the non convective area, on the drift velocities within the flux tubes, on the contrast factor of the flux tubes, and on the contrast of the atmosphere in between the flux tubes. The magnetic field of the

Fig. 1. The solid lines represent the bisectors observed at the quiet disk center. The active regions bisectors have been collected in three groups characterized by a similar shape and shift. For the sake of clarity the three groups have been arbitrarly shifted. The numbers specify the observed shifts in mÅ of the point at 78% of the continuum with respect to the "quiet" bisector.

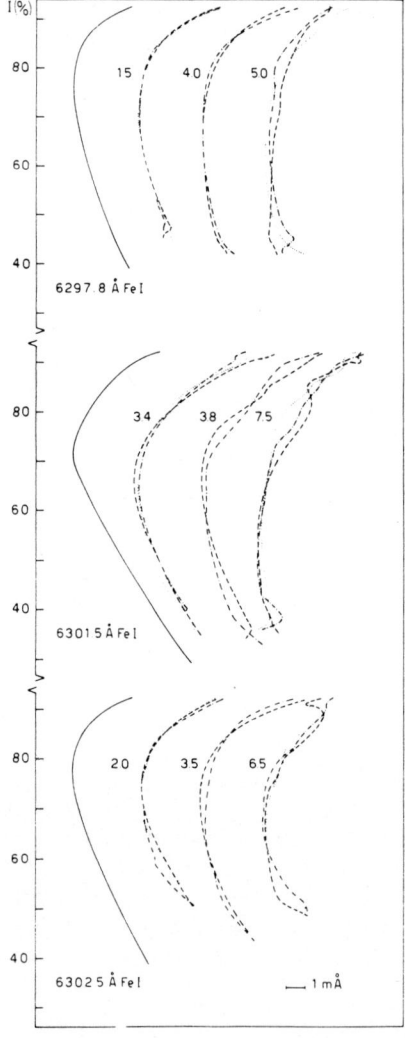

flux tubes is assumed to be 1500 G. We consider a filling factor such to produce an average magnetic field ranging from 100 to 300 G. The contrast factor of the flux tubes is assumed to be 1.4 (Muller and Keil, 1983), and the facular contrast to be 1. This requires that the background in between the facular points in the active region should be darker than the quiet photosphere (Hirayama et al., 1985). In our computations we keep fixed the magnetic field, and the contrast of the flux tubes. We also keep fixed the contrast of the facular area. This condition implies that the contrast of the material in the flux tubes must accordingly change in dependence of the adopted filling factors for the flux tubes and the non convective columns. We consider free parameters the flux tubes drift velocity and the filling factors. The filling factor of the magnetic area determines the average field of the facula region. The final spatially averaged profile results as a blend of the single column profiles weighted accordingly to the contrast and to the filling factors. The non convective column is assumed to give rise to an unshifted profile. The line arising from the magnetic atmosphere will be relatively shifted according to the assumed drift velocity. We recall that this profile is a blend of the Zeeman components since it has been obtained integrating the Stokes V profile. Finally the convective atmosphere is assumed to emit a blue shifted profile due to the convection.

4 RESULTS AND CONCLUSIONS

We note that the observed bisectors in active regions show a global red-shift, a straightening near the line core and an increasing bend near the continuum. The first numerical experiments have shown that, while the suppression of the convection explains the straightening of the lower part of the bisector and the shift, the flux tubes modify the higher part of the bisector. This suggests that in reality both phenomena contribute in changing the line asymmetry and shift into active regions. We have therefore tried different values of the free parameters, i.e. flux tubes drift velocity and filling factors to reproduce the shapes and shifts of the observed bisectors. No fitting was possible supposing an upward velocity or downward velocities larger than 250 m/s. In Fig. 2 we plot the line bisectors obtained by supposing no down draft in the flux tubes. These bisectors (solid lines) are compared with the experimental ones. In this particular case the non convective atmosphere ranges from 15 to 58% of the total surface, and the average magnetic field ranges from 160 to 220 G. These values also fit within ± 1.5 m$\overset{\circ}{A}$ the observed shifts.

This simple heuristic model may satisfactorily account for the behaviour of the bisector shape and shift observed in active regions on some Fe I lines. The main conclusions are that the contribution of the line profile arising from the flux tubes is fundamental to correctly reproduce the observations and that no downdraft in flux tubes must be assumed.

Fig. 2. The leftmost solid lines represent the bisectors observed at the quiet disk center. The others represent (from left to right) the bisectors calculated supposing that the area of the non convective atmosphere is 15%, 30%, 58%, that the average magnetic field is 160, 200, 220 G, and that no downflow is present in the flux tubes. The observed bisectors (the same as in Fig. 1) have been superimposed on the computed ones minimizing the discrepancies in shape. The numbers specify the computed shifts defined as in Fig.1.

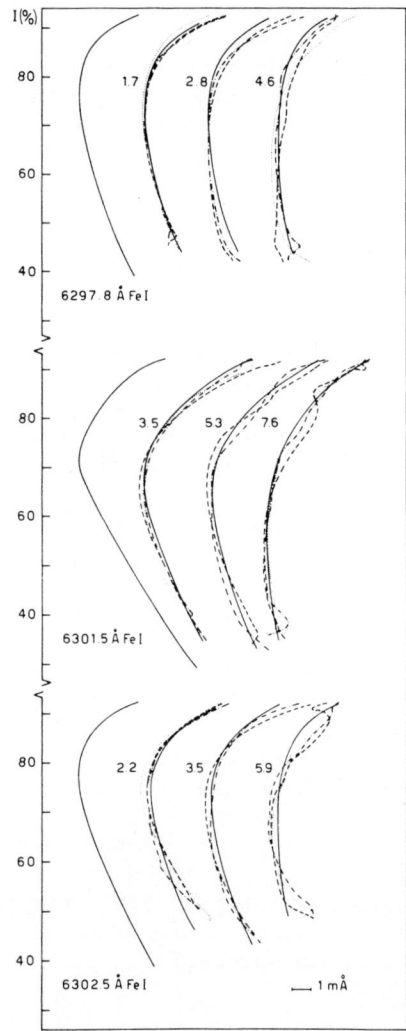

REFERENCES

Beckers, J.M., Schröter, E.H. (1968). Solar Phys. 4, 142.
Cavallini, F., Ceppatelli, G., Righini, A. (1984). In "Small-Scale dynamical processes in quiet stellar atmospheres", S. Keil ed., Sacramento Peak National Observatory, p. 334.
Cavallini, F., Ceppatelli, G., Righini, A. (1985a). Astron. Astrophys. 143, 116.
Cavallini, F., Ceppatelli, G., Righini, A. (1985b). Astron. Astrophys. 150, 256.
Frazier, E.N., Stenflo, O.J. (1972). Solar Phys. 27, 330.
Giovanelli, R., Slaughter, C. (1978). Solar Phys. 57, 255.
Hirayama, T., Hamana, S., Mizugaki, K. (1985). Solar Phys. 99, 43.
Kaisig, M., Schröter, E.H. (1983). Astron. Astrophys. 117, 305.
Koch, A. (1984). Solar Phys. 93, 53.
Koutchmy, S., Stellmacher, G. (1978). Astron. Astrophys. 67, 93.
Livingston, W.C. (1982). Nature 297, 208.
Miller, P., Foukal, P., Keil, S. (1984). Solar Phys. 92, 33.
Muller, R., Keil, S.L. (1983). Solar Phys. 87, 243.
Scholiers, W., Wiehr, E. (1985). Solar Phys. 99, 349.
Solanki, S.K. (1986). Astron. Astrophys. (submitted).
Solanki, S.K., Stenflo, J.O. (1984). Astron. Astrophys. 140, 185.
Solanki, S.K., Stenflo, J.O. (1985). (Private Communication).
Stenflo, J.O. (1973). Solar Phys. 32, 41.
Stenflo, J.O., Harvey, J.W. (1985). Solar Phys. 95, 99.
Stenflo, J.O., Harvey, J.W., Brault, J.W., Solanki, S. (1984). Astron. Astrophys. 131, 333.
Tarbell, T.D., Title, A.M. (1977). Solar Phys. 52, 13.
Wiehr, E. (1978). Astron. Astrophys. 69, 279.
Wiehr, E. (1985). Astron. Astrophys. 149, 217.

JOINT DISCUSSION ON TOPICS OF SESSIONS 3 AND 4, SUMMARIZED BY THE CHAIRMEN

J.O. Stenflo
Institut für Astronomie, Zürich, Switzerland

W. Mattig
Kiepenheuer-Institut für Sonnenphysik, Freiburg, FRG

The Joint Discussion was structured in three parts:
(1) Discussion on the diagnostics of Stokes observations, as reviewed by S.K. Solanki. (2) Discussion of the limits and the optimization of magnetic field observations, as a follow-up of the invited talk by E. Wiehr. (3) Questions relating to the "Short Contributions".

The first part started off by considering some disadvantages of many-line analysis. Thus, the question was raised (Wöhl) whether errors in the atomic parameters and influences of blends would not be a problem. Solanki clarified that blends will contribute to the scatter, and therefore it is safer to use a larger number of lines to reliably bring out the empirical relations. Not much precision in the atomic data is needed for the regression analysis used, except that there is a problem in finding good laboratory wavelengths for the Fe II lines (needed to determine the absolute line shifts). Zwaan commented that so far no calculations exist which conclusively show that atomic orientation (i.e. departures in the populations of the Zeeman states) is unimportant in flux-tubes.

Questions concerning the internal dynamics of fluxtubes were raised by several participants. Wiehr pointed out that although spectral smearing due to insufficient spectral resolution could explain observed apparent downdrafts in fluxtubes, he has occasionally observed blueshifts as well. Solanki agreed that with sufficient resolution you might see both blue- and redshifts from oscillations and waves. Maltby asked if the very high deduced fluxtube velocities of 3.5 km/s had been supported by any other observations. Solanki explained that for single spectral lines the velocities are mixed with other effects and can only be disentangled by using many lines. Schüssler remarked that it cannot be taken for granted that when you average oscillatory dynamics in space and time, the average shift of Stokes V will be zero.

Spruit pointed out that in an ensemble of fluxtubes of various sizes, the smaller fluxtubes would contribute more to an average fluxtube model than the larger fluxtubes at greater heights. Deeper in the atmosphere the larger fluxtubes would dominate. In his reply, Solanki agreed that this effect may be present but it should be checked quantitatively. At present it is unfeasable to construct models based on multiple fluxtubes due to the resulting explosion in the number of free parameters.

Mattig set the tone of the second part of the Joint Discussion by expressing considerable optimism concerning the possibility of improving the spatial resolution of the spectrograph to match that of the preceding telescope, in contrast to the critical view that had been voiced by Wiehr. In the following discussion, various participants contributed to this optimism by suggesting different future improvements.

Thus Wiehr and Schröter commented on the superb seeing they have experienced on Izana. Brandt pointed out the great, still unused potential of the FTS. Righini then mentioned the possibilities with tunable Fabry-Perot filters, while Wöhl stressed the advantages of CCD detectors. Zwaan emphasized the importance of achieving high S/N ratio. Brückner remarked that new technologies open many ways of enhancing the observations, like adaptive optics. Brandt added the importance of correlation trackers.

Finally the discussion relating to the "Short Contributions" is summarized:

Paper by <u>Brandt and Solanki:</u> W. Livingston asked whether blends could influence the results of bisector curvature changes in the line wings. It was answered that both the straightening of the bisector in the lower part and the increasing redshift in the upper part (line wing) is consistently seen in a dozen of medium-strong Fe I lines.

Paper by <u>Pahlke:</u> For calculating the line profiles the Unno theory of line formation in magnetic fields with depth-independent magnetic field and Doppler width is used (clarification in response to question by Mattig).

Paper by <u>Koutchmy and Stellmacher:</u> The question by Wiehr whether the fluxtube inclination reported by the authors fits his 1978 finding of a westward displacement of the magnetic maximum with respect to the Ca+K maximum was answered by pointing out that Ca+K intensities are indeed not well correlated with the magnetic field, but a preferred direction of the displacement has so far not been studied.
The problem of signal-to-noise ratio when using photographic procedures was discussed (question by Brandt). It was pointed out that in special cases, e.g. integration in the spectral direction over 50 mA, a S/N-ratio of 1000 is possible.

Paper by <u>Righini:</u> Brandt pointed out that the FTS observations show (a) a decrease of line depth, and (b) a steepening of the line bisector in the wings, with increasing filling factor. This is not seen in the Arcetri data. It was answered that this cannot be measured with the used bisector program.

Types of Magnetic Flux Emergence

A.M. Title
Lockheed Research Laboratory, Palo Alto, California, USA

ABSTRACT. For the past two decades magnetic field and flow data has been collected at a number of observatories world wide. This data, which spans a complete magnetic solar cycle, has allowed study of collective patterns of flux emergence. It is now clear that a significant fraction of solar flux emerges from subsurface sources, which have their own characteristic rotation rate and which last years. The existence of these sources means that flux emergence can not be studied as an isolated phenomena. Consideration must be given to preexisting flux structures both above and below the surface. Emergence, recombination, and submergence may not be treated as separate processes. New very sensitive video magnetographs have shown that there exists a weak component (1 to 40 gauss) of the solar magnetic field. Perhaps this weak field component is a major factor in the entire picture of flux emergence.

1 INTRODUCTION

There are good theoretical arguments (Schussler 1983) that the flux we see in the surface was generated at the bottom of the convection zone. How the flux ropes are removed from the bottom of the convection zone, how they traverse the convection zone itself, and then emerge through the solar surface is not well understood. Because a significant fraction of flux that emerges does so in complexes of activity (Gauzauskas *et al.* 1983) and nests (Castenmiller *et al.* 1986), emergence, recombination, and submergence may not be treated as separate processes.

In the not too distant past flux emergence was synonymous with the birth of active regions. Virtually the all the flux outside of spots was thought to be in the form of 1500 gauss flux tubes that had been fragmented off and advected from some active region. The Babcock mechanism was responsible for the maintanence of the cycle. Today we have more questions than answers. We must question whether all the flux has emerged from active regions, whether all the flux is strong, and whether the diffusion mechanism is responsible for carrying the flux to the pole and thus renewing the cycle.

2 HOW MANY TYPES OF FLUX ARE THERE?
2.1 *Strong field*

The traditional picture of solar activity has flux

emerging in active regions in the form of pores and sunspots. Active region bipoles have fluxes of 10^{20} to 4×10^{22} Mx and separations of 60,000 to 100,000 km. In the initial stages the emergent flux coalesces into pores which then often merge to form spots. In the decay phase the spots disappear at a rate proportional to their total flux. The flux is finally dispersed by a combination of meridonal flow and diffusion. A tiny fraction makes it to the pole to start the new solar cycle.

Recently Zirin (Zirin 1986) has observed another method of formation of a spot - expansion of pores in place. The magnetic field emerges at its predestined location where the pores begin to form. The pores then expand in place to form the spots. During the process there was little change in the total flux in the area that became the leader spot. It is suggested that plage may be formed by a similar process. That is, that the plage field or at least some of the plage field emerges in place.

Besides the active region flux, the existence of ephemeral regions (Martin and Harvey 1979) has also been recognized although their role in the solar cycle is still unclear. These structures are typically bipoles 3 to 5 arc seconds in diameter which are separated by 5 to 10 arc seconds. Fluxes are in the range of 3×10^{18} to 1×10^{20} Mx. They have lifetimes which range between 8 and 20 hours and a birth rate of one per square arc minute per hour. In the past there has been serious argument that ephemeral regions are the tail end of the distribution of active regions. But distribution functions of total flux versus number for all magnetic structures shows that there are many more ephemeral regions then would occur if the active region (fluxes greater than 10^{21} Mx) distribution is extrapolated down to the ephemeral region values.

The variation in the number of ephemeral regions over the solar cycle has been puzzling because of their association with X Ray bright points (Golub et al. 1979). Ephemeral region number is in phase with the cycle and bright point number is 180 degrees out of phase. Some of the confusion has been resolved by the recognition that a significant fraction of the X Ray bright points are caused by opposite polarity flux elements coming into close association (Harvey 1985).

Work in progress may show that some of the structures previously identified as ephemeral regions are, like a fraction of the bright points, chance encounters of opposite polarities (Martin et al. 1985). Thus, there exists the possibility that ephemeral regions are in fact just the tail end of the active region phenomena. However, it is hard to understand how flux loops emerging from the bottom of the convection zone can have a spacing of only 5 to 10 arc seconds.

The entire active region phenomena is currently under serious examination because of the discovery and the recognition of the importance of active complexes (Gaizauskas et al. 1983) and the related phenomena of nested sunspots (Castenmiller et al. 1986). These observations show that active region flux emergence is not completely random, but a significant fraction of the regions rise from sources at particular longitudes and latitudes in coordinate systems with unique rotation rates. Because of the existence of these persistent sources, the picture of emergence and distribution of flux as evolution from isolated rising flux ropes must be reexamined. We must recognize that under an active complex there may be a tangle of flux ropes through which new flux must travel to emerge. It is very likely, therefore, that there are subsurface reconnections. These reconnections would affect the evolution of the emerged flux and modify the emergence patterns of the second generation into the region.

Evidence for large scale order in the eruption of flux also comes from an analysis of the rotation rate of the source surface fields (2.5 Solar Radii) by Hoeksema and Scherrer (1987). They find that there are two rotation rates at the height of the source surface - 27 and 28 days. Although both rates are present both in the northern and southern hemispheres, the 28 day period has its maximum amplitude in the south and the 27 day in the north. Preliminary analysis (Hoesksema, private communication) shows the same periods are found in the surface magnetic field; and Wilcox et al.(1970) reported this same result from their analysis of Mt. Wilson magnetograms. The Stanford records cover ten years, and the width of the peaks in the fourier spectrum are consistent with an oscillation of at least that duration.

It has been observed that strong field ($B \sim 1200$ gauss) exists nearly everywhere on the sun (Howard and Stenflo, 1972; Tarbell and Title 1977). It has been assumed that isolated high field strength flux tubes have been advected to their positions regardless of how far they are from a possible source active region. Can all isolated strong field structures explained in this manner? Are there any other prototypical types of flux emergence? For example, could magnetic flux emerge in "quasi active regions"? That is, emergence of well separated flux, 30,000 to 100,000 km, that does not have a birthing phase with the majority of the flux in a well collected form.

2.2 Weak field

In addition to active region and ephemeral region flux there seems to be another type of solar magnetic field structure - weak bipoles. In 1975 Livingston and Harvey (1975) recognized an internetwork component in their best magnetograms. Since that time there has been good reason to believe that the internetwork features had a strength on the order of ten gauss. There is now strong evidence from the magnetic measurements of Martin et al.(1984) and Zirin (1985), that there is a pervasive weak field component to the solar flux.

These weak bipoles are 3 to 4 arc seconds in diameter with separation of 4 to 8 arc seconds and have fluxes in the range 10^{16} to 5×10^{17} Mx. Lifetimes are on the order of hours. Unlike ephemeral regions, the weak bipoles are just as visible near the limb as they are at disk center, suggesting a strong horizontal component to their geometry.

The actual field strength of the weak component depends critically on the calibration of the Big Bear Magnetograph (Shi et al. 1986). The pixel size of the magnetograph is one arc second, and an image is acquired every 1/30 second. The published calibration measurements from Big Bear Solar Observatory (BBSO) state that 5 gauss can be detected in 512 integrations in good seeing, and that the signal to noise ratio improves with the square root of the number of magnetograms that are added together. Most weak field studies have been carried out using 4096 integrations, which implies a detectable flux of 1.8 gauss. 4096 integrations requires about 5 minutes and even under good conditions (2 arc second image jitter radius during 5 minutes) a minimum of 1.6×10^{16} Mx might be detected. Averaging over larger areas can reduce the minimum level level of field strength, but it raises the minimum flux level; i.e., averaging over a 4 arc second area could detect about 0.8 gauss and 3.2×10^{16} Mx. Larger area integrations are possible, but at some point the dipoles must overlap.

Very high resolution magnetograms (Ramsey et al. 1977), 0.5 arc second, do not show the weak field structures. A compact flux structure of 100 gauss (2×10^{17} Mx) can be detected on the high resolution magnetograms, so that all weak field structures would not be seen if they were strong and compact; but a significant number would be detectable. The data supports the concept that the intrinsic strength is less than 50 gauss. A best guess might be that the structures seen at Big Bear have a field strength of 5 to 25 gauss. However, without very high resolution magnetograms and/or strength discrimination there is no way of ruling out the idea that the "weak" field is in fact caused by 1/10 arc second diameter dipoles separated by 6 to 10 arc seconds.

The existence of the weak field does not change the fact that the majority of the magnetic energy stored in the solar atmosphere is in kilogauss flux tubes. Using the limiting assumption that a 50 gauss weak field covers all the area that does not have strong field, only 50 percent of the flux in quiet regions is weak (Tarbell et al. 1979). In quiet areas 90 percent of the magnetic energy (E is proportional to B^2) is from the strong field. In plage at most 10 percent of the flux is weak, and virtually all the magnetic energy is in the strong field.

The explanation for the existence of kilogauss structures is convective collapse initiated by the draining of the nearly vertical flux tubes. Another mechanism needs to be found to explain why weak field structures don't collapse, especially when 10^{17} Mx structures are often seen in strong field. Spruit et al. (1986) have suggested that the bipoles are essentially vertical doughnut like structures caused by reconnection below the surface. Because mass is trapped in the doughnut by the field, collapse is not possible.

The weak field is apparently nearly everywhere on the sun. However, it is reasonable to ask whether all of the weak field has the same physical origin. It may be that inter and intra active region weak fields are created by different processes.

3 HOW ARE EMERGENCE AND THE DISSAPEARANCE OF ACTIVE REGION DIPOLE MOMENT RELATED?

A typical 10^{22} Mx active region has a separation of 60,000 km which implies a 6×10^{31} gauss cm^3 dipole moment. By comparison a 10^{19} Mx ephemeral region with a separation of 7000 km has a 7×10^{27} gauss cm^3 dipole moment, and a weak field bipole has a 4×10^{23} gauss cm^3 dipole moment. Dipole moment is not conserved, but to destroy dipole moment requires that opposite polarity flux be moved together. Removal of an active region from the surface requires either interdiffusion of the active region itself, the eruption of another oppositely oriented active region, eruption of nearly 10^5 oppositely oriented ephemeral regions between the poles or about 10^8 oriented weak field bipoles, or some combination of the above.

In order to remove the active region dipole moment with smaller dipoles, the smaller regions must be between the poles of the active region, say in a 60,000 km square box (3.6×10^9 km^2). Since the total solar flux does not grow rapidly during maximum, the active regions must disappear on the time scale they are born on average. Near solar maximum an active region is born about once a day. Suppose most of the dipole moment disappears in four days. To accomplish this task with oriented ephemeral regions, which on average occupy 3×10^7 km^2, would require one to be born every 7 minutes in every 3×10^7 km^2 area ($3.6 \times 10^9 \times 4 \times 24 \times 60/(10^5 \times 3 \times 10^7)$). For weak fields to have a similar effect would require an oriented birth rate of one per 600 km square box (3.6×10^5 km^2) per hour.

From the above, the dipole moment of a typical active region can not be removed by ephemeral regions or the component of the weak field seen by Martin and Zirin. However, an oriented field of weak dipoles with an order of magnitude less flux per dipole emerging every 10 minutes could remove most of an active region in days. An ensemble of oriented weak dipoles, which are mostly below the current detection limit, might be created if flux in the active region reconnected deep below the surface (Spruit et al. 1986).

The Leighton - Mosher (1977) type diffusion model removes most of the flux while the region is young by having the opposite polarity fluxes interdiffuse. However, organized flow fields which carried the opposite polarities together would also make significant contribution to rate of flux disappearance, and perhaps even obviate the need for another mechanism of flux removal.

The currently accepted value of the diffusion coefficient, 300 km^2 s^{-1}, carries flux to the pole in about 60 years. A meridional flow of 10 to 30 m s^{-1} carries flux from the equator to the pole in 3.5 to 1 year. Hence, for all practical purposes it is meridonal flow which carries flux to the pole in diffusion models. All the diffusion constant does is predict something about the shape of active regions early in the decay phase.

4 WHY IS THE SITUATION SO CONFUSED?

The picture drawn above is quite confusing. I think it suggests that we do not understand some of the basic organizations and mechanisms of flux emergence. We do not understand the origin of a rigidly rotating component of magnetic field sources. We do not understand the nature of weak field and its importance in the local redistribution of flux. These shortcomings in understanding are to be expected in the evolution of the comprehension of any complex physical system. It also reflects the fact that we are just beginning to appreciate the implication of the synoptic magnetic and doppler data that has been collected over the last solar cycle. And it points up the need to continue to collect both high spatial and temporal resolution magnetic and flow field data over the next solar cycle.

5 WHERE SHOULD EFFORT BE DEVOTED TO CLARIFY OUR UNDERSTANDING OF EMERGING FLUX?

The critical factor in future observations of emerging flux should be careful calibration of the measurements. The areas of study suggested here are not intended to be inclusive, but rather to be examples of programs that would clarify our understanding.

1) Our understanding of the emergence would be greatly enhanced if we could establish whether weak fields exist or not. A time sequence of magnetograms with 1.5 - 2 arc sescond resolution formed from images in both 5250 and 5247 with sensitivities of 3×10^{16} Mx would answer this question. Forty minutes of observations with a magnetogram pair every 5 minutes would pin down the most critical question of existence.

2) The line ratio technique should establish the field strength

issue; however, it would be very desirable to have a pair of very high resolution (0.5 arc second or better) magnetograms in 5250 and 5247. This might be accomplished by storing the circularly polarized images in both lines and creating the magnetograms after the observing run. The data analysis would require distortion removal procedures similar to those described in Topka et al.(1986).

3) In order to make real progress in the emergence of flux it will be necessary to understand the process by which active regions emerge. An accurate accounting of where the flux emerges and where it goes will have to be obtained. This is a very difficult problem from the ground. From space a 15 cm telescope with a filter to obtain magnetograms in line pairs such as 5250-5247 and a doppplegram in a magnetically insensitive line could, in a year of operation, collect data to answer some of the fundamental questions currently being investigated. Observations every hour would ideal. Observations every two hours would suffice.

Since it is probably unrealistic to expect a free flying spacecraft magnetograph in the next few years it will be necessary to attempt to do such measurements from the ground. This will require a dedicated instrument and an elaborate processing system to produce the magnetograms from images that have the distortion removed.

4) Even more difficult to do from the ground is high resolution magnetograms taken in close time sequence with granulation maps, dopplergrams at several heights, and images in chromospheric lines. Such data sets need to be taken over time scales on the order of hours, and a complete set of images should be taken in less than one minute. With such data sets it should be possible to make significant progress on the relationship between granulation and mesogranulation scale flow fields and the location of the magnetic field.

Initial work with this type of data (Title et al. 1987a) has indicated a relationship between both granule boundaries and somewhat larger doppler cells and the magnetic field. However, there are some serious unresolved problems in relation between the flow and magnetic field. A major difficulty is that the fine scale magnetic field is stable for at least an hour, while the granulation pattern has a measured lifetime of six minutes. Some of the difficulties may be resolved by the SOUP data from Space Lab 2 which has indicated that in magnetic field regions the granule pattern lifetime may be 15 minutes or more (Title et al., 1987b).

High resolution magnetograms interleaved with images in temperature sensitive lines should help to further resolve the confusion between bright structures, higher temperature regions, and the locations of magnetic flux tubes. While it is well known that bright structures and magnetic features are spatially related, the highest resolution magnetograms have poorer detailed correlations with brightness structures, than the same images smeared to resolutions of 1 to 3 arc seconds. That is, relationships that are good at 2 to 3 arc second resolution deteriorate greatly at 0.5 to 1 arc second resolution.

5) The value of magnetograms will be limited by the quality of their interpertation. The work by Wilson and Simon (1983) and Topka et al.(1986) has shown that magnetic field can apparently disappear in place. There have been several suggestions for mechanisms to account for the phenomena. Wilson (1986) believes the explanation is surface currents, while van Ballegooijen (1985) has shown radiation transfer effects can lower the amount of circular polarization signal, if the flux tubes are sufficiently small. In any case, it is required that we develop models that are sophisticated enough that the measurements can be turned into physical parameters.

REFERENCES

Castenmiller, M.J.M., Zwaan, C., van der Zaln, E.B.J. (1986). Solar Phys., 105, 237.
Gaizauskas, V., Harvey, K.L., Harvey, J.W., Zwaan, C. (1983). Ap. J., 265, 1056.
Golub, L., Krieger, A.S., Harvey, J.W.,Viana, G. (1979). Solar Phy., 53, 11.
Harvey, K.L. (1985). Aust. J. Phys., 38, 875.
Hoeksema, T. and Scherrer, P. (1987). Ap. J. in press 1987.
Howard. R. and Stenflo, J.O. (1972). Solar Phys., 22, 402.
Livingston, W.C., Harvey, J.W. (1975). Bull. Amer. Astron. Soc., 7, 346.
Martin, S.F. and Harvey, K.L. (1979). Solar Phys., 40, 87.
Martin, S.F. (1984). In Small Scale Dynamical Processes in Quiet Solar Atmospheres ed. S. Kiel, p. 30. Sunspot, New Mexico: NSO, Sacramento Peak.
Martin, S.F., Livi, S.H.B., Wang, J. (1985). Aust. J. Phys., 38, 929.
Mosher, J. (1977). Thesis, California Institute of Technology.
Ramsey, H.E., Schoolman, S.A., Title, A.M. (1977). Ap. J., 215, L41.
Schussler, M. (1983). In Solar and Stellar Magnetic Fields: Origins and Coronal Effects ed. J.O. Stenflo, p. 213. Dordrecht: Reidel.
Shi, Z., Wang, J., Patterson, A. (1986). BBSO Report 0257.
Spruit, H.C., Title, A.M., van Ballegooijen, A.A. (1986). Solar Phys. submitted.
Tarbell, T.D., Title, A.M. (1977). Solar Phys., 52, 13.
Tarbell, T.D., Title, A.M., Schoolman (1977). Ap. J., 229, 387.
Title, A.M., Tarbell, T.D., Topka, K.P. (1987a). Ap. J., in press.
Title, A.M., Tarbell, T.D., SOUP Team (1987b). Workshop of Theoretical Solar Physics II ed. G. Athay, NCAR Pub. in press.
Topka, K.P., Tarbell, T.D., Title, A.M. (1986). Ap. J., 306, 304.
van Ballegooijen, A.A. (1985). In Measurements of Solar Vector

Magnetic Fields, ed. M. J. Hagyard, p. 322. NASA Conf.
Publ. 2374.
Wilcox, J.M., Schatten,K.H., Tannenbaum,A.S., Howard, R. (1970).
Solar Phys., 14, 255.
Wilson, P.R., Simon, G.W. (1983). Ap. J., 273, 805.
Wilson, P.R. (1986). Solar Phys., 106, 1.
Zirin, H. (1985). Aust. J. Phys., 38, 961.
Zirin, H. (1986). BBSO Report 0266.

OBSERVATIONS OF THE MAGNETIC FINE STRUCTURE OF A FACULA

J.C. del Toro Iniesta[1], M. Semel[2], M. Collados[1]
(1) Instituto de Astrofísica de Canarias, 38200-La Laguna, Tenerife, Spain.
(2) D.A.S.O.P. Observatoire de Meudon, 92195 MEUDON PL CEDEX, France.

ABSTRACT

Simultaneous spectropolarimetric observations of a facula have been carried out in 10 spectral lines with a spatial resolution of 1". Local variations of the magnetic field strength and the filling factor of fluxtubes were obtained. The analysis of the velocities inside fluxtubes shows that positive and negative Doppler shifts are present, at the same time, at different points of the facula.

1 INTRODUCTION

The use of simultaneous spectropolarimetric observations in several lines has become the most reliable method in order to study the non-resolved structure of the solar magnetic fields, where the ability to eliminate the different influences of thermodynamical parameters in the line formation is its most important success (see, e.g. Howard and Stenflo, 1972; Stenflo, 1976; Semel, 1980, 1981; Solanki and Stenflo, 1984, 1985; Scholiers and Wiehr, 1985).

In this paper, we combine the use of several spectral lines with a spatial resolution as high as possible, because only average information can be obtained when low spatial resolution spectra are used (Stenflo et al., 1984). This is an attempt to clarify some controversies concerning the values of magnetic field strength and the velocities in fluxtubes (see, e.g. Stenflo, 1976; Stenflo et al., 1984; Bezanger and Semel, 1986; Giovanelli and Slaughter, 1978; Scholiers and Wiehr, 1985).

We investigate the local variations in different magnitudes which characterize the fluxtubes: magnetic field strength, relative velocities and "filling factor". Finally, the importance of increasing the spatial resolution is briefly discussed.

2 OBSERVATIONS AND DATA REDUCTION

The observations were carried out in October 1974 with the McMath telescope of the Kitt Peak National Observatory (operated by the Association of Universities for Research in Astronomy, Inc., under contract with the National Science Foundation), using a polarimetric analyser as described by Semel (1980). A photographic IIIaJ plate was employed as detector. The observed region which covered 120" -in the direction of the slit-, contained a facula. Two simultaneous spectra were obtained from each point of the region (1"x1" spectrograph slit width x microphotometer slit width): 1/2 (I+V) and 1/2 (I-V) -where I and V are the first and the fourth Stokes parameters-. After digitalisation, the spectral dispersion was 5.6 mA/pixel.

TABLE I

Observed spectral lines

Number	Wavelength (A)	Element	Equiv. width (mA) W_{bi}	Eff. Landè fac
1	5225.5	Fe I	68	2.25
2	5228.4	Fe I	60	0.875
3	5229.85	Fe I	124	1.5
4	5242.5	Fe I	80	1.0
5	5243.8	Fe I	60	1.5
6	5247.06	Fe I	59	2.0
7	5247.57	Cr I	76	2.5
8	5250.2	Fe I	62	3.0
9	5250.65	Fe I	104	1.5
10	5253.46	Fe I	75	1.5

Ten spectral lines were selected (see Table I) in the spectral region (5223-5255 A), with the following characteristics: a) equivalent widths are greater than 50 mA, in order to achieve a high signal to noise ratio; b) the lines belong to neutral atoms because lines in ionic spectra behave differently (Semel, 1981); c) they are blend-free, except the 5250.65 A Fe I line.

For each spectral line and for each of the 60 observed solar points, we determined the equivalent width, the continuum intensity, the "apparent" magnetic field strength -using the "center of gravity method" (Semel, 1967; Rees and Semel, 1979; Semel, 1981)- and the Doppler velocity, relative to the mean position of the line.

3 THE TWO-COMPONENT MODEL

With a two-component model (fluxtubes and background) and using the "center of gravity method" to measure the position of the lines, the following expression can be obtained:

$$\frac{W_i}{W_{b,i}} B_i = \frac{W_i}{W_{b,i}} B_t + (1 - \alpha) \frac{I_{cb}}{I_c} (B_b - B_t)$$

where B_i and W_i are the "apparent" magnetic field strength and the measured equivalent width for the line "i"; I_c the observed continuum intensity; $W_{b,i}$ and I_{cb} the corresponding quantities for the background; and B_t and B_b the magnetic field strength for fluxtube and background respectively. This is a linear relation between $(W_i/W_{b,i})B_i$ and $(W_i/W_{b,i})$, where B_t is the slope and the constant term depends on the partial surface occupied by the fluxtubes, α, leaving B_b as free parameter. Note that, although B_b is a free parameter, its absolute value should not exceed that of the "apparent" magnetic field strength measured in any line. Thus, we have an upper limit for B_b.

An analogous expression holds for the velocities:

$$\frac{W_i}{W_{b,i}} v_i = \frac{W_i}{W_{b,i}} v_t + (1-\alpha) \frac{l_{cb}}{l_c} (v_b - v_t)$$

where the v's have the same significance as the corresponding B's.

Fig 1: Two observed facular points with "good" (a) and "bad(b) fits. Crosses indicated lines not included in the fit.

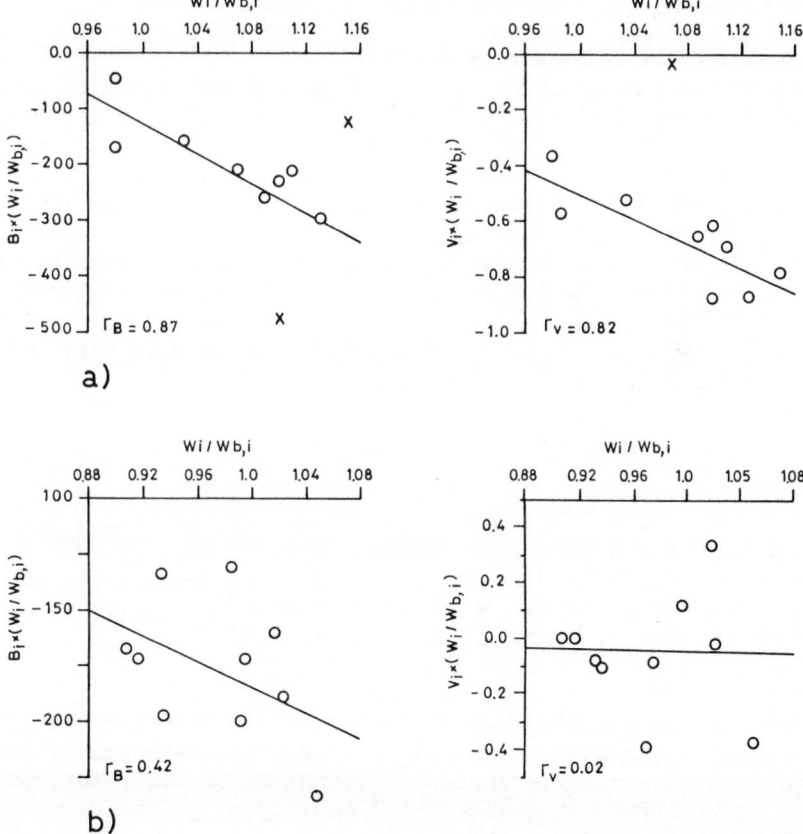

Therefore, we can calculate B_t, α, v_t, and v_b with only one free parameter, B_b, fitting the observational data to the above relationships. However, accurate estimates are needed for the equivalent widths, $W_{b,i}$, and continuum intensity, I_{cb}, for the background. We have adopted average values for these quantities over the observed quiet zone, which is defined by the absence of signal in the V-profiles. In each studied point of the facula, the spectral lines whose data present a deviation greater than 2 σ are eliminated from the fit. Although this procedure is necessary in almost all the points, it may be argued that there is no solar origin because these eliminated lines are different from one point to another. As examples, Figure 1 shows the fits corresponding to two points in the facula. While the data fit well enough to the straight line in part (a), the correlation coefficients in part (b) are small, may be because of the noise in the spectra, may be because the two-component model is not valid everywhere in the facula.

4 RESULTS AND CONCLUSIONS

A large range of values for magnetic field strength in fluxtubes (350-1650 G) and α (0.67-0.02) have been obtained, indicating that both the magnetic field strength and the surface density of fluxtubes may vary inside the same facula.

Values for the velocity of the material inside fluxtubes relative to that of the background ($|v_t - v_b|$) as high as 6 km/s, together with "normal" values of 1 km/s (see, e.g. Giovanelli and Slaughter, 1978; Scholiers and Wiehr, 1985) have been found. Moreover, both positive and negative velocities are present. Thus, it seems that fluxtubes with material downflow and upflow coexist within the same facula.

Finally, using all the data as a whole, average values of B_t=600 ± 50G, α =0.14±0.07 and v_t=-0.5 ± 1.0 km/s have been obtained, showing the simulated effect of a loss of spatial resolution. This study suggests that the physical conditions of fluxtubes vary markedly across one single facula.

REFERENCES

Bezanger, C., Semel, M. (1986). In preparation.
Giovanelli, R.G. & Slaughter, C. (1978). Solar Phys., 57, 255.
Howard, R. & Stenflo, J.O. (1972). Solar Phys., 22, 402.
Rees, D.E. & Semel, M. (1979). Astron. Astrophys., 74, 1.
Scholiers, W. & Wiehr, E. (1985). Solar Phys., 99, 349.
Semel, M. (1967). Ann. d'Astrophys., T. 30, n° 3.
Semel, M. (1980). Astron. Astrophys., 91, 369.
Semel, M. (1981). Astron. Astrophys., 97, 75.
Solanki, S. & Stenflo, J.O. (1984). Astron. Astrophys., 140, 185.
Solanki, S. & Stenflo, J.O. (1985). Astron. Astrophys., 148, 123.
Stenflo, J.O. (1976). In Basic Mechanisms of Solar Activity, eds. V. Bumba & J. Kleczek, I.A.U. Symp. 11, 69.
Stenflo, J.O., Harvey, J.W., Brault, J.W. & Solanki, S. (1984). Astron. Astrophys., 131, 333.

ASYMMETRY OF STOKES PROFILES ACROSS A SUNSPOT-MEASUREMENTS

K.S.Balasubramaniam [*]
Joint Astronomy Program
Indian Institute of Astrophysics
Bangalore 560034, India

[*]Dept. of Physics
Indian Institute of Science
Bangalore 560012

Introduction

Measurements of asymmetries of the polarization content of Zeeman broadened spectral lines either in the broad-band (Illing, Landman, Mickey 1974a, 1974b, 1975) or of the total line profile (Solanki, Stenflo 1984, Solanki 1986a, Solanki 1986b) have yielded information on the velocity, temperature and magnetic fields in flux tubes. Theoretical approaches to the problem have also been attempted in terms of the radiative transfer of polarized light (Landi degl Innocenti & Landofi 1983, Auer & Heasley 1978). We have attempted to measure the symmetries in Stokes V profiles across a sunspot using the lines FeI λ 6301.5A (g=1.667) and FeI λ 6302.5A (g=2.5) and present some preliminary results.

Instrument

The Kodaikanal Solar Tower/Tunnel Telescope and Spectrograph were used in the observations (Bappu 1967). The instrument polarization cause by the coelostat mirrors had been earlier worked out and it depends on the geometry of the telescope and the time of observation (Balasubramaniam, Venkatakrishnan & Bhattacharyya 1985). The accuracy of the measurements which depend on the orientation of the polarization optics has also been calculated (Bhattacharyya & Balasubramaniam 1986).

Observations & Reductions

The observations were carried out on March 6, 1986 on a sunspot KKL 18265 (B=1°, L=13°E), located approximately at the centre of the solar disc. The image scale is 5.5 arc-seconds per mm. The spectral lines used were the FeI $\lambda\lambda$ 6301.5 A, 6302.5 A, at a dispersion of 6.5 mm/A. A 10 angstrom narrow band filter, centered at λ 6302A was used to reduce scattered light in the spectrograph. The slit width correspond to .55 arc-seconds.

All the four Stokes parameters were recorded in six frames I+Q,I-Q,I+V,I-V,I+U,I-U successively. The observations reported here were completed in five minutes centered at $3^h 57^m$ GMT. The spectra were recorded on 103-aE emulsion. The seeing was about 2 arc-seconds.

The spectra were digitised using the PDS microdensitometer.

The sampling rate along the dispersion direction was 4 mA and along the orthogonal direction corresponded to 2.2 arc-seconds.

The digitised spectra was further reduced using the VAX 11/780 computer. The software package RESPECT (Prabhu, Anupama & Sunetra 1986) was used in the reductions. The noise in the data was filtered in the Fourier domain between 0.075-0.1 cycles $(mA)^{-1}$.

If $[I] = (I,Q,U,V)^T$ is the stokes vector of a beam of light incident on an optical system, then after passing through the system the resultant Stokes Vector is given by $[I'] = (I',Q',U',V')^T$.

$$[I'] = [M][I]$$

where [M] is the combined Mueller matrix of the optical path. The observed Stokes profiles had to be premultiplied by the inverse of the matrix [M] relevant during the time of observation. The inverted matrix is given below:

$$[M]^{-1} = \begin{pmatrix} 1.1169 & 0.0265 & 0.0383 & 0.0011 \\ -0.0460 & 0.484 & -1.0054 & -0.0457 \\ -0.0073 & 0.9783 & 0.4580 & 0.2081 \\ 0.0011 & -0.2334 & -0.1614 & 1.0792 \end{pmatrix}$$

Since only the Stokes I and V profiles were strong, the profiles corrected for instrumental polarization are not significantly distorted (figure 1).

The wavelength calibration was done using the three terrestrial oxygen lines $\lambda\lambda$ 6209.2 A, 6302.0 A, 6302.8 A. These lines are stable within 5 m/s (Caccin et al 1985). The solar wavelengths so obtained were then corrected for the doppler shifts due to solar rotation, earths rotation, earths orbital velocity and for the gravity shift (Bhatnagar 1964, Kawakami 1983). The laboratory wavelengths used for the comparison of the two Fe I lines were $\lambda\lambda$ 6301.515 A, 6302.507 A (Moore 1945).

Results

The following quantities were determined (see e.g. Solanki 1986b) using the Stokes V profile

$$\text{Area asymmetry} = \frac{A_B - A_R}{A_B + A_R}$$

$$\text{Amplitude Asymmetry} = \frac{a_b - a_r}{a_b + a_r}$$

$$\text{Velocity} \quad v = c \frac{\lambda_{6302.5} - \lambda_{lab}}{\lambda_{lab}}$$

where c is the velocity of light and $\lambda_{6305.5}$ is the zero crossing wavelength and, the magnetic field strengths from the Zeeman separation of the centroid wavelengths of the two circularly polarized components.

These quantities are given in Table I. It may be noted that the quantities along the column vary from the centre of the umbra to the penumbra closer to the photospheric edge. The relative broadening parameter $(\lambda b - \lambda r) 6301.5/(\lambda b - \lambda r) 6302.5$ also showed a systematic drop in values ranging from about 0.9 to 0.75 as one moved from the umbra to the photospheric edge of the penumbra.

Discussion and conclusion

Although we are yet to attempt at any modelling, we are certain that gradients in both the velocity and the longitudinal magnetic field can result in area and amplitude asymmetries.

The zero crossing velocity of the V profile, the velocities within the flux tubes, does not show any systematic variations in our observations. Their signs are positive showing that there are indeed downdrafts of the order of 1-2 kms/sec, which is in contradiction to the results of Solanki (1986a) although such velocities have been reported earlier in the literature.

These results are very preliminary. We wish to repeat such observations for a variety of sunspots as a function of the heliocentric angle and include the asymmetries of the Q profiles and, model the results.

Acknowledgements

The author is indebted to Prof.J.C.Bhattacharyya and Dr. P.Venkatakrishnan for encouragement and help. Mr.P.Devendran helped in the observations.

The RESPECT package was of immense help in the reductions. The author is thankful to Ms.G.C.Anupama for help in the course of the reductions.

Thanks are also due to Prof.Arne Wyller and The Royal Swedish Academy of Sciences, The Department of Science and Technology, Govt. of india, The Director, Indian Institute of Astrophysics and Mr.S.Gopalkrishnan for the financial support to attend the conference.

References

1. Auer,L.H., and Heasley,J.N., 1978 Astron. Astrophys. 35, 327.
2. Balasubramaniam,K.S., Venkatakrishnan,P., and Bhattacharyya,J.C., 1985 Solar Physics 99, 333.
3. Bhattacharyya,J.C., and Balasubramaniam,K.S., 1986 Kodaikanal Observatory Bull. 6, 30.
4. Bhatnagar,A., 1964 Ph.D., Thesis, Agra University.
5. Bappu,M.K.V., 1967 Solar Physics 1, 151.
6. Caccin,B., Cavallini,F., Capatelli,G., Righini,A., and Sambuco,A.M., 1985 Astron. Astrophys. 149, 357.
7. Illing,R.M.E., Landman,D.E., Mickey,D.L., 1974a Astron. Astrophys. 35, 327.

8. Illing,R.M.E., Landman,D.E., Mickey,D.L., 1974b Astron. Astrophys. 37, 97.
9. Illing,R.M.E., Landman,D.E., Mickey,D.L., 1975 Astron. Astrophys. 41, 183.
10. Kawakami,H., 1983 Publ. Astron. Soc. Japan 35, 459.
11. Landi degl Innocenti,E., and Landofi,M., 1983 Solar Physics 87, 221.
12. Moore,C.E., 1945 Contr. Princeton Univ. Observatory No.20.
13. Solanki,S.K., and Stenflo,J.O., 1984 Astron. Astrophys. 140, 185.
14. Solanki,S.K., 1986a. Preprint-Velocity in Solar Magnetic Flux Tubes.
15. Solanki,S.K., 1986b in these Proceedings.
16. Prabhu,T.P., Anupama,G.C., Giridhar,S., (in preparation).

Table I: The various quantities mentioned in the text are shown for the Fe I λ 6302.5A line. The results are similar for the Fe I λ 6301.5 A line. The spectrograph is limited to measuring field strengths greater 280 gauss and velocities greater 0.6 Km s^{-1}.

Position	area asymmetry	amplitude asymmetry	velocity kms sec^{-1}	field strength Gauss
umbra near centre	-0.82	-0.21	1.3	3235
umbra	-0.61	-0.04	2.33	3127
penumbra	-0.73	0.08	2.28	2912
penumbra	-0.77	-0.23	2.32	2695
penumbra near photosp.	-0.79	-0.17	2.26	2693

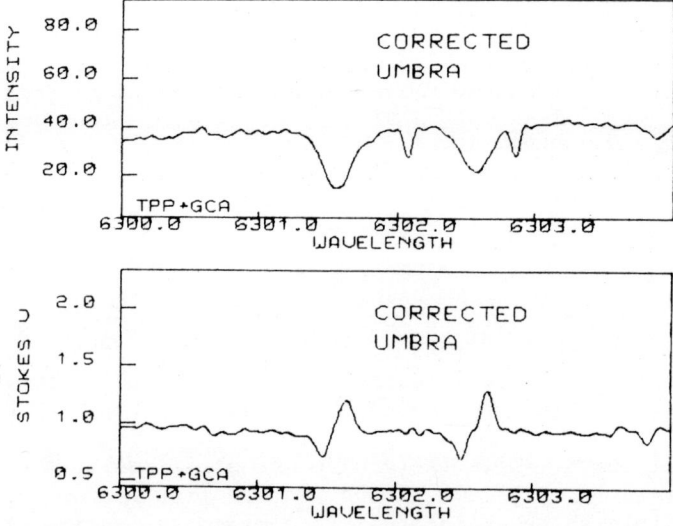

Figure 1. Sample Stokes I and V profiles recorded in the umbra. These profiles have been corrected for the instrumental polarization caused by coelostat mirrors.

SUNSPOT PROPER MOTIONS AS A TOOL FOR THE STUDY OF THE BEHAVIOUR OF RISING MAGNETIC FLUX TUBES

F. Mazzucconi
Osservatorio Astrofisico di Arcetri, Firenze, Italy

G. Godoli
Istituto di Astronomia, Università degli Studi, Firenze, Italy

Proper motions in longitude of the preceding (p) and following (f) components of sunspot groups, if interpreted as the result of ascending subphotospheric magnetic flux tubes being subject to a buoyancy force, can give information on the behaviour of these tubes.

As is well known, while the groups are still small, the p components move forward in longitude to the Western limb for several consecutive days at diminishing speed; the f components, on the contrary, retrograde or remain stationary in longitude.

Greenwich sunspot data for the years 1917-1955 (cycles No. 15-19) have been examined.

In the catalogues of the yearly Greenwich photo-heliographic results following the tables of positions and areas of sunspots and faculae for each day of the year and the general catalogue of groups of sunspots for each transit, two ledgers are reported respectively for recurrent and non-recurrent groups. In these ledgers the daily time of observation, the area and the heliographic coordinates (latitude ϕ and Carrington longitude L) are given for each sunspot group during its whole life-time. These data are also given separately for important p and f components of the principal groups. In these ledgers proper motions of the p and f components have been published for the years 1917-1955. The set of data includes 335 sunspot groups for which we have deduced plots such as those in Fig. 1.

The representation of the observed points by arcs of ellisses is so good that the authors felt justified in extrapolating the arcs of ellipses in order to evaluate when the sunspot group first appeared, as has been done in Fig. 1: actually this method has been proved to be reliable when the sunspot group first appeared on the disc.

The possibility of representing the observed points with ellipses strongly suggests that the axis of the ascending parts of the subphotospheric magnetic flux tubes can also be represented by arcs of ellipses and that the ascent velocity of a flux tube is constant.

Let us indicate with q the axis (in meters) parallel to the solar surface of one of these ellipses and with n = kq the axis normal to the solar surface. The axis n is related to the ascent velocity v by the relationship:

$$n = kq = 2 \Delta t\, v \qquad (1)$$

where Δt is the time interval between the (calculated) time of the first appearance of the group and the time in which the maximum separation of the p and f component is observed.
In Fig. 2 q values are plotted versus $2\Delta t$ values for each sunspot group. Only groups for which the q axis can be observed directly have been considered.

Since relationship (1) gives

$$q/2\, \Delta t = v/k$$

if we assume that v/k is constant for all the magnetic flux tubes (as

Fig. 1. Left representations: positions in longitude given separately for the p and f components of sunspot groups Nos. 16752, 16775, 16651, 15961 (abscissae) for consecutive days (ordinates - time increasing from above to below). Right representations: observed positions in longitude of the p and f components referred to their average positions (dots) and calculated position by arcs of ellipses (crosses).

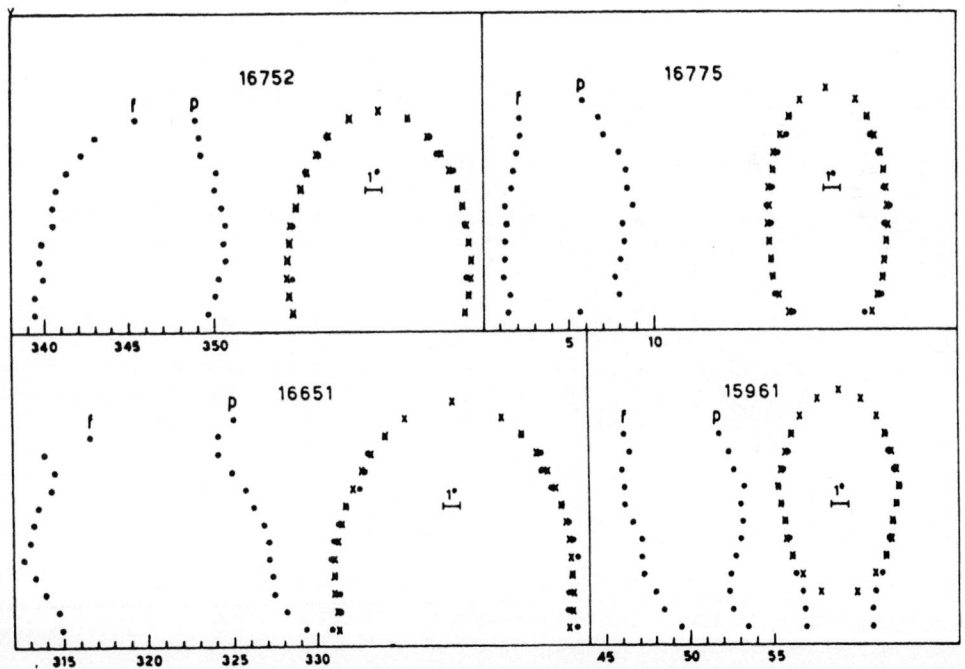

would happen if the ascent velocity and the shape were the same), we should interpret the scattering of the dots in Fig. 2 as a normal distribution around the average straight line $v/k = 105$ ms^{-1}. Extreme values for v/k are 185 and 49 ms^{-1}.

These values give the ascent velocity of a circular shape ($k = 1$). In general we can assume, taking into account X-ray, EUV, radio and optical observations, that

$$0.75 < k < 1.25$$

so that the values found for the ratio v/k can be considered within an approximation of 25% to represent the ascent velocity also.

We emphasize that these deductions must be considered only as a first approximation: certainly at least part of the scattering of the dots in Fig. 2 must be due to a coupling between v and k different than that assumed.

Fig. 2. Relationships between q values (in 10^8 meters) and $2\Delta t$ values (in days for each sunpost group).

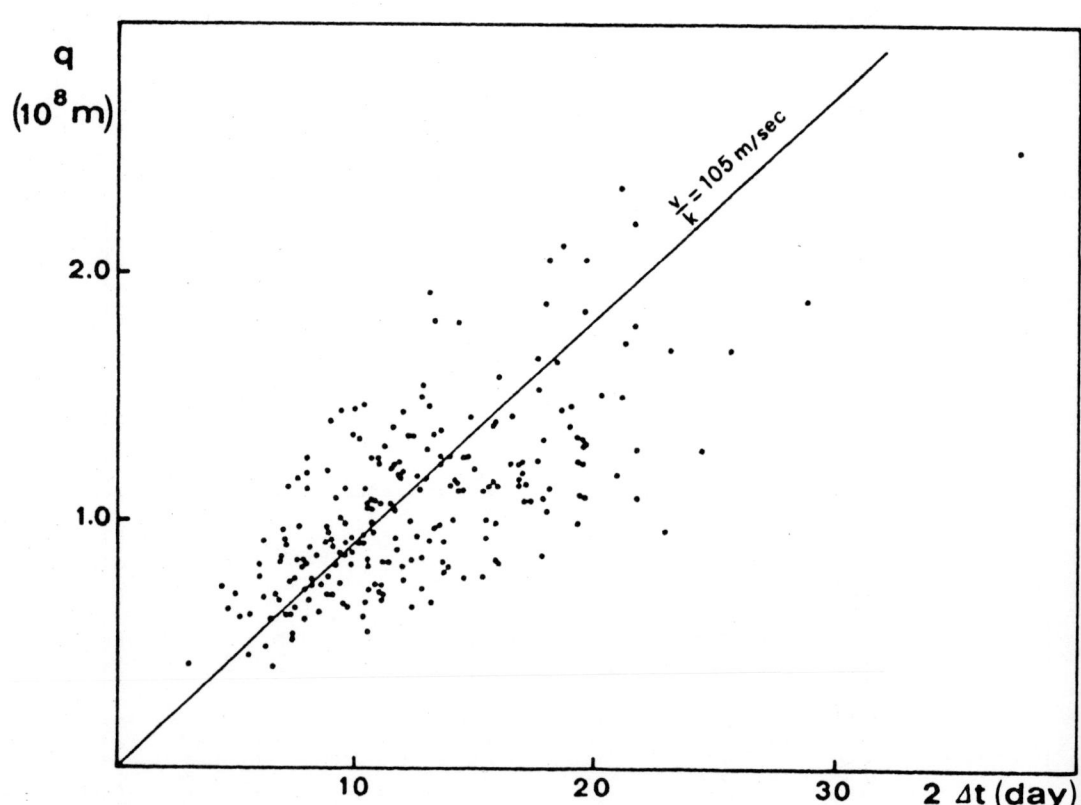

In any case our q values seem to agree with those deduced from the observations of Krieger et al. (1971) and Foukal (1975). On the contrary our results are about an order of magnitude greater than those of Kundu et al. (1980) and of Fang and Martres (1986): we must yet observe that the footpoints of the archshaped structures studied by Fang and Martres (1986) avoid the local maxima of the photospheric line-of-sight magnetic fields.

Our ascent velocities seem to be greater by an order of magnitude than that generally assumed (Schroeter and Woehl, 1978; Parker, 1979).

References

Fang, C. and Martres, M.J.: 1986, Solar Phys. <u>105</u>, 51.
Foukal, P.: 1975, Solar Phys. <u>43</u>, 327.
Greenwich Photo-Heliographic Results: 1917, 1918, ..., 1955, Her Majesty's Stationary Office, London.
Krieger, A.S., Vaiana, G.S. and Van Speybroeck, L.P.: 1971, in R. Howard (ed.), Solar Magnetic Fields, IAU Symp. <u>43</u>, 397.
Kundu, M.R., Schmahl, E.J., and Gerassimenko, M.: 1980, Astron. Astrophys. <u>82</u>, 256.
Parker, E.N.: 1979, Cosmical Magnetic Fields, Carendon Press, Oxford.
Schroeter, E.H. and Woehl, H.: 1978, Proceedings of the Workshop on Solar Rotation, Oss. Astorf. Catania <u>162</u>, 35.

ESTABLISHED AND NON-ESTABLISHED PROPERTIES OF UMBRAL AND
PENUMBRAL FINE STRUCTURES

J.I. García de la Rosa
Instituto de Astrofísica de Canarias, 38071 La Laguna,
Tenerife, Spain

1 INTRODUCTION

Three planets like the Earth, set side by side, can easily fit in the diameter of a typical large sunspot. To give an impression of the required resolution, the structures that we are going to consider in this review are the size of Tenerife island. The most important features of that size are umbral dots and penumbral filaments.

In this review we should like to investigate whether the presently available observations suffice to discriminate between the alternative ideas on the nature of those fine structures. Are the umbral dots non-thermal phenomena? (Kitai 1986), Joule-heated material? (Bruzek 1977), penetrative convection? (Parker 1979b) or overstable convection? (Knobloch & Weiss 1984) (Cowling 1976). Are the penumbral filaments convective rolls? (Danielson 1961), elevated filaments? (Moore 1981), alignment of grains? (Mu

Figure 1.- After Grossmann-Doerth et al. (1986). At left umbra with dots

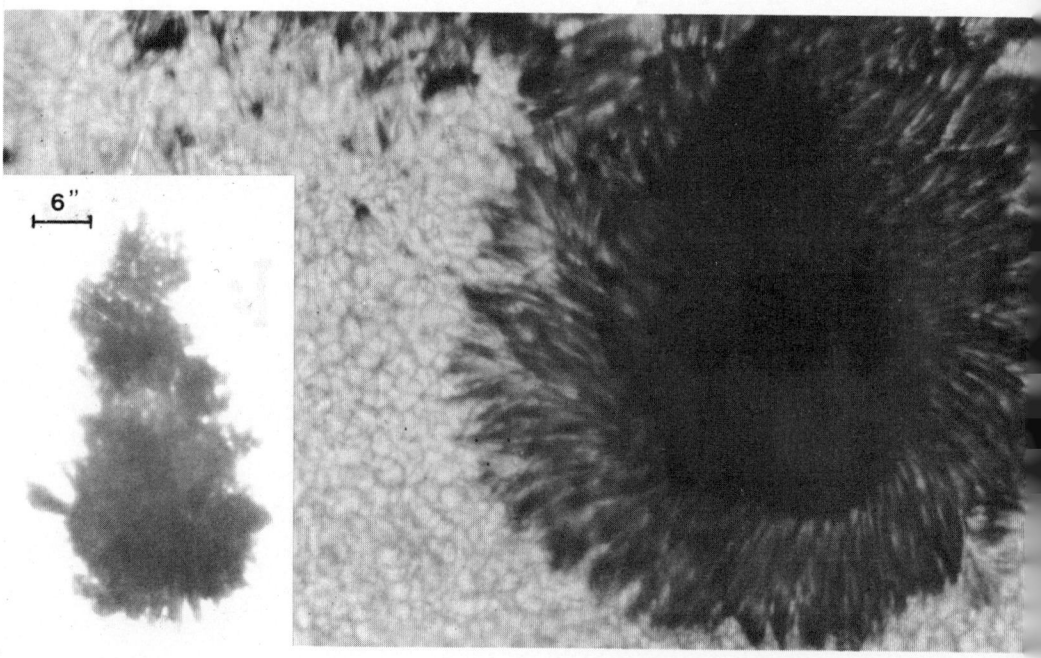

ller 1973a) or floating filaments? (Schmidt et al. 1986).

Four main aspects of the fine structures are considered in turn: brightness and morphology (Section 2), magnetic field (Section 3), velocity (Section 4) and oscillations (Section 5).

2 BRIGHTNESS AND MORPHOLOGY
2.1 Umbral dots

Although the umbral dots are spread out all over the whole umbra (Bumba & Suda 1980), their distribution is inhomogeneous, with brightness and contrast depending on their location (Loughhead et al. 1979). The distribution is generally far from random and a characteristic behaviour can be established: the brightest umbral dots are located both at the periphery and the interior of the umbrae; but in the last case they form chains and rings encircling darker nuclei (Peripheral Umbral Dots: PUDs) (Krat et al. 1972). In the interior of those dark nuclei the faintest and less contrasted dots are found (Central Umbral Dots: CUDs). The reported difficulties, even for the photographic detection of the CUDs, suggest that all the reviewed observations only refer to the PUDs.

Our own observations (García de la Rosa 1986) suggest, that the sunspots (umbrae) form after the coalescence of several fragments with typically 10^{21} Mx, which apparently keep their identity until the sunspot decay. In the frame of this picture, the bright PUDs are located at the junctions of the fragments, while the dark nuclei with the CUDs correspond to the fragments themselves. In Table 1 the observations on size/temperature of the umbral dots are summarized. Direct measurements are limited by

Table 1.- Direct and indirect observations of size/temperature of umbral dots.

DIRECT OBSERVATIONS			
Authors	Observation	Telescope diameter	u.d. size
* Danielson (1964)	Stratoscope	30 cm	≲300 km
* Krat et al. (1972)	Stratoscope	50 cm	≲160 km
* Harvey & Breckinridge (1973)	Speckle Int.	150 cm	<180 km
* Adjabshirzadeh & Koutchmy (1981)	High resol. pictures	76 cm	165 km

INDIRECT OBSERVATIONS			
Authors	Observation	Temperature	u.d.size
* Beckers & Schröter (1968)	Colour Temp. (simultaneous)	$T_{ud} \simeq T_{ph}$	160 km
* Kneer (1973)	Photosph. line in umbra	$T_{ud} = T_{ph}$	------
* Koutchmy & Adjabshirzadeh (1981)	Colour Temp. (not simult.)	$T_{ud} \simeq T_{ph}$	190 km
* Grossmann-Doerth et al. (1986)	Brightness T. (peak I)	$T_{ud} \ll T_{ph}$	290/660 km

the telescope resolution and therefore lower values are always expected; but there is agreement on typical sizes of the order 100-200 km. Indirect observations mainly rely upon the determination of the colour (or brightness) temperature. The comparison of the corresponding blackbody intensity with the measured flux, allows us to deduce the emitting area. Almost photospheric temperatures and 100-200 km sizes are suggested from the observations; although variance is shown with the work by Grossmann-Doerth et al. (1986). Their result can reflect real temperature fluctuations, although caution should be taken with this peak intensity observation (to measure brightness temperature), likely to suffer from atmospheric and telescope degradation.

It is a sensible idea to check whether our interpretation of the high resolution observations fits to the more reliable low resolution ones:

i) Wiehr (1985) tested different distributions of umbral dots with photospheric temperatures in a dark background and showed, that they were unable to explain the contrast variation with wavelength, for several umbrae. The contradiction with the fine structure model is however relieved after noticing, that his umbrae observations were restricted to the dark umbral nuclei, where no PUDs are generally present. The result consequently indicates, that the CUDs do not show photospheric temperatures.

ii) Van Ballegooijen (1984) after the low resolution (5") observation of a umbra with evident presence of umbral dots, concluded, that a two-component (hot and cool) model was needed to explain the behaviour of spectral lines with different temperature sensitivity.

iii) The observation of the independence on size of the umbral intensity (Albregtsen & Maltby 1981) seems to favour the idea of a fragmented umbra instead of a monolithic tube.

2.2 Penumbral filaments

The morphology of the penumbra filamentary structure is better described if, at least, three regions are considered: inner, medium and outer penumbra.

i) Inner penumbra ($\leq 0.2\ L_P$: penumbral length): Filaments are comet-like with a bright nucleus (penumbral grains) and a tail of attenuating brightness pointing towards the photosphere (Muller 1973a). They usually overlap the dark umbra producing a ragged edge formed by grey ($I_{umb} < I_{fil} < I_{ph}$) filaments and dark ($I = I_{umb}$) interstices (e.g. see Figure 1).

ii) Medium penumbra (~ 0.2-$0.8\ L_P$): A mixture and superposition of comet-like filaments is the predominant feature of this complex part of the penumbra. Our own high resolution observations indicate that the penumbra filaments are not the result of the alignment of bright grains, as suggested by Muller (1973a), although some fortuitous examples are occasionally found.

iii) Outer penumbra ($\geq 0,8\ L_p$): Several observations suggest,that this part is formed by the dark tails of the filaments,which apparently overlay the photosphere (Moore 1981) (unpublished observations by several authors). Although this idea is disputed by Muller (1985), our extensive program on the study of penumbra formation has provided several examples of apparently elevated dark filaments (Collados et al. 1985). These first stages of the penumbra development are particularly enlightening for the morphological studies because, in that time,the penumbra is just a thin filamentary layer,lacking the later complexity.

Links between umbral dots and penumbral filaments have been reported by several authors: Soltau (1982), for instance, provides some examples and Muller (1973a) reports the frequent transition from penumbral to umbral bright dots. These observations suggest a continous, rather than abrupt, umbra-penumbra transition, at least from the morphological point of view.

Photometry establishes a quantitative base on the study of the penumbral filaments, although the results are not completely independent on the assumptions. Muller (1973b) carried out photometric studies of the penumbra fine structure,under the assumption of only two components: bright filaments and dark background. He separately measured the mean brightness of both components. Grossmann-Doerth & Schmidt (1981) without any previous assumption, photometrically digitized several penumbrae obtaining a single peaked distribution of values, instead of the expected double peak of the two-component assumption. The single peak distribution can be easily explained after noticing,that in any typical photometric tracing (Figure 2 b) some of the "bright" structures show lower brightness than the "dark" ones. Representing their peak brightness, separating "bright" and "dark" structures, two curves are obtained, but with a very strong overlapping (single peak) (Wiehr et al. 1984). It can be concluded,that there is a variety of intensities in the penumbra, instead of dark and bright filaments. These can only be locally defined (i.e. a filament brighter than its adjacent one)

Figure 2.- (a) Single peak distribution of penumbral brightness, obtained by Grossmann-Doerth & Schmidt (1981). (b) In a typical photometric tracing of the penumbra (after Wiehr et al. 1984), some "dark" structures are in fact brighter than the "bright" ones.

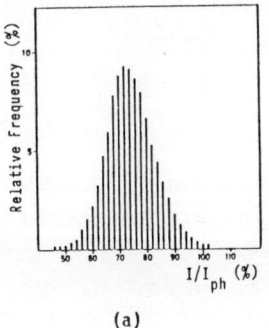
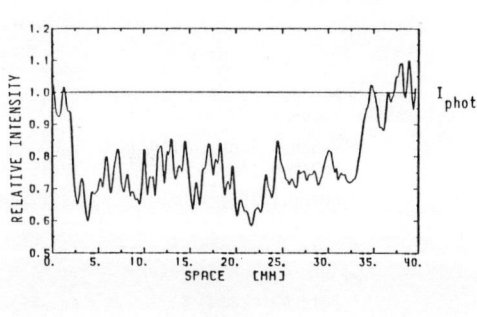

(a) (b)

Collados et al. (1986) have found double peaked and asymmetric brightness distributions during penumbra formation, which suggest a two-component structure. This finding reinforces our previous morphological view of the young penumbrae, as a thin layer of dark tails of filaments overlaying the photosphere. Equally, we forsee that this same double peak behaviour is to be expected, if photometry were restricted to the outer penumbra. In fact, this same idea is suggested after the comparison with low resolution observations. Kjeldseth-Moe & Maltby (1974) compared the contrast (I_{pen}/I_{phot}) variation with wavelength to the predictions of the fine structure models. They concluded that a good fit between model and observations was only obtained with a two-component model of the outer penumbra defined by Muller (1973b).

3 MAGNETIC FIELDS
3.1 Umbral dots

Indirect observations are usually accomplished by means of different temperature sensitivity lines. Pure photospheric lines (high excitation) provide information from only the hot umbral regions, whereas pure umbral lines only represent the cool and dark background. Provided that due allowance is made for the stray light, one is able to spectroscopically study the fine structures without even needing to resolve them. A probably incomplete, but representative list of observations is shown in Table 2.

The varied umbral dot distributions observed in different spots could give account of the quantitative differences; but they qualitatively coincide in suggesting that the magnetic field is generally weaker in the bright dots than in the dark background, in agreement with the magnetostatic equilibrium expectations.

Direct measurements of magnetic fields in fine structures require very high resolution spectroscopic observations and are consequently scarce. In a high resolution spectroscopic study, Zwaan et al. (1985) failed to observe a clear relation between bright umbral structures and magnetic field variations. However, the tendency for a few cases is to be weaker

Table 2.- Magnetic fields measured in the umbra, with lines of opposite temperature sensitivity.

Authors	Lines	Results
* Mogilevsky et al. (1968)	FeII and TiI	$B_d = 1.8 \times B_b$
* Zwaan & Buurman (1971) Buurman (1973)	High and low excit. lines	$B_d = 3120 \pm 40$ G $B_b = 2850 \pm 100$ G
* Kneer (1973)	FeII and CaI	$B_d = 2600 \pm 150$ G $B_b = 1200 \pm 300$ G
* Guseinov (1983)	FeII and FeI	$B_b = 1.4-1.7$ KG
* Adjabshirzadeh & Koutchmy (1983)	FeI 6302.5 Å	$B_d = 2850$ G $B_b = 2600$ G

in bright structures. This result introduces some doubts on the indirect results. Several authors contrarily propose an extremely weak field in bright dots, as a consequence of the observations discussed in Section 3.2.

3.2 π-component splitting

According to the standard model of the sunspot magnetic field (Bray & Louhhead 1964), the observation of an umbra in the disk center should not present any unshifted Zeeman component. Observations clearly show the opposite. Several explanations for the presence of the " π " component have been developed in order to preserve the model:
i) Scattered photospheric light; ii) presence of molecular blends, as found, for instance, in the widely used 5250.2 Å line (Kjeldseth-Moe et al. 1970); iii) complex σ-profile of saturated lines (Henoux 1968). The calculation of the Stokes V profile discards those three effects, leaving behind two oppositely circularly polarized components (i.e. the " π -component splitting"). Generally, those small σ-components show displacements from the λ_0 position, which are respectively opposite and one order of magnitude smaller than the Zeeman displacements of the main σ-components (Mogilevsky et al. 1968) (Beckers and Schröter 1969a) (Figure 3). These authors accordingly attributed the effect to very weak fields (300G) with opposite polarity to the background and probably associated with the brightness inhomogeneities (umbral dots).

Figure 3.- (a) Ideal observation of a sunspot in the disk center, without stray light, molecular blends and saturation effects (π -component splitting). (b) The Stokes V profile, as observed by Beckers & Schröter (1969a).

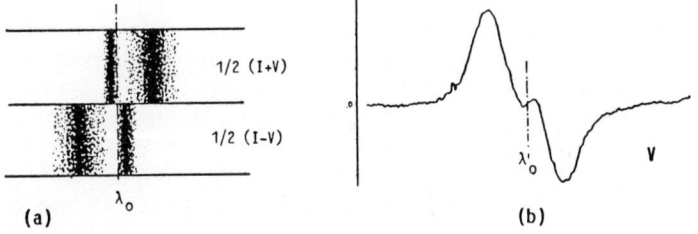

Figure 4.- (a) Idea proposed by Moore (1981b) to reproduce the observed Stokes V profile (see Figure 3b): weak magnetic field of same polarity in structures with upward motion (blueshifts). (b) Although V profile is "similar" to the observations, the 1/2 (I+V) and 1/2 (I-V) profiles (c) show disagreement (see Figure 3a).

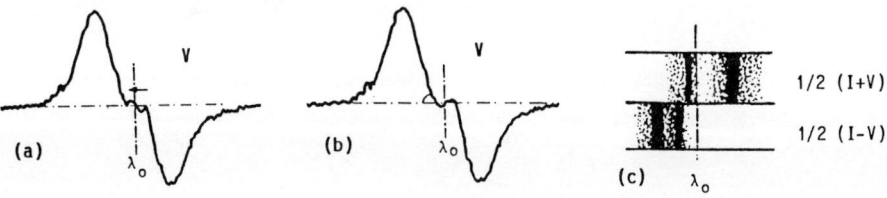

A new version of this interpretation has been proposed by Moore (1981b), after considering the presence of upward motions in umbral dots (see Section 4). The combination of blueshifts and weak magnetic fields of the same polarity, reproduce the observed V profile (Figure 4 a,b). This interpretation however, contradicts the observed I profiles as shown in Figure 4 c.

The idea of very weak fields in umbral dots can be criticized, at least, with three examples: i) Pure umbral line 6435 Å shows the π-splitting (Harvey 1971); ii) Photospheric lines with special Zeeman patterns, without unshifted components, fail to show the presence of very weak fields (Mehltretter 1969) (Harvey 1971); iii) Guseinov (1983) observed the same behaviour of the magnetic field in bright dots (measured with ionized lines), for a group of 11 spots, with 4 cases lacking π-splitting

An apparently more likely alternative explanation of the π-component splitting is based in magneto-optical effects, in the presence of twisted or strongly inclined magnetic fields (Wittmann 1971) (Kunzel & Staude 197 Although we are not going into the details of this explanation, it should be mentioned, that the somewhat restrictive conditions required (inclination $> 45°$) are easier accomodated in a fragmented umbral model, than in a monolithic one.

3.3 Penumbral filaments

Measurements to determine the possible correlation between magnetic fields and brightness fluctuations have been carried out mainly using direct procedures: i.e. correlating continuum fluctuations with magnetic line shifts. The spatial resolution of the spectroscopic observations is generally poor (1"-2"), compared to the typical filament width (0"367) (Bonet et al. 1982), indicating that those measurements represent bundles of filaments with a variety of intensities. This resolution handicap, probably explains the small magnetic differences found between dark and bright spectroscopic streaks (unresolved bundles of filaments) shown in Table 3. The apparent contradiction between the first two result of Table 3, has probably a simple geometric explanation, as shown in Figure 5. In conclusion, the magnetic field in the dark streaks seems to be stronger than in the bright ones, although certainly there are also strong fields in the bright filaments (Harvey 1971)

In the picture preferred by the present author (see Section 6), the

Table 3.- Penumbral magnetic field measurements.

Authors	Results
* Mattig & Mehltretter (1968)	Bright elements show strong $B_{//}$ (Spot at 45°)
* Beckers & Schröter (1969b)	More horizontal and stronger $|B|$ in dark filaments.
* Abdussamatov (1976)	$B_d = B_b + 225$ G
* Stellmacher & Wiehr (1981)	No systematic B fluctuations in b and d streaks.

penumbral bright streaks can correspond either to almost vertical filaments connected to umbral dots or to the underlying photosphere. On the other side, dark penumbral streaks can correspond to: i) underlying umbra ii) dark tails of filaments; iii) interfilamentary lanes where the magnetic field is concentrated, after being pushed aside by the moving material; iv) photospheric intergranular lanes. In the two possible bright structures the magnetic field $|B|$ is expected to be lower, than in the four possible dark ones (except perhaps ii). In the frame of this picture, it is extremely important to know, which part of the penumbra has been observed, in order to understand the local nature of bright and dark structures (Section 2.2).

Figure 5.- Although the magnetic field is more intense in the dark penumbral structures, observations of its longitudinal component ($B_{\parallel}$) in inclined sunspots (Mattig & Mehltretter 1968), can give a larger value in bright structures. In fact, bright structures are likely to correspond to the almost vertical parts of the filaments. (Also to the non-magnetic photosphere in the outer penumbra).

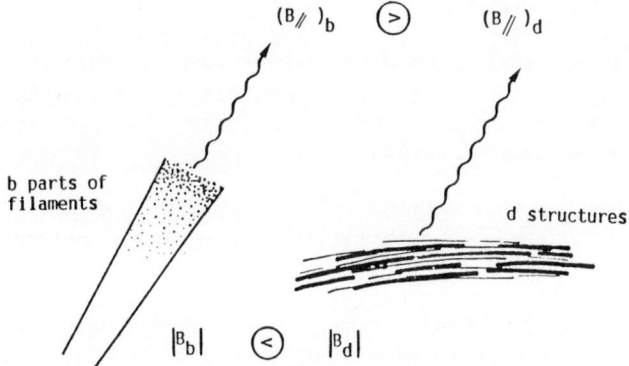

Figure 6.- Correlation diagram (after Beckers 1969) produced by the careful comparison (point by point) between simultaneous 2-D images and Dopplergrams of a sunspot near the limb.

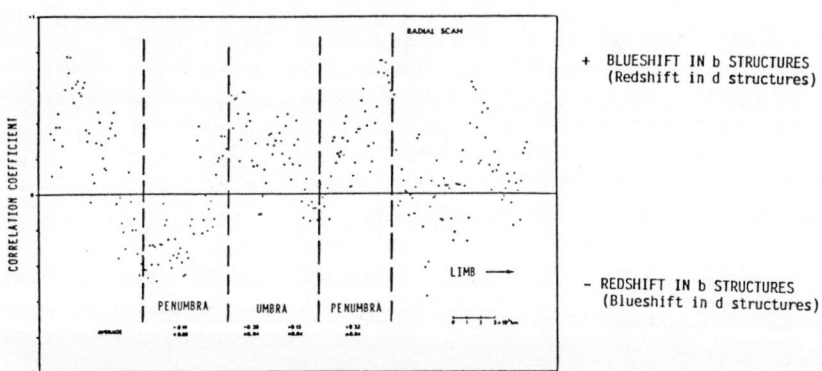

4 VELOCITY FIELD
4.1 Umbral dots

Velocity inhomogeneities in the umbra have been scarcely measured and generally, only qualitative results are available. One of th highest resolution studies is that by Beckers (1969) using two-dimensiona Dopplergrams of a sunspot near the solar limb. The results are represente in a correlation diagram (Figure 6), where positive values of the correlation coefficient correspond to bright structures showing blueshifts (or dark structures with redshift) and negative values of the coefficient correspond to dark structures showing blueshifts (or bright structures with redshifts). It is clearly seen that umbral bright structures show bl shifts which correspond to upward motions in umbral dots. This result has been confirmed by Beckers & Schröter (1969b) and Kneer (1973) who measure upward flows of $3\pm0,5$ km s^1 in bright dots (probably a cluster of umbral dots). The dark background velocity field is contrarily very small, obser ved to be smaller than $0,025$ km s^1 (Beckers 1977).

4.2 Penumbral filaments

At the photospheric level, a displacement of the center of gr vity in spectral lines at the penumbra, indicates an almost horizontal ou flow from the spot. This so called Evershed effect is really the result c two phenomena: a line core shift and an asymmetry in the same direction. (See figure 2 in Wiehr et al. 1984).

Although the Evershed effect has been extensively studied (see the reviev by Schröter 1967), fundamental questions remain to be answered about its relation to the penumbral fine structure. According to Becker's (1969) observations (Figure 6), the radially outward Evershed motion must be concentrated in the dark penumbral structures. Positive and negative cor lation coefficients are shown, respectively in the limb and centerside penumbrae. Other observations confirm this result; e.g.: Mattig & Mehltr tter (1968); Abdussamatov & Krat (1970); Stellmacher & Wiehr (1971); Mamadazimov (1972).

This result is certainly insufficient after the finding by Grossmann-Doerth & Schmidt (1981), that the penumbra is characterized by a continu variety of intensities rather than two components. In fact Stellmacher & Wiehr (1980) found that a two-component model (Figure 7 a) was unable to describe the Evershed effect. The goal of an updated multi-component model (Figure 7 b) is to look for a correspondence (if there is any) between the degree of brightness in the penumbral filaments and the velocity parameters of the Evershed effect: core shift ($\Delta\lambda$) and asymmetr Wiehr et al. (1984) with very highly resolved spectroscopic observations attempted to find that correspondence. Their results present a large scatter of values around a mean line which shows a larger core shift and smaller asymmetry in the darker structures (Figure 7 c,d). This large scatter probably reflects, both the resolution handicap of the spectrosco pic observations and the weak links between brightness and velocity para meters.

5 OSCILLATIONS

The purpose of the present section, rather than to review the observations on sunspot oscillations (see review by Moore 1981b), is to emphasize several of their characteristics, which suggest a promising future for the high resolution sunspot seismology.

The various oscillations observed in sunspots can be divided into two groups: i) those with periods around 5 min, interpreted as a passive response of the spot to the atmospheric p-modes; ii) those with periods around 3 min, interpreted as a resonant mode of the sunspot itself (Thomas et al. 1982). The second group can inform us about the physical conditions in the resonant cavity, provided we know exactly where it is. In fact, an active debate is taking place to decide whether the resonator is chromospheric (Zhugzhda et al. 1984) or photospheric (Thomas & Scheuer 1982).

The correlation between the oscillations (period and amplitude) and the sunspot structure (not yet fine structure) is evidenced by the changes

Figure 7.- (a) Two-component model with b and d streaks formed by bundles of DRs and BRs. (b) Multi-component model with b and d streaks formed by regions with a variety of intensities and unknown velocity parameters. (c) & (d) represent the result of the attempt by Wiehr et al. 1984, to correlate brightness and velocity parameters.

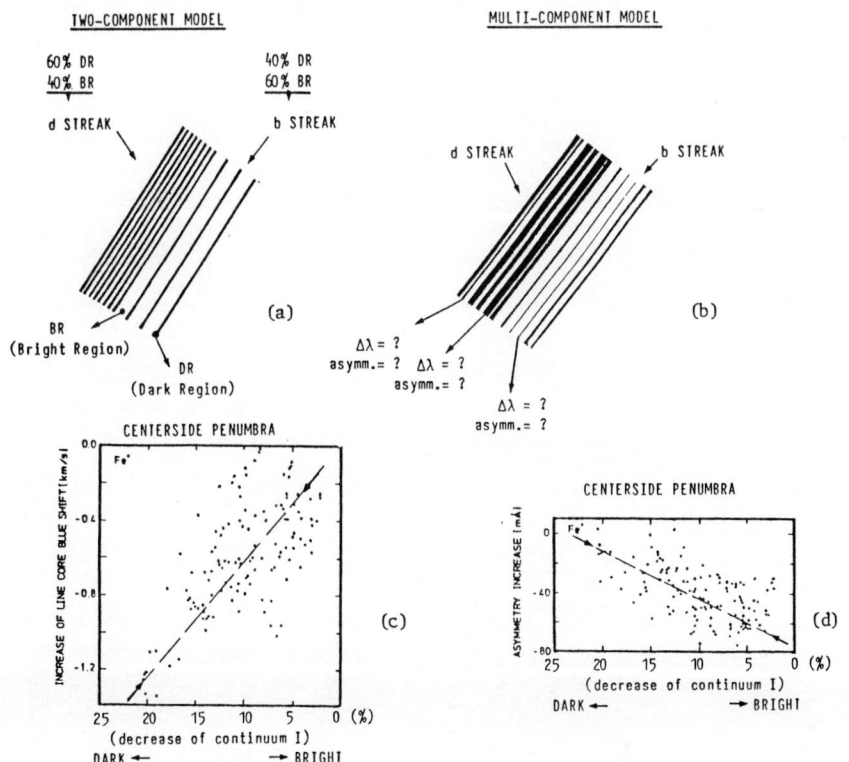

Table 4 .- Check of several umbral dot (UD) models by some observations.
+: Possible agreement model/observation. - : Difficult agreement model/
observation or observation out of the reach of the model.

	NON-THERMAL	JOULE HEATING	CONVECTION IN MONOLITHIC TUBE	PENETRATIVE CONVECTION
UDs around dark nuclei	+	+	-	+
Fragments in sunspots	+	+	-	+
No B opposite polarity	+	-	+	+
Weaker B in UDs	-	-	+	+
Upward motions in UDs	-	-	+	+

Figure 8.- Some proposed models of the umbral dots. (a) Non-thermal heating (Kitai 1986); (b) Joule heating (Bruzek 1977); (c) Overstable convection in monolithic flux tubes (Cowling 1976) (Knobloch & Weiss 1984); (d) Penetrative convection (Parker 1979b).

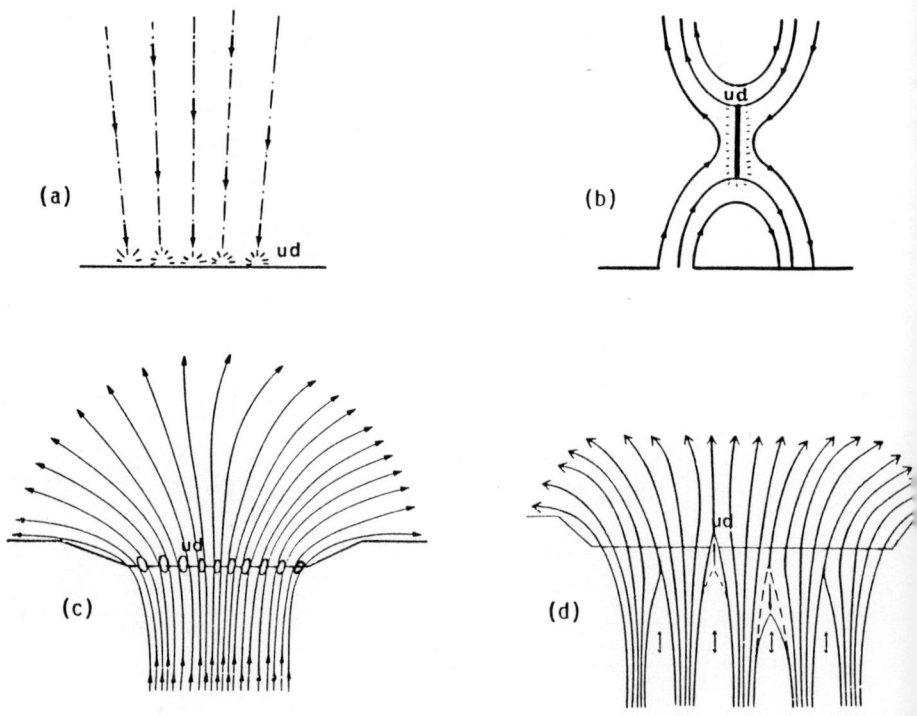

Table 5. - Check of several penumbral filament (PF) models by some observations. + : Possible agreement model/observation. - : Difficult agreement model/observation or observation out of the reach of the model.

	CONVECTIVE ROLLS	ALIGNMENT OF GRAINS	ELEVATED CHANNELS	FLOATING FILAMENTS
* Connection UDs/PFs	-	+	+	-
* Attenuating bright. (comet-like)	-	-	+	+
* Filaments overlaying umbra	-	-	+	-
* Strong B also in b structures	-	-	+	+
* Magnetic canopies (Giovanelli 1982)	-	-	+	-

Figure 9.- Some proposed model for the penumbral filaments. (a) Convective rolls (Danielson 1961); (b) Alignment of grains (Muller 1973a); (c) Channels of flowing matter (Cram et al. 1981); (d) Floating filaments (Schmidt et al. 1986).

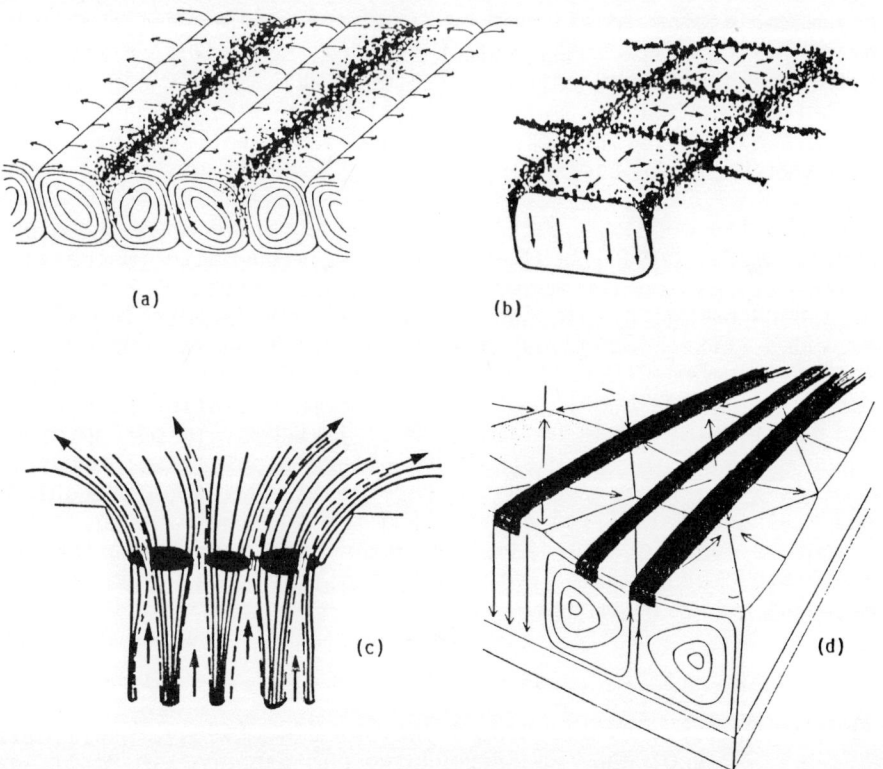

with position and time.

Several examples of changes in oscillations with position have been reported by Beckers & Schultz (1972); Abdelatif et al. (1984); Lites & Thomas (1985) and Balthasar et al. (1986). For instance, Abdelatif et al (1984) report an intensification of the 3 min oscillation in the umbral dark nuclei, with respect to the whole umbral structure. This suggests once again, different atmospheric conditions inside and outside the dark nuclei.

Both fast (hours) and slow (days) changes with the umbral oscillations have been reported. Fast changes (Schröter & Soltau 1976) (Zhugzhda & Makarov 1982) are difficult to explain; but the slow ones (Thomas et al. 1984) (Lites & Thomas 1985) are easier to understand, as structural changes usually accompany them. For instance, Lites & Thomas (1985) report a clear example where a sunspot in the process of formation (still a group of fragments) oscillates with a 5 min period, probably due to the lack of a proper resonant cavity. Four days later, after attaining a regular shape, its predominant period of oscillation became 3 min.

6 DISCUSSION

The attempt to answer the questions raised in the introduction of this review is a risky task, because a single author cannot probably make justice to all models. In fact, the present author has "unfortunately" a favorite model and, therefore, some caution should be taken with the results presented in Tables 4 and 5. They can, nevertheless, contribute to stimulate the discussion. In Tables 4 and 5 the predictions of several models are compared to some observations, excluding those which agree with all of them.

Our favorite model is that in which penetrative convection produces peripheric umbral dots (PUDs) and upward flowing matter which is channeled outwards, pushing the magnetic field aside (Figure 9 c). The possibility of a non-thermal origin for the central umbral dots (CUDs) cannot be excluded (Kitai 1986), because downward motions are observed in the chromospheric umbrae (Dialetis et al. 1985). This model is a mixture of Parker's (1979a) model (modified for a much smaller number of fragments) and the model by Cram et al. (1981). It shows, in our opinion, the best agreement with the fine structure observations. Its theoretical development is still poor, due to the lack of a levitation mechanism for dense material and a detailed explanation for the contrast changes along the filaments (Cram et al. 1981). Nevertheless, other examples of elevated dense structures exist in the Sun, e.g.: the Arch Filamentary Systems in emerging active regions.

REFERENCES

Abdelatif, T.E.; Lites, B.W. & Thomas, J.H. (1984). Oscillations in a sunspot and the sorrounding photosphere. In Small Scale Dynamic Processes in Quiet Stellar Atmospheres, ed. S.L. Keil, pp. 47. Sunspot.

Abdussamatov, H.I. (1976). On fine structure of the magnetic field and brightness in the penumbrae of sunspots. Solar Phys., **48**, 117-9.

Abdussamatov, H.I. & Krat, V.A. (1970). On the motion of penumbra filaments in sunspots. Solar Phys.,**14**, 132-5.

Adjabshirzadeh, A. & Koutchmy, S. (1983). Photometric analysis of sunspot umbral dots. Astron. Astrophys., **122**, 1-8.

Albregtsen, F. & Maltby, P. (1981). Solar Cycle variation of sunspot intensity. Solar Phys., **71**, 269-83.

Balthasar,H.; Fangmeier, E.; Küveler, G. & Wiehr, E. (1986). Oscillations in sunspot umbra-penumbra and the sorrounding photosphere. To be published in Proc. of the IAU Symp. n. 123.

Beckers, J.M. (1969). The microstructure of sunspots. In Plasma Instabilities in Astrophysics, eds. D.G. Wentzel & D.A. Tidman, pp. 139-51. Gordon and Breach Publications.

Beckers, J.M. (1977). Material motions in sunspot umbrae. Astrophys. J., **213**, 900-5.

Beckers, J.M. & Schröter, E.H. (1968). The intensity, velocity and magnetic structure of a sunspot region.II., Solar Phys.,**4**, 303-14.

Beckers, J.M. & Schröter, E.H. (1969a). The intensity, velocity and magnetic structure of a sunspot region.III., Solar Phys., **7**, 22-25.

Beckers, J.M. & Schröter, E.H. (1969b). The intensity, velocity and magnetic structure of a sunspot region.IV., Solar Phys., **10**, 384-403.

Beckers, J.M. & Schultz, R.B. (1972). Oscillatory motions in sunspots. Solar Phys., **27**, 61-70.

Bonet, J.A.; Ponz, D. & Vázquez, M. (1982). On the width distribution of penumbral filaments in sunspots. Solar Phys., **77**, 69-75.

Bray, R.J. & Loughhead, R.E. (1964). Sunspots. Chapman & Hall Ltd. London.

Bruzek, A. (1977). Spots and Faculae. In Illustrated Glossary for Solar and Solar-Terrestrial Physics, eds. A. Bruzek & C.J. Durrant, pp. 71-9. D. Reidel Pub. Co.

Bumba, V. & Suda, J. (1980). Internal structure of sunspot umbrae. Bull. Astron. Inst. Czech. **31**, 101-11.

Buurman, J. (1973). The spectrum of a large umbra.I. Astron. Astrophys. **29**, 329-34.

Collados, M.; García de la Rosa, J.I.; Moreno-Insertis, F. & Vázquez, M. (1985). Observations of the birth and fine structure of sunspot penumbrae. In High Resolution in Solar Physics, ed. R.Muller pp. 133-37. Springer Verlag.

Collados, M.; del Toro Iniesta, J.C. & Vázquez, M. (1986). The intensity distribution in sunspot penumbrae. These Proceedings.

Cowling, T.G. (1976). On the thermal structure of sunspots. Monthly Notices Roy. Astron. Soc., **177**, 409-14.

Cram, L.E.; Nye, A.H. & Thomas, J.H. (1981). Conjetures regarding the structure of a sunspot penumbra. In The Physics of Sunspots, eds. L.E. Cram & J.H. Thomas pp. 384-8. Sacramento Peak Observatory.

Danielson, R.E. (1961). The structure of sunspot penumbras.II. Theoretical. Astrophys. J. **134**, 289-311.

Danielson, R.E. (1964). The structure of sunspot umbras.I. Observations. Astrophys. J. **139**, 45-7.

Dialetis, D.; Mein, P. & Alissandrakis, C.E. (1985). The Evershed effect as a steady-state homogeneous phenomenon. Astron. Astrophys., **147**, 93-102.

García de la Rosa, J.I. (1986) in preparation.
Giovanelli, R.G. (1982) Sunspot geometry and pressure balance. Solar Phys. 80, 21-31.
Grossmann-Doerth, U. & Schmidt, W. (1981). The brightness distribution in sunspot penumbrae. Astron. Astrophys., 95, 366-72.
Grossmann-Doerth, U.; Schmidt, W. & Schröter, E.H. (1986). Size and temperature of umbral dots. Astron. Astrophys., 156, 347-53.
Guseinov, M.D. (1983). Magnetic field polarity in umbral bright dots in sunspots. Izv. Krymsk. Astrof. Obs., 68, 36-8.
Harvey, J.W. (1971). Solar magnetic fields-small scale. Publ. Astron. Soc. Pacific, 83, 539-49.
Harvey, J.W. & Breckinridge, J.B. (1973). Solar speckle interferometry. Astrophys. J., 182, L137-39.
Hénoux, J.C. (1968). Sur une particularité de la composante du triplet normal dans l'ombre d'une tache. Solar Phys., 4, 315-17.
Kitai, R. (1986). Solar Phys., 104, 287.
Kjeldseth-Moe, O.; Brückner, G.E. & Hagyard, M.J. (1970). Bull. A.A.Soc., 2, 331.
Kjeldseth-Moe, O. & Maltby, P. (1974). The temperature of penumbral filament Solar Phys., 36, 101-8.
Kneer, F. (1973). On some characteristics of umbral fine structure. Solar Phys., 28, 361-7.
Knobloch, E. & Weiss, N.O. (1984). Convection in sunspots and the origin of umbral dots. Monthly Notices Roy. Astron. Soc., 207, 203-14.
Koutchmy, S. & Adjabshirzadeh, A. (1981). Photometric analysis of the sunspot umbral dots: 2- Size, shape and temperature. Astron. Astrophys., 99, 111-9.
Krat, V.A.; Karpinsky, V.N. & Pravdjuk, L.M. (1972). On the sunspot structure. Solar Phys., 26, 305-17.
Kunzel H. & Staude, J. (1975). The anomalous splitting of the π-component of a Zeeman triplet in sunspot umbrae and suggestions for its interpretation. Astron. Nachr., 296, 171-6.
Lites, B.W. & Thomas, J.H. (1985). Sunspot umbral oscillations in the photosphere and low chromosphere. Astrophys. J., 294, 682-8.
Loughhead, R.E.; Bray, R.U. & Tappere, E.J. (1979). Improved observations of sunspot umbral dots. Astron. Astrophys., 79, 128.
Mamadazimov, M. (1972). On the fine structure of the Evershed effect. Solar Phys., 22, 129-36.
Mattig, W. & Mehltretter, J.P. (1968). Fine structure of brightness, velocity and magnetic field in the penumbra. In Structure and Development of Solar Active Regions, ed. K.O. Kiepenheuer, pp. 187-92. IAU Symp. 35. D.Reidel.
Mehltretter, J.P. (1969). On π-components in Zeeman-split lines of the umbral spectrum. Solar Phys., 9, 387-90.
Mogilevsky, E.I.; Demnika, L.B.; Ioshpa, B.A. & Obridko, V.N. (1968). On the structure of the magnetic field of sunspots. In Structure and Development of Solar Active Regions, ed. K.O. Kiepenheuer IAU Symp. 35,215-29. D. Reidel.
Moore,R.L. (1981). Structure of the sunspot penumbra. Astrophys. J., 249,39
Moore, R.L. (1981b). Dynamic phenomena in sunspots. In The Physics of Sunspots, ed. L.E. Cram & J.H. Thomas, pp. 259-311. Sacramento Peak Observatory.

Muller, R. (1973a). Etude morphologique et cinématique des structures fines d'une tache solaire. Solar Phys., 29, 55-73.
Muller, R. (1973b). Etude photométrique des structures fines de la pénombre d'une tache solaire. Solar Phys., 32, 409.
Muller, R. (1985). On the structure of sunspot penumbra. Solar Phys.,98,51-2.
Parker, E.N. (1979a). Sunspots and the physics of magnetic flux tubes.I. The general nature of the sunspots. Astrophys. J., 230, 905-13.
Parker, E.N. (1979b). Sunspots and the physics of magnetic flux tubes. IX. Umbral dots and longitudinal overstability. Astrophys. J., **234**, 333-47.
Schmidt, H.U.; Spruit, H.C. & Weiss, N.O. (1986). Energy transport in sunspot penumbrae. Astron. Astrophys., **158**, 351-60.
Schröter, E.H. (1967). The Evershed Effect in Sunspots. In Solar Physics, ed. J.N. Xanthakis. Proc. of NATO Advanced Study Institute on Solar Physics. Lagonissi, Athens. pp. 325-51. Interscience Publ.
Schröter, E.H. & Soltau, D (1976).On the time behaviour of oscillations in sunspot umbrae. Astron. Astrophys., **49**, 463-5.
Soltau, D. (1982). A morphological study of some umbral fine structures. Astron. Astrophys., 107, 211-3.
Stellmacher, G. & Wiehr, E. (1971). Magnetically non split lines in penumbae. Solar Phys., **17**, 21-30.
Stellmacher, G. & Wiehr, E. (1980). Line shifts and asymmetries in sunspot penumbrae. Astron. Astrophys., **82**, 157-62.
Stellmacher, G. & Wiehr, E. (1981). Line profiles and magnetic field in penumbral fine structures. Astron. Astrophys., **103**, 211-15.
Thomas, J.H.; Cram, L.E. & Nye, A.H. (1982). Five minute oscillations as a subsurface probe of sunspot structure. Nature, **297**, 485.
Thomas, J.H.; Cram, L.E. & Nye, A.H. (1984). Dynamical phenomena in sunspots. I. Observing procedures and oscillatory phenomena. Astrophys. J., **285**, 368-80.
Thomas, J.H. & Scheuer, M.A. (1982). Umbral oscillations in a detailed model umbra. Solar Phys., **79**, 19-29.
Van Ballegooijen, A.A. (1984). On the temperature structure of sunspot umbrae. Solar Phys., **91**, 195-217.
Wiehr, E. (1985). Influence of umbral dots in sunspot models. In High resolution in Solar Physics, ed. R. Muller. pp.254-5. Springer Verlag.
Wiehr, E.; Koch, A.; Knölker, M.; Küveler, G. & Stellmacher, G. (1984). The influence of penumbral fine structures on line profiles. Astron. Astrophys. 140, 352-6.
Wittmann, A. (1971). On magneto-optical effects in sunspots. Solar Phys., **20**, 365.
Zhugzhda, Y.D. & Makarov, V.I. (1982). On the dynamics of the chromosphere above sunspots. Solar Phys., **81**, 245-51.
Zhugzhda, Y.D.; Staude, J. & Locans, V. (1984). A model of the oscillations in the chromosphere and transition region above sunspot umbrae. Solar Phys., **91**, 219-34.
Zwaan, C.; Brants, J.J. & Cram, L.E. (1985). High resolution spectroscopy of active regions. Solar Phys., **95**, 3-14.
Zwaan, C. & Buurman, J. (1971). Magnetic field strengths derived from various lines in the umbral spectrum. In Solar Magnetic Fields, ed. R. Howard. IAU Symp. **43**, 220-2. D. Reidel.

Photospheric Fine Structure Close to a Sunspot

Oskar von der Lühe

National Solar Observatory
Sunspot, NM 88349

1. Introduction

Over the last decade, several investigators (Mehltretter, 1974; Koutchmy, 1977; Muller, 1983; Muller and Keil, 1983) have studied photospheric faculae, which were first described by Dunn and Zirker (1973). Facular points, which typically have angular scales smaller than one arc second, are believed to represent the footpoints of magnetic flux tubes in the photosphere. They have been studied using single frames or time series of photographs covering periods as long as one hour (Dunn and Zirker, 1973; Muller, 1983). Except for Koutchmy's (1978) study, no attempt has been made to deconvolve the atmospheric-telescopic point spread function from the observation in order to obtain intensity profiles of facular points.

Recent theoretical models (Deinzer et. al., 1984) of very small flux tubes predict the appearance of bright footpoints (i. e., facular points) in the continuum at solar disk center having a diameter of some 300 km and a maximum brightness approximately 1.4 times the average photospheric intensity. The centers of the structures are less bright, so the features actually have a ring-like appearance. Features of this scale should be observable with existing solar telescopes (entrance pupil diameters between 50 .. 100 cm), provided that disturbing effects caused by seeing and telescope aberrations can be corrected.

Recent developments and improvements of speckle imaging techniques (Knox and Thompson, 1974; Weigelt, 1977; Stachnik et al., 1983; von der Lühe, 1985; see Dainty, 1984, for a review) should permit this goal to be attained. I made observations of facular points in the continuum and tried to to resolve them using the Knox-Thompson speckle imaging technique (Knox and Thompson, 1974; Stachnik et al., 1983). The objective was to measure the spatial extent of facular points and, if possible, to resolve their internal structure. Also, by using a time series of images, the lifetime of facular points can be determined and their interaction with the surrounding medium can be examined.

2. Observations and data reduction

Two 35mm film cameras and an RCA 501 "thinned" 512 by 360 pixel CCD were used to record data at the NSO-Sacramento Peak vacuum tower telescope in July 1984 (see fig. 1). One of the film cameras received light through a 15 Å bandpass filter centered at the 3933 Ca II K line, with exposures of 0.015 s. The other film camera received light through a 1 Å Halle $H\alpha$ filter tuned 1 Å towards the red, with exposures of 0.075 s. The CCD camera received light through a 50 Å wide interference filter centered at 5890 Å. In order to move the bandpass away from the sodium D lines into the continuum towards shorter wavelengths, this filter was slightly tilted. Additional K3 infrared blocking filters and neutral density filters were used to avoid saturating the CCD. The CCD exposures were 0.004 s. The spatial scale was 0.043 arc sec per pixel; thus the diffraction limit of the vacuum tower telescope was twofold oversampled. An area 128 × 128 pixels (5.5 × 5.5 arcsec) was actually recorded.

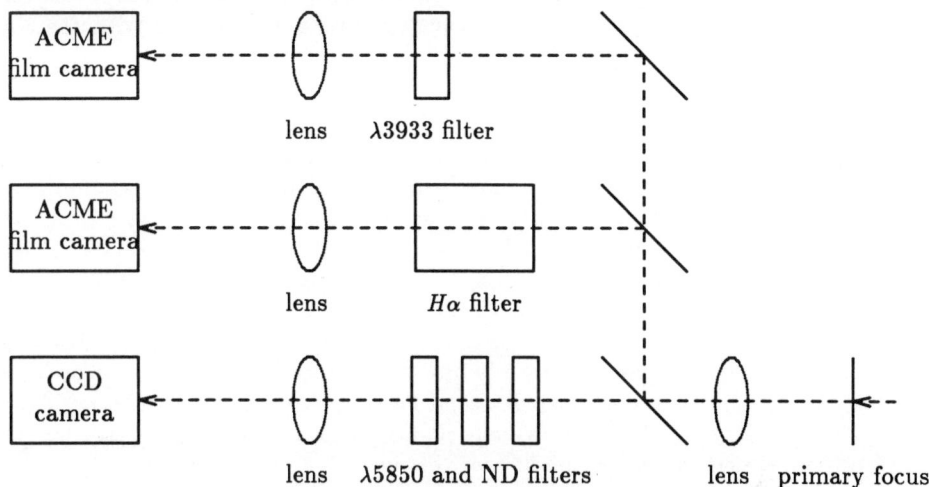

Fig. 1 Experimental setup. See text.

Pictures were taken with all three cameras simultaneously. Bursts of 100 frames were taken, and the time between consecutive exposures within each burst was 0.6 s. The CCD data were written to magnetic tape. A three-bar target at the prime focus of the telescope was used to establish the spatial scale for all three cameras and to determine the position of the field of the CCD with respect to the film cameras.

The data taken on the morning of July 22 are of particularly good quality, since the seeing was good ($r_o \approx 13 \cdots 15\,cm$). Seven bursts, taken very close to the penumbra of a sunspot during a period of good seeing, have been analyzed to date. It has been possible to recover small scale structure information approaching the diffraction limit of the telescope. However, because of the small observed field, it was extremely difficult to view a particular position on the sun for an extended period. Subsequent bursts do not always show the same region. Also, because it was necessary to correct for solar rotation, bursts were taken several minutes apart, and the appearance of small scale structure varies considerably over this time scale. Unfortunately, only one of the analyzed bursts shows a feature that can be clearly identified as a facular point.

The Ca and the Hα films were visually inspected to select the sharpest picture from each of the 100 frame bursts. The CCD pictures were processed using the von der Lühe (1985) implementation of the Knox-Thompson (Knox and Thompson, 1974) speckle imaging algorithm. The Fourier transform of each frame of the burst is calculated and averages are computed in a manner that preserves the high spatial frequency information of both the Fourier amplitudes and the Fourier phases. The Fourier amplitudes are reweighted to correct for atmospheric attenuation (von der Lühe, 1984), the amplitudes and phases are recombined, and the reconstructed image is obtained through an inverse Fourier transformation. The reconstructions were obtained from the central 100 × 100 pixel (4.3 x 4.3 arc sec) subfields in order to reduce artifacts due to frame recentering. Sets of 50 frames, corresponding to 30s elapsed time, were processed. It is possible to recover amplitude information approaching the diffraction limit of the telescope (0.16 arc sec). The major error source is inaccuracies of the speckle transfer function models used to correct the Fourier amplitudes at high spatial frequencies, which limits resolution to slightly less than 0.2 arc sec. The reconstructions are compared with the Ca- and $H\alpha$-prints which serve to identify filigree and other, presumably magnetic regions.

3. Preliminary results and discussion

The power of the speckle imaging technique is best demonstrated in the run taken at 14:19 UT. Very fine structure appeared close to the center of the field of view of the CCD camera. Since this feature was also seen as brightening in the Ca- and $H\alpha$-pictures, I conclude that a facular point or a collection of facular points was observed. Figs. 2 a, b, c and d show the Ca image, $H\alpha$ image, a sample frame from the CCD camera, and the reconstruction, respectively. The structure is situated in an intergranular lane and is elongated; it is approximately 0.8 arc sec long and 0.2 arc sec wide. The arkest part of the surrounding intergranular lane is at 90% average photospheric intensity; its average intensity is 93%. There are two distinguishable maxima in the structure. One maximum is approximately 107% of the average photospheric intensity, and the broader and more elongated second maximum is110% of the average photospheric intensity. Fig. 3 a to c show the intensity profile along three cuts through the feature. Fig. 3a shows a cut along the structure; fig. 3b and fig. 3c show cuts across the two maxima of the feature.

The size of the feature is at the resolution limit of the reconstruction technique and the telescope. Fig. 3d shows the point spread function (PSF) of the reconstruction process. The full width at half maximum (FWHM) of the structure is only slightly larger than the FWHM of the point spread function of the reconstruction process (see table I). This means that the structure is essentially a line source or a collection of unresolved point sources. The actual size of the structure has to be much smaller than the dimension of the point spread function, i. e., significantly smaller than 100 km. It is impossible to infer accurate actual sizes or brightnesses from the reconstructions, since errors in the reconstruction (of order 10% .. 20% of the recovered fluctuation) would generate orders-of-magnitude errors in such inferences.

The small scale structure studied here did not exist when the previous burst was taken 4 min earlier. Judging from the subsequent Ca- and $H\alpha$-pictures, the structure seemed to persist for at least 11 min. Unfortunately, it was not within the field of the CCD camera during that period, so no additional reconstructions were obtained. The feature is significantly smaller than the models of Deinzer et al. (1984) would predict. If the internal structure of such features is to be studied, much larger baselines (larger telescopes) will be required.

Table I reconstructed feature dimensions				
	min	max	FWHM (arcsec)	FWHM (km)
cut 1	0.938	1.072	0.159	117
cut 2	0.933	1.093	0.142	104
PSF	-	-	0.136	100

Literature

Dainty, J. C. "Stellar Speckle Interferometry", in "Laser Speckle and Related Phenomena", 2nd ed., Vol. 9 of "Topics in Applied Physics", Ed. J. C. Dainty, Springer Verlag Berlin Heidelberg New York Tokio

Deinzer, W., Henseler, G., Schüssler, M., Weisshaar, E. (1984), Astron. Astrophys. **139**, 435-449

Dunn, R. B., Zirker, J. B. (1973), Solar Physics **33**, 281-304

Knox, K. T, Thompson, B. J. (1974), Astrophys. J. **193**, L45-48

Koutchmy, S. (1978), Astron. Astrophys. **61**, 397-404

von der Lühe, O. (1985) "Speckle Image Reconstruction of Solar Small Scale Features", PhD thesis, University of Freiburg i. Br., West Germany

von der Lühe, O. (1984), J. Opt. Soc. Am. A **1**, 510-519

Mehltretter, J. P. (1974), Solar Physics **38**, 43-57

Muller, R. (1983), Solar Physics **85**, 113-121

Muller, R., Keil, S. L. (1983), Solar Physics **87**, 243-250

Stachnik, R. V., Nisenson, P., Noyes, R. W. (1983), Astrophys. J. **271**, L37-40

Weigelt, G. P. (1977), Opt. Commun. **21**, 55ff

Fig. 2 The observed region. **a**: Caλ3933 image. **b**: $H\alpha + 1$Å image. **c**: a representative λ5850 4ms exposure from the series of 100. **d**: the Knox-Thompson reconstruction from the first 50 frames of the series of 100. The field size of **2a** and **2b** is 14.1 arc sec, The field size of **2c** and **2d** is 4.3 arc sec. The arrows in **2a**, **2b** and **2c** point at the small structure that appears slightly below the center in **d**.

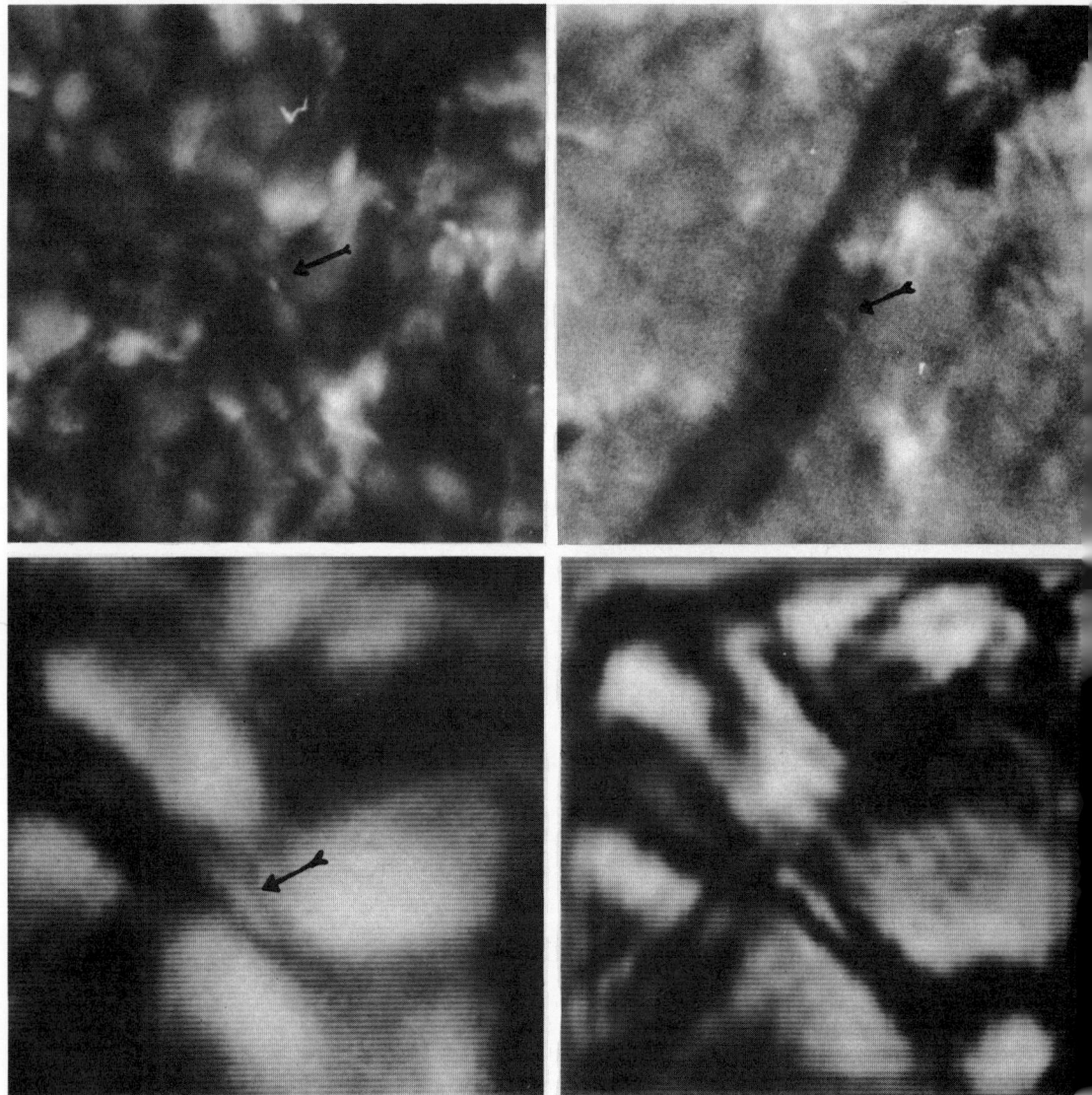

Fig. 3 Intensity profiles along (**a**) and across (**b** and **c**) the small feature seen slightly below the center in fig. **2d**. Profile **3b** cuts the feature across the upper left maximum, profile **3c** cuts through the lower right maximum. Both transverse cuts go from lower left to upper right. **d**: the point spread function of the reconstruction process. The effective modulation transfer function is a circular box that cuts off at 6 Lp/arcsec (91% of the resolution limit), which explains why the minima of the PSF are negative.

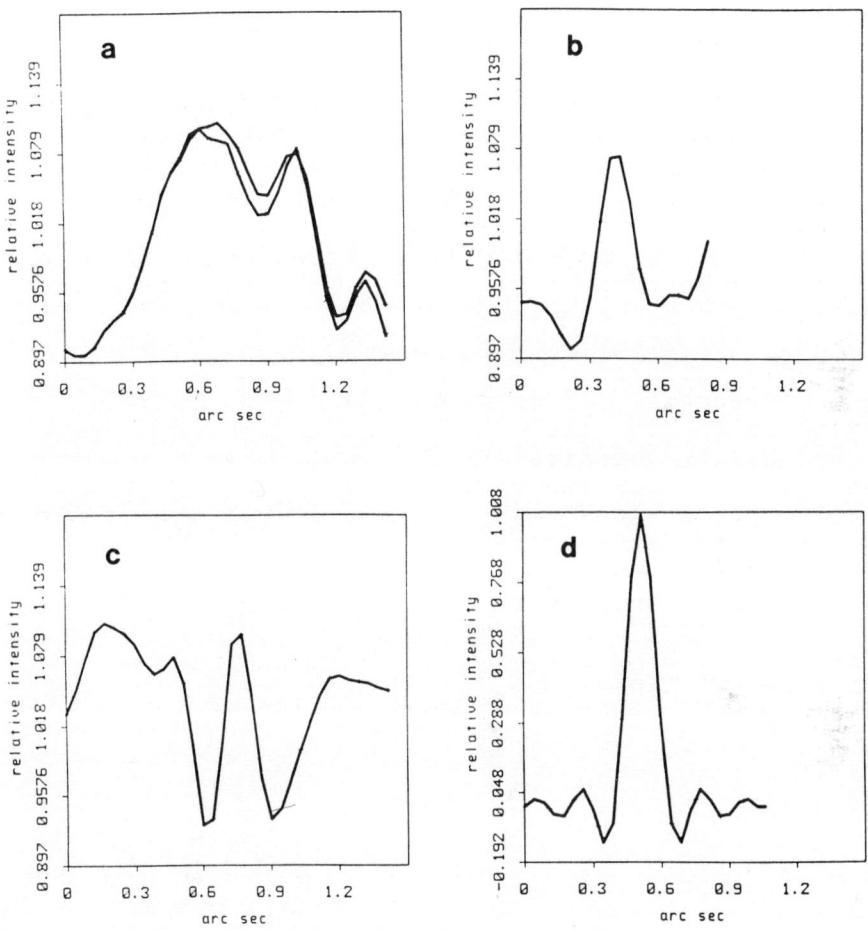

HIGH RESOLUTION SPECTROSCOPY OF SUNSPOT PENUMBRAE

E.Wiehr, M.Knölker, H.Grosser
Universitäts-Sternwarte; D-3400 Göttingen

G.Stellmacher
Inst.d'Astrophysique; F-75014 Paris

The spatial variation of velocity- and magnetic field within penumbral finestructures is investigated from two very highly resolved spectra obtained on Oct.8, 1978, at θ = 30 deg, and on Aug.21, 1980, at θ = 12.7 deg. These two spectra (analyzed by Wiehr et al., 1984, and, resp., by Stellmacher and Wiehr, 1981) have been re-investigated applying new photometry and improved line fitting methods.

The θ = 30 deg spectrum is used to study the horizontal component of the Evershed flow (Fig.1) which is found to be highly cospatial with dark continuum streaks; this confirms the findings by Beckers and Schröter (1969) and by Stellmacher and Wiehr (1981). Those dark streaks are also locations of line-width enhancements, dFWHM, which originate almost entirely from enhanced line asymmetries,

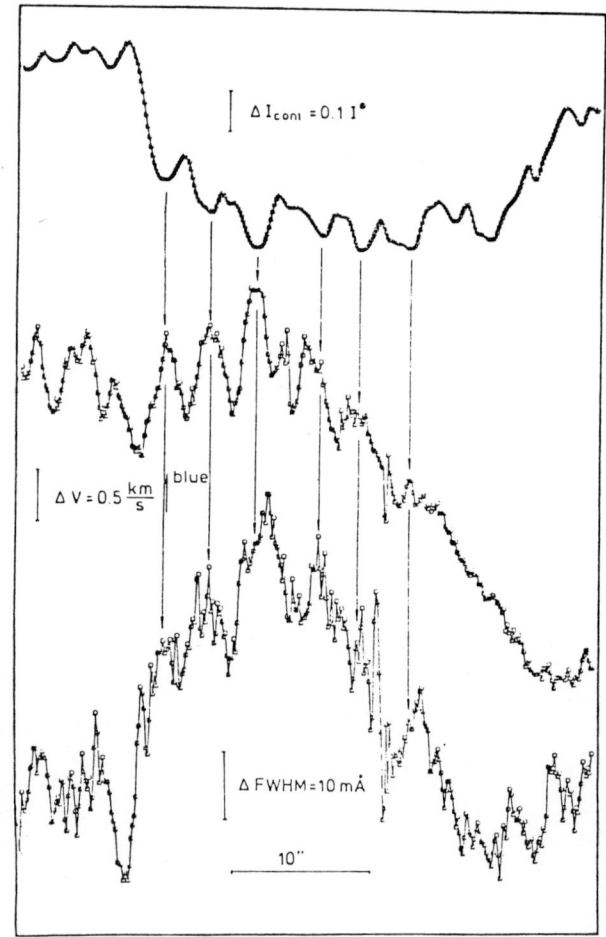

Fig.1: The very small split line Fe+ 5264.8 (g=0.1; 2mÅ/1500 Gs) in a center-side penumbra at θ = 30 deg yielding enhanced line core shifts (central curve) and line widths (lower curve) in dark continuum streaks (upper curve).

(possible influences from turbulence would be much smaller). This latter result supports Maltby (1964), Beckers and Schröter (1969), Stellmacher and Wiehr (1980, Fig.5) but it conflicts with Wiehr et al. (1984) who report enhanced line asymmetries in bright continuum streaks.

The spectrum near disc center is used to study the magnetic field fluctuation within continuum brightenings and -darkenings: for a strong Zeeman splitting (e.g. Fe6302.5), the cospatial occurrence of line- broadening and -flattening indicates an enhancement of the magnetic field. Nummerical calculations (Stellmacher and Wiehr, 1981) yield for dB = 100 Gs, fluctuations of dFWHM = 10 mÅ and dR = 2.5%. The field enhancements in dark streaks Fig.2, confirm Beckers and Schröter (1969) and Stellmacher and Wiehr (1981) but additionally quantify the field fluctuations to 100 Gs.

The spectrum near disc center is furthermore used to investigate the existence of a vertical velocity component. In order to avoid any influence of the (much stronger) main component of the Evershed flow, the line shifts must be determined at a heliocentric angle, θ, equal to the angle of inclination, i, of the main flow (see Fig.3) which Schröter (1965) determines to 15 deg.

Hence, for the limb-side penumbra of our θ=12.7 deg sunspot observation the line-of-sight is perpendicular to the main Evershed flow.

From the small split line Fe 6344.2 we find a marginal tendency for blue shifts in bright continuum streaks.

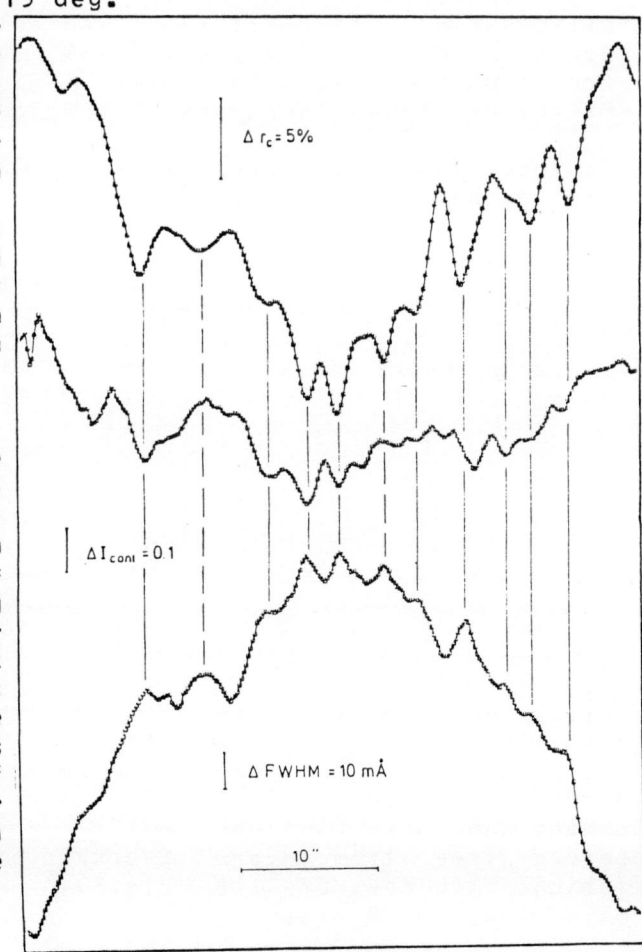

Fig.2: The large split line Fe 6302.5 (g=2.5; 70 mÅ/1500 Gs) in a sunspot penumbra near disc center (θ = 12.7 deg) yielding magnetic field enhancements in dark continuum streaks (upper curve): the cospatial fluctuations (vertical bars) of line width (lower curve) and line depth originate from the Zeeman effect.

A slight indication for such a tendency has already been claimed by Beckers and Schröter (1969). This poor relation between fluctuations of vertical velocities and continuum streaks might be due to influences from oscillations in the penumbra (Balthasar and Wiehr, 1986), if these would exclusively be visible e.g. in bright structures. Systematic velocity fluctuations could then only be seen in temporal averaged data, but not in single spectra.

Our spectra show a minimum distance of neighbouring continuum streaks of 1.2 arcsec = 870 km; they are still far from reflecting the true penumbra-structures of about 270km widths (Muller, 1973). However, even if one were able to observe spectra of individual penumbral filaments, one could not expect to obtain two families of magnetic field- or velocity values: highly resolved white light pictures do not show two distinct brightness distributions (Grossmann-Doerth and Schmidt, 1981). Instead the denotations 'bright' and 'dark' are only _locally_ defined (Stellmacher and Wiehr, 1980), since the intensity of locally dark filaments often exceeds that of (non-neighbouring) locally bright filaments (Wiehr et al., 1984).

REFERENCES

Balthasar,H., Wiehr,E.: 1986; IAU collquium Århus/Denmark
Beckers,J.M., Schröter,E.H.: 1969, Sol.Phys.$\underline{10}$, 284
Grossmann-Doerth,U.,Schmidt,W.: 1981,Astr.Astrophys.$\underline{95}$, 366
Maltby,P.: 1964, Astrophys. Norwegia $\underline{8}$, No.8
Muller,R.: 1973, Sol.Phys.$\underline{32}$, 409
Schröter,E.H.: Z.f.Astrophysik $\underline{62}$, 228
Stellmacher,G., Wiehr,E.: 1980, Astr.Astrophys.$\underline{82}$, 157
Stellmacher,G., Wiehr,E.: 1981, Astr.Astrophys.$\underline{103}$, 211
Wiehr,E.,Koch,A.,Knölker,M.,Küveler,G.,Stellmacher,G.:
1984, Astron.Astrophys.$\underline{140}$, 352

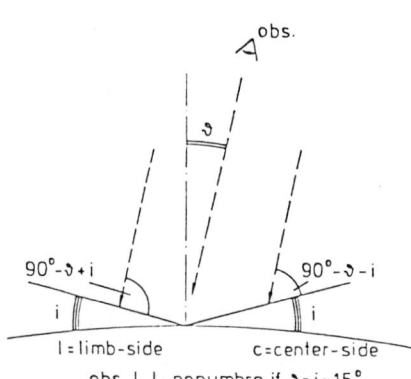

Fig.3: Due to the inclination, i, of the Evershed flow, a perpendicular velocity component can only be observed free from interfering influences of the main flow, for θ = i.

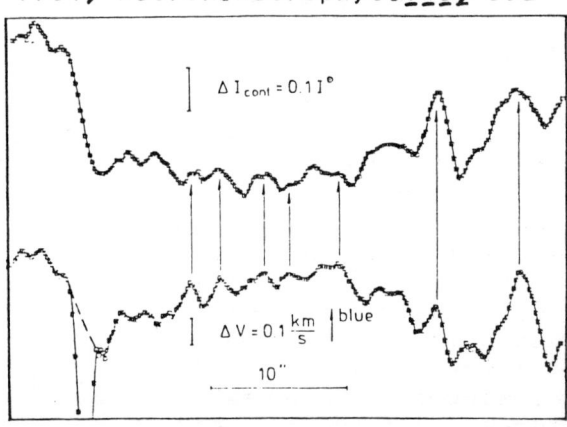

Fig.4: The moderately split line Fe 6344.2 (g=1.17) in the limb-side penumbra from Fig.2; θ=12.7 deg) where the main Evershed flow is invisible (see Fig.3).

JOINT DISCUSSION ON TOPICS OF SESSIONS 5 AND 6, SUMMARIZED
BY THE CHAIRMEN

C. Zwaan
Astronomical Institute
Utrecht, The Netherlands
and
P. Maltby
Institute of Theoretical Astrophysics
Oslo, Norway

Kneer commented on the search for correlations between velocity, intensity and magnetic field, as discussed by Garcia de la Rosa, that one should not reject observational data because they do not show high (anti-)correlations. This would narrow our view. Low correlation may tell us something about the structure and the dynamics as well: it may reveal turbulence, or breaking up of structures, for instance.

On Zirkers's question Garcia de la Rosa confirmed that sunspots do form from fragments that remain independent during the lifetime of the spot, and that reappear during the decay of the spot.

Moreno-Insertis asked Title whether he sees in his data evidence of a process of convective collapse in emerging flux. The answer was no, but Title added that for next spring they are planning to collect alternate 5250-5247 magnetograms to look for such collapse.

Schröter noticed on a superposed filtergram (granulation-mesh-system, plus magnetic fields), shown by Title, an almost ideal co-spacing between magnetic fields and intergranular lanes. But the Spacelab 2 data show a large-scale flow pattern with sources and sinks, and more than 60% of the granules are exploding, which should affect the surrounding structure. How do these two facts fit together? Title answered that the observed magnetic field has pores of 3 to 5", which suggests that the field is <u>not</u> typically concentrated in granulation boundaries. The agreement is better with the size scale of "exploding" granules which also have diameters of 3 to 5".

At a question by Zwaan, Title answered that there are exploding granules in plages, but their number is smaller there.

Deubner asked whether observations showing that concentrations of (longitudinal) magnetic field are not exactly cospatial with the line center brightenings (or "line gaps") may be reconciled with the assumption that part of the brightenings occurs in regions with horizontal field lines. Title agreed but he pointed out that we have good reasons to believe that the strong field is very nearly vertical because of buoyancy. Moreover, there is another observational fact: the brightenings are transient with a time scale of 10 minutes while the field changes very little on the scale of an hour.

Priest asked von der Lühe what he did see in Hα above the filigree - are there spicules? Von der Lühe had used a 1 Å Halle Hα filter tuned 1 Å towards the red. There is a small brightening above the fine structure in the continuum, which has approximately the same extent and shape, but he does not know whether there were spicules to be seen in the core of Hα.

Von der Lühe does not yet know the answer to Knölker's question whether in the data there is a filigree structure of larger sizes than those discussed in the paper.

THE SUBSURFACE STRUCTURE OF SUNSPOTS AND THE ORIGIN OF SOLAR ACTIVE REGIONS

F. Moreno-Insertis
Instituto de Astrofisica de Canarias,
38200 La Laguna (Tenerife), Spain

Abstract
Several topics concerning the subsurface structure and evolution of sunspots are discussed. These include the morphology and stability of the magnetized plasma below a sunspot as well as the heat blockage in it. In the second part of this paper, current ideas on the origin and physical processes preceding the formation of an active region are reviewed.

1 INTRODUCTION

The study of the structure, origin and evolution of active regions involves several fundamental topics in solar physics, like dynamo theory or the theory of convection. An active region is the observable consequence of magnetohydrodynamic processes taking place in deep layers of the Sun; it therefore contains valuable information about the solar interior. However, it has proved very difficult to draw inferences about the hidden parts of a magnetic region from the ever improving observations of its visible layers. In fact, there are still basic issues under discussion concerning, e.g., the morphology of sunspots below the surface or the depth of origin of active regions.

With a view to a necessary selection, we can consider 4 groups of problems which are characteristic of the research activity in this area. The first concerns the morphology and instabilities of the magnetic region below a sunspot. The dichotomy of 'monolithic' and 'cluster' models and the depth to which the spot extends as a coherent structure are the main issues debated in this context.

The second group deals with the energy balance and energy transport in the spot. It is unclear how the spot manages to transport 20 per cent of the photospheric radiative flux despite the inhibition of convection by the magnetic field. Nor is it easy to explain where the remaining 80 per cent is diverted to or how this is done. The solution of these problems requires a better understanding of the nature of convection in and around the spot and of the modes of oscillation supported by the umbral plasma.

Another class of problems relates to the 'prehistory' of active regions (A.R.'s), i.e., the origin and evolution of the A.R. fields before it appears at the surface. At issue here are, in particular, the depth of origin of the A.R. and the physical phenomena leading to the formation and rise to the surface of its flux tubes. Progress in this area will require a deeper insight into the structure and dynamics of the convection zone, specially its cellular structure, differential rotation and magnetic fields.

Finally, the further destiny of the magnetic field after sunspot decay is also
an open and exciting topic. It is still unclear what causes the sunspot's loss
of stability and dissolution. Does the magnetic flux submerge or does it diffuse
on the surface? Which reconnection processes take place at this stage? What
is the extent of the subsurface control of A.R.'s?

Each of these groups of problems is worth a review on its own. The first two
have been treated extensively in several excellent review papers in the past
(Spruit 1981a,b; Thomas 1981; Maltby 1981; Wilson 1981). We will summarize
some major points concerning these questions in Sec.2. The second part of
the review (Sec.3) will be devoted to recent work regarding the evolution of
the active region field prior to its emergence at the surface. This paper
does not cover the late stages of A.R.'s, which will be treated by E. Priest
(Session 11); nor, due to lack of space, will it deal with self-similar sunspot
models (Schlüter & Temesvary 1958; Deinzer 1965; Yun 1970), current sheet
models (e.g.Schmidt & Wegmann 1980;cf.also Pizzo 1986) or empirical umbra
models   (cf. refs. in Tandberg-Hassen 1967, Zwaan 1984).

2   THE SUBSURFACE STRUCTURE OF A SUNSPOT.
2.1   Morphology and stability

Two main alternatives are currently under discussion. The sunspot field may be contained either in a single magnetic column or in several,
perhaps many, magnetic flux tubes. The first case has been advocated by
Meyer et al (1974, 1977), Cowling (1976) and, more recently, by Knobloch &
Weiss (1984). In the model by Meyer et al (1974), the magnetic field is held
together by two "collars" at different heights. The gas pressure compresses
the spot at the photosphere, constituting a sort of straight jacket for the
field, as Gokhale & Zwaan (1972) describe it. The second collar operates at
depth and is provided by the ram pressure of the converging flow in the
moat, the latter being the annular convective cell surrounding most mature
spots (cf. Vrabec 1974, Pardon et al 1979).

As to the origin of this structure there are two extreme possibilities. The
single flux tube shape can just reflect the monolithic structure of the
original magnetized region, say a flux tube in the deep convection zone
(cf.Sect. 3), which has not been disrupted during its rise to the surface. Or
it could be the result of local concentration of preexisting, more elementary
magnetic units, perhaps thinner tubes of high field strength, by the supergranulation. This second possibility is not generally accepted nowadays,
since it seems to conflict with observations (Zwaan 1985).

The alternative is that the sunspot field separates into many individual
strands below the photosphere (Parker 1979; Severny 1965). A model of this
sort is referred to in the literature as 'cluster model' or 'spaghetti model'.
The main argument in favour of this structure is the lack of stability of a
magnetic field column against fluting below the surface (cf. below). Another
possibility is that overturning convective motions in the deep spot layers
split the field column into more or less separate flux tubes. To hold together
the field strands below the surface, Parker introduced a material flow which
converges towards the spot and down between the individual flux tubes. This
flow would also dispose of the excess heat accumulated below the spot. This

'countermoat' cell is a controversial item in the model; first, no trace of this flow has been observed. Second, the heat pile-up below the spot is likely to generate a moat cell and not a convective cell flowing down along the spot.

A necessary test for any sunspot model is the stability of the proposed configuration. The fluting instability, in particular, is characteristic of the interface between a magnetized region and the unmagnetized plasma, when the field lines are concave towards the latter. In their 1977 paper Meyer et al proved that a magnetic structure whose field lines fan out towards higher levels can be stabilized against fluting through the buoyancy force: if the magnetic region is lighter than its surroundings, then work has to be done to raise the unmagnetized matter in the crests of the perturbation and to lower the magnetized plasma in the valleys. This is the case for the surface layers of the spot, which are much less dense than the matter outside. By using Bernstein et al' energy principle (1958), they obtained the following condition for stability:

$$\sin \chi > 2 H_p/R_c \tag{1}$$

with H_p pressure scale-height, R_c radius of curvature and χ the angle between the field lines and the vertical. Accordingly, (1) a greater inclination, (2) a shorter scale-height and (3) less curvature favour stability. Meyer et al estimated that sunspots with total flux above 10^{19} Mx ought to be stable against fluting at or near to the photospheric levels. Yet, the stabilizing effect does not work in lower levels where H_p increases and the spot boundary becomes less inclined. Therefore, it is difficult for the spot to keep a monolithic structure along its life. The instability is especially obvious in field configurations with a 'wasp waist' below the surface (e.g. Parker 1979a. Cf. also Pizzo 1986), which, however, are not confirmed by current sheet models (Schmidt & Wegmann, 1980).

If the fluting instability indeed splits the magnetic structure below the photosphere, then one should ascertain how far the instability proceeds, i.e., at which stage the growing instability saturates, as well as the timescale of

Fig.1 A monolithic and a multi-flux-tube spot. Adapted from Meyer et al (1977) and Parker (1979a) respectively.

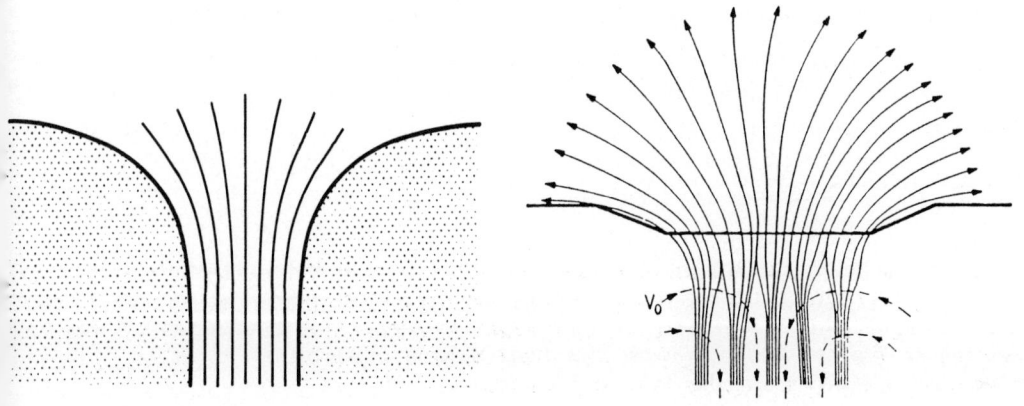

the process. Although these are difficult questions, two simple arguments can readily be given. The resulting strands should have high field strength, since it is difficult to accommodate a 'sea' of diffuse field beneath a concentrated structure of about 3000 G and total flux up to a few times 10^{22} Mx. Besides, the subsurface control of the active region, apparent at least during its first evolutionary stages, indicates that the flux tubes keep their coherence individually and as a bundle to deep levels.

Spruit (1981c) has proposed a sunspot model in which the tubes of the cluster are rooted in the deep convection zone and are pressed together at the surface owing to their strong buoyancy, as if they were a bunch of tethered balloons. Although still not worked out in detail, this model possesses several interesting features. For instance, the field lines' radius of curvature should be greater than the scale-height at all levels except, perhaps, near the surface; also,the inclination angle in (1) need not go through zero to negative values. Therefore, the instability problem is not so acute as in wasp waisted spot models. On the other hand, a countermoat cell is not necessary here to keep the flux tubes in a bundle since they are held together in deep layers.

2.2 Heat flux blockage in the spot.

As is well known, it is not an easy task to explain the diversion of 80 per cent of the normal photospheric flux from the spot; nor is it easy to explain how the strongly magnetized plasma manages to transport the remaining 20 per cent to the photosphere.

Biermann (1941) was the first to propose that the inhibition of convection through the strong magnetic field (cf. Chandrasekhar 1961) was the cause of the reduced heat flux and temperature in the umbra. Meyer et al (1974), in fact, calculated that overturning convection was not possible in about the first 2000 km below the umbral photosphere; a limited convective heat transport could be nevertheless present through overstable convection. Therefore, the heat flux must be blocked in the spot below that level.

The blocked heat cannot pile up in the magnetized region, since this would lead to the destruction of the spot. Assuming that the spot cross section does not diminish drastically with depth, we can easily compare the accumulated heat with the heat contents of the spot or with its magnetic energy. If, for instance, the heat would be deposited in one scale-height H at 2000 km depth

$$\frac{|\text{div } \vec{F}|}{p/(\gamma-1)} \simeq \frac{|\vec{F}|}{H\,p}(\gamma-1) \simeq 0.5 \text{ day}^{-1} \qquad (2)$$

and

$$\frac{|\text{div } \vec{F}|}{B^2/(8\pi)} \simeq 10^2 \text{ day}^{-1}, \qquad (3)$$

so that the heat blocked in one day would amount to half the thermal energy or about 100 times the magnetic energy of that stretch of the sunspot. Such a pile-up of heat would lead to a pressure increase in the spot which would destroy it (Parker 1979a). Thus, the heat excess must be efficiently transported away from the blocking region.

Cooling by Alfvén waves (Parker 1974a,b) is not a clear solution of the problem, since Alfvén waves are not generated efficiently (e.g. Cowling 1976) and are strongly reflected downwards in the photospheric levels (Beckers and Schneeberger 1977; Parker 1979c; Nye and Hollweg 1980). Other MHD modes, e.g. resonant fast modes, are trapped in the umbra (Scheuer & Thomas 1981); they cannot transport heat away from the spot.

Cooling by turbulent heat transport outside the magnetic region (Spruit 1977b, 1982a,b) or a systematic flow circulating adjacent to the spot are better possibilities. In the monolithic model heat blockage occurs within the spot column, specially at the level above which overturning convection is not possible; the lateral transport of heat is also reduced in the magnetized column. The spot should therefore be cooled very efficiently from outside. Perhaps the moat circulation pattern is a direct consequence of the heat blocking process (Meyer et al 1974).

In a multi-flux-tube spot the interface between magnetized and unmagnetized plasma increases like $n^{1/2}$, n being the number of individual flux tubes Cooling of the spot from outside is therefore easier in this model. Here, again, a material flow would help to cool the spot; yet, as remarked above, the heat pile-up tends to generate a flow circulating in the sense of the moat cell rather than the contrary. On the other hand, if the flux tubes of the cluster are thin enough, then the convective transport would be severely hampered in them. The heat flux blocked would then be only that of the intertubular plasma.

As to the further fate of the diverted heat flux, the best founded theoretical possibility is that it be stored in the convection zone (Spruit 1982a,b), where the turbulent heat transport coefficient greatly increases with depth. Following Spruit two timescales are characteristic for the thermal adjustment of a star after the sudden appearance of a spot. These are a diffusive timescale, $t_D = D_{cz}^2/\kappa$, and the Kelvin-Helmholtz time of the convection zone, $t_{KH} = U/L_\odot$. Here κ is the turbulent thermal diffusivity, D_{cz} and U are the depth and thermal contents. respectively, of the convection zone. Since the spot's lifetime is much shorter than those timescales, the solar luminosity must decrease during the spot's life by an amount roughly corresponding to the blocked heat flux. This indeed is in agreement with the appearance of minima in solar irradiance measurements coinciding with the passage of the spot across the disk (Willson et al 1981).

More detailed studies of the irradiance curves, however, seem to indicate that an amount of radiative energy of the same order of magnitude is lost by the Sun in the form of excess radiation by active region faculae in a timescale up to a few times the spot's life (Oster et al 1982; Chapman et al 1984). Although this is a suggestive possibility, it is difficult to imagine a direct physical relationship between the heat flux blocked by the spot and that emitted in excess by the faculae. For that, the radiant energy diverted from the spot should be stored somehow in the surroundings and then be reradiated; yet, it is not easy to devise a mechanism that could possibly do this job (but see Wilson 1981). The excess radiation of the faculae will probably cancel the perturbation caused by the spot only in an average sense over timescales longer than the lifetimes of both types of objects (Chiang & Foukal 1985).

2.3 Convection in umbrae, umbral dots.

Basic problems in this respect are the nature of umbral dots and the amount of radiant energy they transport; of interest is, in particular, whether they can transport the 20 per cent of the photospheric flux seen to come out of the umbra. Research into these matters has progressed along two main avenues. One can carry out a MHD study, trying to determine the possibility and characteristics of oscillatory convection in the umbra. The alternative is to set up empirical umbra models which reproduce the observed spectral features of the umbral radiation (cf. refs. in Sec.1), in particular 2-component models (e.g. Wiehr & Stellmacher 1985, Adjabshirzadeh & Koutchmy 1983).

Umbral dots bear information about the subsurface layers of the sunspot. Up to now not much information could be extracted from them because of observational uncertainties and limitations of theory. Concentrating on the latter, the MHD studies of convection in umbrae afford different results depending on the basic sunspot model used:

(1) Monolithic model. According to linear theory, monotonously increasing convection cannot take place where the magnetic Prandtl number is smaller than 1. Thus, only oscillatory convection is possible in the uppermost 2000 km of the umbra, which could give rise to umbral dots (Zwaan 1968; Knobloch & Weiss 1984). Using nonlinear Boussinesq theory of magnetoconvection, the latter authors propose that umbral dots could be the external manifestation of supercritical (Ra>Ra_{cr}) nonlinear oscillatory convection in vertically elongated cells, with periods much longer than obtained through linear theory. They suggest that these cells could have diameters up to 300 km and a depth of some 1500 km. Alternatively, umbral dots could be the result of overshooting from the deeper spot layers where overturning convection is possible.

Until recently the stumbling block for the application of magnetoconvection studies to solar problems was the fact that convection in the Sun is clearly non-Boussinesq. In the case of sunspots, for instance, one need only compare the photospheric scale-height, O(100 km), with the values just mentioned for the diameter and depth of umbral dots. Only recently fully compressible calculations of magnetoconvective processes are becoming possible thanks to the advances in computer science (cf. N.Weiss & N. Hurlburt´ presentations at this meeting).

(2) Cluster models. In this case overstable convection can occur in the interstices between the individual flux tubes. In fact Parker (1979d) showed through a Boussinesq calculation that overstable convection was possible in a vertical field-free fluid sheet separated from the magnetized plasma by two vertical planes. He suggested that umbral dots are intrusions of the field-free plasma between the magnetic tubes to visible layers. If so, umbral dots would have to show the following observational signatures: first, they should be largely field free. Second, they should have a tendency to appear forming a network around the flux tubes, although only if the latter have larger cross sections than the dots. The observations of umbral dots are as yet inconclusive about these extremes. In particular the recent observations of umbrae covered by dots (refs. in Garcia de la Rosa, this volume) can be compatible with both overstable convection in the spot and oscillations in the intertubular lanes.

Progress in this field is necessary both in observation, in particular to ascertain occupation factors, field strength, etc, as well as in theory (Non-Boussinesq models, studies with realistic sunspot stratification). As regards these topics, however, both observation and theory are likely to advance rapidly in the next years, as can be judged from this meeting.

3 IDEAS CONCERNING THE ORIGIN OF ACTIVE REGIONS
3.1 Depth of origin.

Nowadays it is widely believed that the field concentrations in active regions originate in deep layers of the convective zone. This idea has repeatedly appeared in the literature along the years (Parker 1955; Babcock 1961; Schmidt 1968; Ponomarenko 1970; Zwaan 1978; Spiegel & Weiss 1980; Golub et al 1981; Parker 1984a). There are both observational and theoretical grounds for this assumption. Observationally, we find striking regularities in the appearance of active regions, like Hale's polarity law, their approximate east-west orientation or the presence of active longitudes. All these seem to be indications of the existence of a toroidal system of magnetic field submerged in deep layers of the convective envelope and in interaction with the material flows of the latter. Also, the typical dimensions of active regions are comparable to the length scales, say the density scale-height, of the deep convection zone rather than to those of the photosphere. Finally, the overall pattern of emergence of active regions (see reviews by Zwaan 1978, 1985 and Mc Intosh 1981) can be interpreted as the arrival at the surface of magnetic flux which was already concentrated in the form of flux tubes in deeper layers.

Several theoretical arguments are also available. The magnetic field appears in concentrated structures with high field strength not only in the photosphere but probably also in the convection zone, where turbulent convective motions tend to expel the magnetic flux from the interior of the turbulent eddies and concentrate it at their boundaries (cf Galloway & Weiss 1981). The concentrated field should be located preferentially near the base of the convection zone: on the one hand, magnetic buoyancy tends to deprive the convection zone of its concentrated magnetic flux on short timescales (Parker 1955,1975; Schüssler 1977 Moreno-Insertis 1983). On the other hand, the region of overshooting convection below the convection zone proper seems to be an adequate site for the operation of the solar dynamo (cf. Van Ballegooijen 1982b; Schmitt et al 1984; Schmitt 1985; Pidatella & Stix 1986). One reason for this is, in particular, that the magnetic flux may be stored there for longer periods of time than in the convection zone.

The formation of active regions starting from the magnetic field concentrated near the base of the convection zone must be due to processes which cause a loop of field to rise to the surface. If so, an active region would be a direct manifestation of the flux system hidden deep in the Sun. On the other hand, the study of the evolution of the magnetic flux before its emergence at the surface can provide valuable clues as regards the structure of sunspots in the subphotospheric layers.

The depth of origin of active regions must be determined mainly through theoretical arguments. Golub et al (1981) have suggested the possibility of different origins for ephemeral A.R.'s as for normal, longer lived A.R.'s. The latter would originate from a magnetized layer near the base of the convection zone. The instabilities of this layer would produce rising flux ropes, which are thick enough to withstand the disruptive effect of the turbulent convection on their way to the surface. Flux tubes with weaker field strength and less total flux will be shredded by convection. The (re)concentration in shallower layers of this flux would produce magnetic tubes which emerge as E.A.R.´s.

More recently, Parker (1984a) has proposed a simple way of estimating the depth of the anchor points of an A.R. His argument is based on the assumption of the subsurface control of the latter along the spot life. The estimate is obtained by calculating the equilibrium path of an arch-shaped flux tube with fixed anchor points at a depth h. The maximum possible separation of the foot points, L, such that magnetic tension and buoyancy force can still compensate each other along the tube can be tentatively identified with the separation of polarities at the surface. L in turn, depends on the depth h; in the calculation carried out in Parker´s paper the result is $h \cong L$. Since L is of order 10^5 km, this can be taken as a further indication that active regions are rooted in the deep convection zone.

3.2 The rise of magnetic flux from the base of the convection zone.

Assuming a horizontally magnetized region near the base of the convection zone, the two main mechanisms which may produce loops of field rising to the surface are:

1. an instability started by a perturbation that bends the field lines in a vertical plane, like a modified form of the magnetic Rayleigh-Taylor instability or the kink instability, and

2. an ordered, large-scale convective flow that may pick up magnetic flux in its updraft and lift it to higher layers.

Both kinds of phenomena can be studied in one of the following ways. One can study the equilibrium and possible instabilities of a continuum of field, in particular doubly- or multiply-diffusive instabilities (Acheson 1979; Schmitt & Rosner 1983; Hughes 1985a,b). Some of these instabilities may give rise to concentrated, flux-tube like structures. Alternatively, one can consider from the outset an ´intermittent´ structure in the form of individual tubes and study the evolution of one of them. This possibility has the advantage of being amenable to simplified treatment by means of the thin flux tube approximation (Defouw 1976; Roberts & Webb 1978; Spruit 1981d), allowing fully compressible, nonlinear numerical computations.

In the following we will concentrate on the second alternative. A basic calculation in this context is the rise of an initially horizontal flux tube through the convection zone caused by the development of the kink instability. The adequate candidates for this calculation are neutrally buoyant flux tubes, which are known to be liable to the kink instability (Spruit & Van Ballegooijen 1982). The flux tubes in initial thermal equilibrium with their surroundings are not of interest in this study: they float up to the surface as a whole so rapidly that the instability hardly has time to grow during the journey.

The results of a calculation of this sort allowing for the motion of the magnetized matter both in the direction parallel and perpendicular to the field lines can be seen in Fig.2 (Moreno-Insertis 1986). Here a cartesian geometry and a numerical background stratification (taken from Spruit 1977) were used; on the other hand, the influence of a large-scale, ordered convective flow, a giant cell, was ignored. The initial field strength used, B_o, was above the equipartition value (about 10^4 G at the bottom of the convection zone) but well below the field value for which the magnetic pressure equals the gas pressure outside (about 10^7 G). In the case represented in the figure, $B_o = 10^5$ G; such initial field strengths are not unreasonable for the deep convection zone (Parker 1984b, Roberts & Campbell 1986; cf. also Van Ballegooijen 1982a). On the other hand fields weaker than equipartition do not yield results of interest in this context (cf. below)

As can be seen in the figure, the development of the instability starting from a small sinusoidal perturbation leads to the rise of the upper loop of the flux tube to the photosphere; the lowermost portion of the tube leaves the convection zone through its lower boundary, entering the subadiabatic region underneath. Concerning the rising loop, two clear results emerge from this calculation; first, a material downflow sets in along the field lines. Second, the field strength at the top of the tube, B_{top}, becomes weaker by a large factor along the rise.

The latter result is a direct consequence of mass and magnetic flux conservation in the tube. In fact, in the absence of any stretching of the matter along the field lines, B should decrease as ρ. The downflow along the tube prevents such extreme decline but, anyhow, it cannot avert the rapid decrease of B_{top} in the uppermost 10^4 km of the convection zone For example, the top of a flux tube initially at $2\ 10^5$ km depth with 10^5 G had a field strength of 1600 G when crossing a level 8000 km below the photosphere and that field value was rapidly decreasing. In this computation the field strength in the last few thousand km beneath the surface would diminish below the local equipartition value. The top of the tube would thus resemble a rising bubble of weakly magnetized matter.

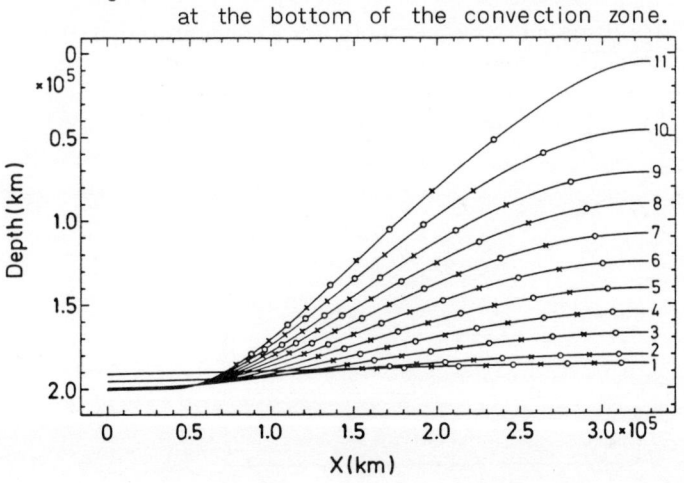

Fig.2. Time evolution of a kink-unstable flux tube initially at the bottom of the convection zone.

However, the assumptions of this calculation are inadequate to explain the stages of the rise inmediately before the emergence at the surface. In fact, the observations of the birth of active regions tell us that the field reaches the photosphere in a concentrated fashion (cf. Zwaan 1985, Title, this meeting). Hence, some additional physical processes must take place in the rising tube to keep the field concentrated. The timescale of the whole rise to the surface for values of the magnetic flux typical of active regions is several weeks to a few months (Fig.3).

3.3 Further physical processes during the rise

1. Convective flows. From a theoretical point of view there is increasing evidence of the existence of convective cells extending over several scale-heights from the bottom to the upper layers of the convection zone (Simon & Weiss 1968; Hurlburt et al 1984). If the coupling of the flux tube to the external medium through the drag term is strong enough, then the convective cell will pick up the flux tube in the updraft and hold it down in the downdraft (Fig.4). Therefore, the convective cell will impose its horizontal scale, possibly of order D_{cz}, on the evolving flux tube. Similarly, the timescale of rise of the top of the tube will be related to the cell turnover time (Moreno-Insertis 1984). Note also that this mechanism works for initially isothermal ($T_i = T_e$) flux tubes as well as for initially non-buoyant ones.

It is easy to estimate the relative strength of the drag and magnetic forces in a thin flux tube. Calling r the radius of the tube cross section, R_c its radius of curvature and using the aerodynamic drag formula, we obtain:

$$\frac{\text{magnetic force}}{\text{drag}} = \frac{B^2}{4\pi\rho} \frac{1}{R_c} \frac{r\pi}{v_r^2} = \left(\frac{B}{B_{eq}}\right)^2 \frac{\pi r}{R_c}, \qquad (4)$$

where B_{eq} stands for the equipartition field strength, $B_{eq}^2 = 4\pi\rho v_c^2$, v_c is a typical value of the convective speed and v_r is the speed of the flux tube relative to the surrounding medium. In (4) we have set $v_r = v_c$, the maximum value of v_r of interest in what follows.

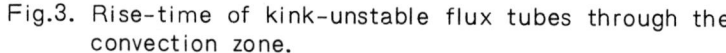

Fig.3. Rise-time of kink-unstable flux tubes through the convection zone.

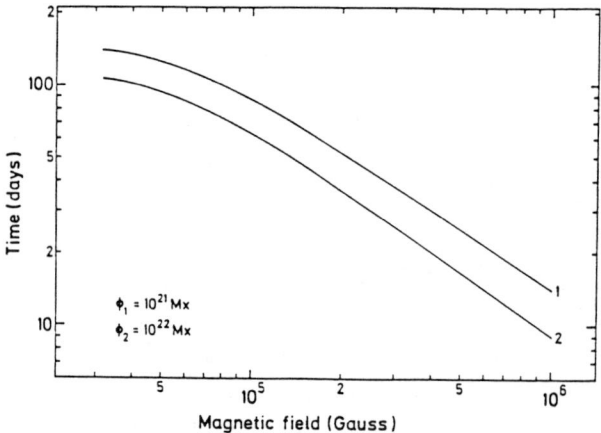

Assuming now for simplicity a tube with a sinusoidal shape of amplitude A and wavelength λ, the ratio (4) will be smaller than 1 if

$$\left(\frac{B}{B_{eq}}\right)^2 \frac{rA}{\lambda^2} < (4\pi^3)^{-1} . \tag{5}$$

Using (4) and (5) we can estimate the minimum field strength, B_m, that can resist deformation of the horizontal shape by the flow. Setting $\lambda = 2\ D_{cz}$, $A = \lambda/20$, $\Phi = 10^{21}$ Mx and the equipartition field at the bottom of the convection zone, we obtain $B_m = O(10^5\ G)$. Weaker (or thinner) tubes will be correspondingly more deformed; more precisely: $B_m \propto \Phi^{-1/3}$. Therefore, we can expect the convective flows in the lower convection zone to play a non negligible role in the dynamics of the flux tubes.

Similarly, we can compare the buoyancy force with the drag in the case of a flux tube with the same temperature as its surroundings. The result is

$$\frac{\text{buoyancy force}}{\text{drag}} = \frac{g}{\beta} \frac{\pi r}{v_r^2} , \quad \beta = \frac{8\pi p}{B^2} , \tag{6}$$

which is less than one if

$$\left(\frac{B}{B_{eq}}\right)^2 \lesssim \frac{2}{\pi} \frac{H_p}{r} . \tag{7}$$

Using the values corresponding to a depth of $1.7\ 10^5$km in the convection zone and $\Phi = 10^{21}$Mx, one obtains $B \lesssim O(10^5 G)$. This, again, shows that the convective flows may noticeably modify the evolution of the flux tubes.

Fig.4. Flux tube in a bidimensional convective cell model. (From Moreno-Insertis 1984)

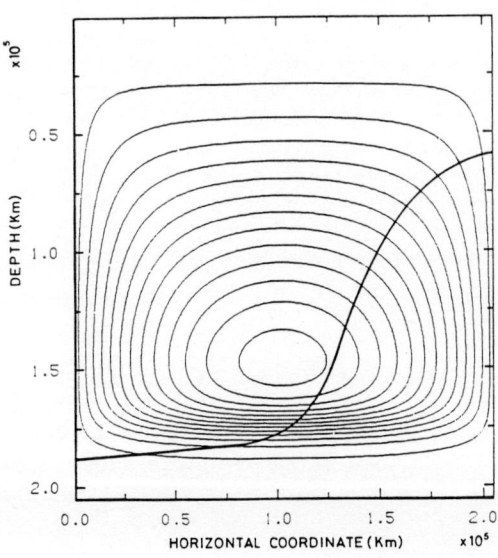

2. Fragmentation of the flux tube during the rise. This may happen, e.g., through the action of MHD instabilities like the fluting instability (Schüssler 1984,1987). In fact Meyer et al´ mechanism for stabilization through buoyancy does not work in the deep convection zone since there $\rho_i \simeq \rho_e$. The instability may therefore tear up the rising flux tube into thinner strands. Using

$$\tau = O\left(\frac{R_c\, r}{v_A^2}\right)^{\frac{1}{2}} \qquad (8)$$

as the growth-time of the instability (Schüssler 1987), with v_A = Alfvén speed and representative values for the quantities on the right ($\Phi = 10^{21}$Mx, $R_c = 1/2\ D_{cz}$, $B = 10^4$G, $\rho = \rho(z=10^5\text{km})$), one obtains $\tau \simeq 20$ d, which is of the same order as the expected rise-times. Still smaller values of τ would ensue for higher B ($\tau \propto B^{-3/4}$ for fixed Φ).

Another possibility, disruption of a rising buoyant flux tube by the flow system adjacent to it, has been discussed by Schüssler (1979).

It is very uncertain how far these processes may affect the rising flux tube. If a fragmentation process takes place, then the flux tube would become a bundle of thinner strands and the resulting sunspot will probably consist of a cluster of individual tubes, like in Spruit´s model (1981c). But then the rise would deviate in several respects from the simple image given in previous paragraphs. First, it ought to occur at a slower pace, since the drag force is inversely proportional to the radius of the cross section. Second, the surrounding turbulent convection could perhaps further shred the flux strands. Nonetheless, a limit to this process of disruption can be set a posteriori: the rising flux tubes do arrive at the surface as a bundle, so they cannot be torn up so far as to be passively carried around by the turbulent flows outside. Also, the subsurface control of active regions, as far as it occurs, is an indication that the individual strands keep a certain degree of coherence to deep layers.

3. Coriolis force. The Coriolis force has always been thought to have a perceptible effect on the rising flux tube, for instance causing the tilt with respect to the azimuthal direction with which opposite polarities of an active region are seen to emerge (cf. Schmidt 1968). Yet, only now it is beginning to be included in actual calculations of the rise of the flux tubes. Choudhuri & Gilman (1986) have calculated the effect of the Coriolis force on a purely toroidal flux ring rising without bending of the field lines through the convection zone. They conclude that, owing to the Coriolis force, the rise may significantly deviate from just a radial expansion. For instance, a flux ring starting from low latitudes may be deviated to latitudes poleward of the activity belt.

The Coriolis force will also affect a kink-unstable flux tube. Nevertheless, during the rise of the latter a material downflow along the field lines takes place, which will give rise to a non-zero contribution to the term $2\vec{\Omega} \times \vec{v}$. The extent of the effect of the Coriolis force on these tubes is still unknown; it will have to be determined with the help of numerical calculations.

3.4 Additional remarks.

Any theory of the evolution of the A.R. fields prior to their emergence at the surface can be subjected to various tests. For instance, it has to fit in with the observations of A.R. development; on the other hand, it has to be understood within the framework of the solar dynamo. Concentrating on the former, we will briefly comment on the proper motion of sunpots.

A direct indication of the dynamics of the magnetic tubes below the surface is the separation of the opposite polarities in young A.R.'s. According to Van Ballegooijen (1982a), this behaviour corresponds to the lateral drift of the almost vertical flux tubes underlying the growing active region; in the stationary situation the drag due to this motion is balanced by the horizontal component of the magnetic force. In need of explanation is, still, why the separation of the opposite polarities comes to a halt after several hours. Two possibilities could be investigated in this respect. The progressive splitting of the spot below the surface could cause the increase of the drag force and the reduction of the drift velocity to low values. A different possibility would be that the opposite polarities stop at locations corresponding to selected positions in an underlying large-scale convective cell, for instance near the cell's downdraft. A behaviour of this kind is found in numerical simulations of the motion of flux tubes in granules or supergranules (Meyer et al 1979, Schmidt et al 1985)

The rotation rate of the spot, Ω_s, does not coincide with that of the photospheric plasma, Ω_{ph}. The usual interpretation of this fact is that the motion of the spot reflects the rotation rate of those layers where it is rooted. This coupling of the spot to the surroundings must be understood in an average sense over a range of depths; observations should then be used to extract information about those layers. Yet, the inversion problem, namely to draw conclusions about the latter from the existing data on sunspot motion, is difficult; its solution requires, e.g., detailed knowledge of the differential rotation of the convection zone, which is not yet available. Schüssler (1987) has pointed out that the drag force possibly plays a subordinate role in the flux tube dynamics near the photosphere, where the buoyancy force is very strong. However, in deeper layers it may become predominant with respect to the buoyancy and Lorentz forces; the observed proper motion of the spot would then correspond to the rotation rate of those levels. The fact that $\Omega_s \to \Omega_{ph}$ along the spot's life (cf. Balthasar et al 1986) would thus reflect, according to Schüssler, the coupling of the flux tube to its surroundings at increasingly shallower layers, possibly as a result of the further splitting of the tube.

Given the current uncertainties in our knowledge of the actual profile $\Omega(r)$ in the convection zone as well as about the possible fragmentation processes in the flux tube, we cannot determine the actual range of depths in which the tubes are coupled to the plasma. Nevertheless the foregoing arguments, if applicable to actual sunspots, favour a multi-flux-tube model, at least at the advanced stages of development.

Acknowledgments

The author is grateful to Drs. H.C.Spruit, H.U.Schmidt and M. Schüssler for discussions concerning the subject of this review.

Reference list

Acheson, D.J. (1979) Instability by magnetic buoyancy. Solar Phys. 62, 23
Adjabshirzadeh, A. & Koutchmy, S. (1983) Photometric analysis of sunspot umbral dots. III Spectrophotometry and preliminary model of a 2-component umbra. Astron. Astroph. 122, 1-8.
Babcock, H.W. (1961) The topology of the Sun's magnetic field and the 22 year solar cycle. Astroph.J. 133, 572
Balthasar, H., Vázquez, M. & Wöhl, H. (1986) Differ.rotation of sunspot groups in the period from 1874 through 1976. Astron. Astroph. 155, 87
Beckers, J.M. & Schneeberger, T.J. (1977) Alfvén waves in the corona above sunspots. Astroph.J. 215, 356-363.
Bernstein, I.B., Friemann, E.A., Kruskal, M.D. & Kulsrud, R.M.(1958) An energy principle for hydromagn.stability problems.Proc.Roy.Soc.A244,17
Biermann, L. (1941) Vierteljahrschr. Astr.Ges. 76, 194
Chandrasekhar, S. (1961) Hydrodynamic & Hydromagnetic Stability.Oxford U.P.
Chapman, G.A., Herzog, A.D., Lawrence J.K. & Shelton, J.C. (1984) Solar luminosity fluctuations & active region photometry. Astroph.J. 282,L99
Chiang, W.-H. & Foukal, R. (1985) The influence of faculae on sunspot heat blocking. Solar Phys. 97, 9-20.
Choudhuri, A.R. & Gilman, P.A. (1986) The influence of the Coriolis forces on flux tubes rising through the solar convection zone. Preprint. Nat.Center Atm. Res.
Cowling, T.G. (1976) On the thermal structure of sunspots.Mon.Not.R.astr. Soc. 177, 409-414.
Defouw, R.J. (1976) Wave propagation along a magnetic tube.Astroph.J.209,266
Deinzer, W.(1965) On the magnetohydrost.theory of sunspots.Astroph.J.141,548
Galloway, D.J. & Weiss, N.O. (1981) Convection and magnetic fields in stars Astroph. J. 243, 945-953.
Gokhale, M.H. & Zwaan, C. (1972) The structure of sunspots. Solar Phys.26, 52
Golub, L., Rosner R., Vaiana, G.S. & Weiss, N.O. (1981) Solar magnetic fields: the generation of emerging flux.Astroph.J.243,309-316
Hughes, D.W. (1985a) Magnetic buoyancy instabilities for a static plane layer. Geophys. Astroph. Fluid Dynamics 34, 273-316.
Hughes, D.W. (1985b) Magnetic buoyancy instabilities incorporating rotation. Geophys. Astroph. Fluid Dynamics 34, 99-142.
Hurlburt, N.E., Toomre, J. & Massaguer, J.M. (1984) 2-dim. compressible convection extending over multiple scale-heights. Astroph.J. 282, 557
Knobloch, E. & Weiss, N.O. (1984) Convection in sunspots and the origin of umbral dots. Mon.Not.R.astr.Soc. 207, 202-214.
Mc Intosh, P.S. (1981) The birth and evolution of sunspots: observations.In The Physics of Sunspots,eds.L.Cram & J.Thomas, Sacramento Peak Obs.
Maltby, P. (1981) Sunspot theories confronted with observational data. In Solar Activity, C.Jordan ed., 3d european solar meeting.
Meyer, F., Schmidt H.U. Weiss N.O. & Wilson, P.R. (1974) The growth and decay of sunspots, Mon.Not.R.astr.Soc. 169, 35-57.
Meyer, F., Schmidt, H.U. & Weiss, N.O. (1977) The stability of sunspots. Mon.Not.R.astr.Soc. 179, 741-761.
Meyer, F., Schmidt, H.U., Simon, G.W. & Weiss, N.O. (1979) Buoyant magnetic flux tubes in supergranules. Astron.Astroph.76, 35.
Moreno-Insertis, F. (1983) Rise-times of magnetic flux tubes in the convection zone of the Sun. Astron. Astroph. 122, 241.

Moreno-Insertis, F. (1984) Ph.D.thesis, Univ. of Munich (Germany)
Moreno-Insertis, F. (1986) Nonlinear time-evolution of kink-unstable magnetic
    flux tubes in the convect.zone of the Sun. Astron.Astroph.166,291
Nye, A.H. & Hollweg, J.V. (1980) Alfvén waves in sunspots. Solar Phys. 68,279
Oster, L. Schatten, K.H. & Sofia, S. (1982) Solar irradiance variations due to
    active regions. Astroph. J. 256, 768-773.
Pardon, L., Worden, S.P. & Schneeberger, T.J. (1979) The lifetimes of sunspot
    moats. Solar Phys. 63, 247-250.
Parker, E.N. (1955) The formation of sunspots from the solar toroidal field.
    Astroph. J. 121, 491.
Parker, E.N. (1974a) The nature of the sunspot phenomenon: I. Solutions of
    the heat transport equation. Solar Phys. 36, 249.
Parker, E.N. (1974b) The nature of the sunspot phenomenon: II. Internal
    overstable modes. Solar Phys. 37, 127.
Parker, E.N. (1975) The generation of magnetic fields in astrophysical bodies,
    X.Magnetic buoyancy and the solar dynamo. Astroph. J. 198, 205.
Parker, E.N. (1979a) Sunspots and the physics of magnetic flux tubes:
    I. The general nature of the spot. Astroph. J. 230, 905-913.
Parker, E.N. (1979b) Sunspots and the physics of magnetic flux tubes: VII.
    Heat flow in a convective downdraft. Astroph. J. 232, 291
Parker, E.N. (1979c) Sunspots and the physics of magnetic flux tubes: VIII.
    Overstability in a magn.field in a downdraft.Astroph.J.233,1005
Parker, E.N. (1979d) Sunspots and the physics of magnetic flux tubes: IX.
    Umbral dots and longitudinal overstability. Astroph.J. 234, 333
Parker, E.N. (1984a) Depth of origin of solar active regions.Astroph.J.280,423
Parker, E.N. (1984b) Magnetic fields in the radiative interior of stars: I.
    Thermal shadows and forced convection. Astroph. J. 286, 666.
Pidatella, R.M. & Stix, M. (1986) Convective overshoot at the base of
    the Sun's convection zone. Astron. Astroph. 157, 338.
Pizzo, V.J. (1986) Numerical solution of the magnetostatic equations for thick
    flux tubes. Astroph. J. 302, 785-808.
Ponomarenko, Y.B. (1970) Astr.Zh. 47, 98 (Sov.Astr. 14, 78).
Roberts, B. & Campbell, W.R. (1986) Magnetic field corrections to solar
    oscillation frequencies Nature 323, 603-605.
Roberts, B. & Webb, A.R. (1978) Vertical motions in an intense magnetic flux
    tube. Solar Phys. 56, 5-35.
Scheuer, M.A. & Thomas, J.H. (1981) Umbral oscillations as resonant modes of
    magneto-atmospheric waves. Solar Phys. 71, 21.
Schlüter, A. & Temesvary, S. (1958) The internal constitution of sunspots. In
    electromagnetic phenomena in cosmical physics, ed. B. Lehnert,
    p.263. Cambridge University Press.
Schmidt, H.U. (1968) Magnetohydrodynamics of an active region. In Structure
    and developments of active regions, ed. Kiepenheuer, p.95.
Schmidt, H.U., Simon G.W. & Weiss N.O. (1985) Buoyant magnetic flux tubes.
    II.3-dimensional behaviour in granules and supergranules.
    Astron. Astroph. 148, 191.
Schmidt, H.U. & Wegmann, R. (1980) A free boundary value problem for
    sunspots. In B.Brosowski. E Martensen, eds. Methoden und
    Verfahren der mathematischen Physik, Oberwolfbach.
Schmitt, D. (1985) Ph.D.thesis Univ. of Gottingen (Germany).
Schmitt, J.H.M.M. & Rosner,R. (1983) Doubly diffusive magnetic buoyancy
    instability in the solar interior. Astroph. J. 265, 901

Schmitt, J.H.M.M., Rosner,R.& Bohn,H.U. (1984) The overshoot region at the bottom of the solar convection zone. Astroph. J. 282, 316

Schüssler, M. (1977) On buoyant magnetic flux tubes in the solar convection zone. Astron. Astroph. 56, 439-442.

Schüssler, M. (1979) Magnetic buoyancy revisited. Analytical & numerical results for rising flux tubes. Astron. Astroph. 71, 79-91.

Schüssler, M. (1984) The interchange instability of small flux tubes. Astron. Astroph. 140, 453-458.

Schüssler, M (1987) Magnetic fields and rotation of the solar convection zone. In Procs. Nat.Solar Obs. Summer Meeting. Sacramento Peak.Obs.

Severny, A.B. (1965) Astr. Zh. 42, 217. (Soviet Astron. 9, 171.)

Simon, G.W. & Weiss, N.O. (1968) Supergranules and the Hydrogen convection zone. Zeitschrift für Astrophysik, 69, 435.

Spiegel, E.A. & Weiss, N.O. (1980) Magnetic activity and variations in solar luminosity. Nature 287, 616.

Spruit, H.C. (1977a) Ph.D.Thesis, Univ.of Utrecht (The Netherlands).

Spruit, H.C. (1977b) Heat flow near obstacles in the solar convection zone. Solar Phys. 55, 3.

Spruit, H.C. (1981a) MHD of sunspots. Spa.Sci.Revs. 28, 434.

Spruit, H.C. (1981b) Small Scale Phenomena in umbras and penumbras. The role of convective processes. In The Physics of Sunspots, L.E.Cram & J.H.Thomas eds, Sacramento Peak Observatory.

Spruit, H.C. (1981c) A cluster model for sunspots. Same procs.as previous ref.

Spruit, H.C. (1981d) Motion of magnetic flux tubes in the solar convection zone and chromosphere. Astron. Astroph. 98, 155-160.

Spruit, H.C. (1982a,b) Effect of spots on a star's radius & luminosity Astron. Astroph. 108, 348-355 and 356-360.

Spruit, H.C. & Van Ballegooijen, A.A. (1982) Stability of toroidal flux tubes in stars. Astron. Astroph. 106, 58-66.

Tandberg-Hassen, E. (1967) Solar Activity. Blaisdell Pub.Comp..Waltham.

Thomas, J.H. (1981) Theories of dynamical phenomena in sunspots. In The Physics of Sunspots, eds L.E.Cram & J.H.Thomas, Sac.Peak Obs.

Van Ballegooijen, A.A. (1982a) The structure of the solar magnetic field below the photosphere. Astron. Astroph. 106, 43-52.

Van Ballegooijen, A.A. (1982b) The overshoot layer at the base of the solar convective zone and the problem of magnetic flux storage. Astron. Astroph. 113, 99-112.

Vrabec, D. (1974) Streaming magnetic features near sunspots. In chromospheric fine structure. R.G.Athay, ed. Reidel. p.201.

Wiehr, E. & Stellmacher, G. (1985) Influence of umbral dots on sunspot models In High resolution in solar physics, ed. R.Muller, Springer. p254.

Wilson, P.R. (1981) Theories of sunspot structure and evolution. In The Physics of Sunspots, eds L.E.Cram & J.H.Thomas, Sac.Peak Obs.

Willson, R.C.. Gulkis, S., Janssen. M.. Hudson. H.S. & Chapman. G.A. (1981) Observations of solar irradiance variability. Science 211, 700.

Yun, H.S. (1970) Theoretical models of sunspots. Astroph.J. 162, 975.

Zwaan, C. (1968) The structure of sunspots. Ann. rev. Astron. Astroph. 6, 135

Zwaan, C. (1978) On the appearance of magnetic flux in the solar photosphere Solar Phys. 60, 213-240.

Zwaan, C. (1984) Ths structure of sunspots. II: a continuum model atmosphere for dark umbral cores. Solar Phys. 37, 99-111.

Zwaan, C. (1985) The emergence of magnetic flux. Solar Phys. 100, 397-414

THE WILSON EFFECT IN SUNSPOTS

M. Collados, J.C. del Toro, M. Vazquez
Instituto de Astrofísica de Canarias, 38200-La Laguna,
Tenerife (Canary Isles), Spain

ABSTRACT

An analysis of the center to limb variation of the geometrical properties of spots has been carried out. It has revealed that spots do not have a symmetrical behaviour with respect to the centre of solar disk. Thus, the Wilson effect is not zero at $\vartheta = 0°$, but at $\vartheta \sim 45°W$. Moreover, the inverse Wilson effect is the general rule in that interval, while the normal phenomenon is maximum at a heliocentric angle of 40°-50°E.

1 INTRODUCTION

Recent studies have found that the behaviour of sunspots is different in the two hemispheres. Thus, the effect seems to be more pronounced at the eastern hemisphere (Prokakis (1974), Obashev et al. (1982)). Recently, Hejna and Solov'ev (1985) have proposed a simple explanation based on a slightly shorter western penumbra. Moreover, those cases which do not present a different foreshortening of the limb-side and center-side penumbras, or even those which present the inverse Wilson effect, appear preferentially at the western hemisphere. The number of anomalies is great enough to consider them as the rule rather than the exception.

In the last few years, we have carried out systematic observations of those regular, more or less round-shaped and long-lived spots which may be reliably used to study the phenomenon, which convert the present analysis into the one with a larger statistical sample and a higher spatial resolution. As a result, the study has confirmed the penumbral asymmetry. Thus, the inverse Wilson effect has proved to be not the exception, but, on the contrary, it is the rule.

2 OBSERVATIONS

The observations were carried out during the period 1982-84, at the Observatorio del Teide, in Izaña (Tenerife), with the 40 cm vacuum Newton telescope, property of the Kiepenheuer Institute. AHU Kodak film ($\gamma \sim 3$) was used to record the spot images, with a scale on the negative of 5."5/mm. The wavelength of observation was 500 nm, in 1982, and 550 nm, in 1983-84.

We selected, among our collected material, the passage across the solar disk of 17 spots. The selection criterion was based on their special symmetry and stability, and the good seeing existent during the observations (most of the pictures having a resolution better than 1"). For each spot, one frame per day of observation was chosen, except when very near to the solar limb, in which case we selected two or even three photographs per day, if the seeing allowed it, giving rise to a total of 145 images.

The negatives were digitized by placing them on the microdensitometer plate in such a way that the radial direction on the Sun coincided with the X-axis of the scanning. This allowed an easy determination of the radial and perpendicular umbral, penumbral and spot sizes. A square slit, with a side of 20 μm long (0."11), and a digitation step of 20 μm in both directions were used.

The films were calibrated by the PFD process, based on the half-filter method (Collados and Bonet (1984)), and the characteristic curves thus obtained were used to convert the images from densities to relative intensities. Restoration was not applied because of the big size of spots, much larger than the resolution achieved. The uncertainties are only significant very close to the limb, but there, even restoration will not avoid them.

The umbral and penumbral limits, in both radial and perpendicular directions were determined visually with the aid of a high resolution graphics display, the uncertainties lying around 5 pixels (0."55) in most cases, which agrees with the resolution of the pictures.

3 ANALYSIS

Regarding the Wilson effect, i.e., the behaviour of the radial penumbras, the variation from east to west of the ratios d_e/A_p and d_w/A_p (the size of the eastern and western penumbras in the radial direction expressed in units of the perpendicular penumbra) can be seen in fig. 1. While the first parameter is rather well described by the expected centre to limb variation laws, the same does not occur for the second one, specially at the eastern hemisphere. Most of the observational data lie below the predicted curve, although their scatter is rather large. The only reasonable explanation we have found is that both penumbras do not have the same size at the center of the disk. A rather good description (fig. 2) is found if the western penumbra is about 8% shorter than the rest of the penumbra (agreeing with Hejna and Solov'ev (1985)). As seen in fig. 3, this single hypothesis is able to explain the inverse Wilson effect which is present up to a heliocentric angle of 40° west, agreeing with previous results published by other authors (Prokakis (1974), Obashev et al. (1982)). Certainly, this asymmetry in the penumbra must be a consequence of the spatial configuration of the magnetic field in old spots, which usually are the remainder of the the leader part of the groups.

It must be emphasized that for a spot with a diameter of 30000 km the western penumbra is just some 700 km shorter than the eastern one. It seems very difficult that this asymmetry can be detected visually. Only a statistical analysis like the one we present here is able to find such a result.

Fig. 1: East to west variation of the eastern (d_e) and western (d_w) penumbral sizes, expressed in units of the perpendicular penumbral size (A_p). Although the first parameter is rather well described by the expected center to limb variation laws, the same does not occur for the second one, specially at the eastern hemisphere, where most of the observational data lie below the predicted curve.

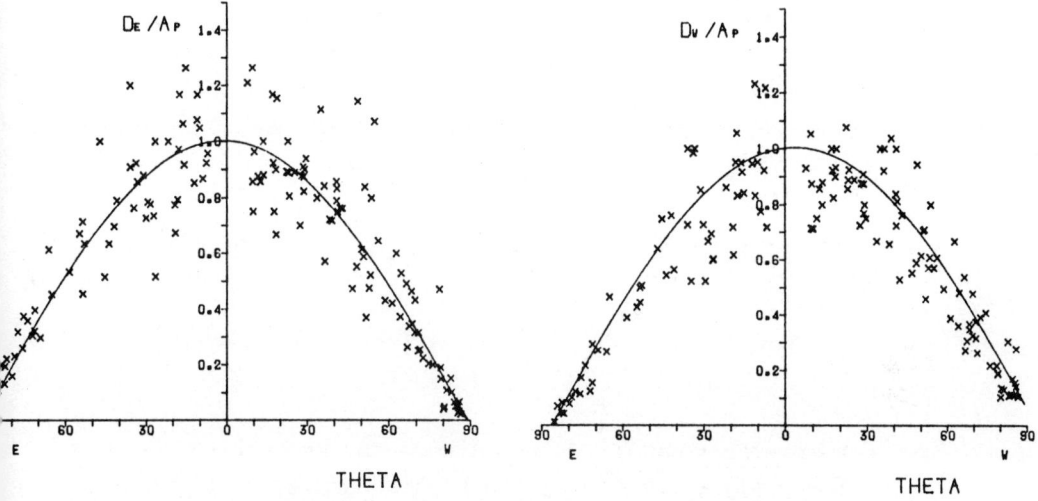

Fig. 2: New description of the east to west variation of the relative size of the western penumbra. A rather good fit is obtained if it is about 8% shorter than the rest of the penumbra. This result agrees with the explanation given by Hejna and Solov'ev (1985).

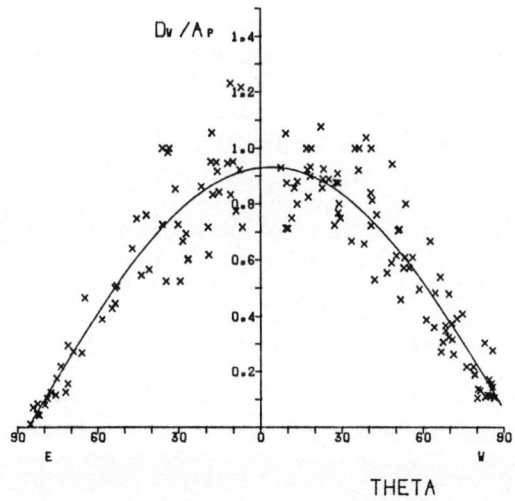

Fig. 3: East to west variation of the parameter ($\Delta = (d_e - d_w)/2A_p$). The inverse Wilson effect is present from the center of the disk up to 40°W. The normal phenomenon is maximum at 40°-50°E.

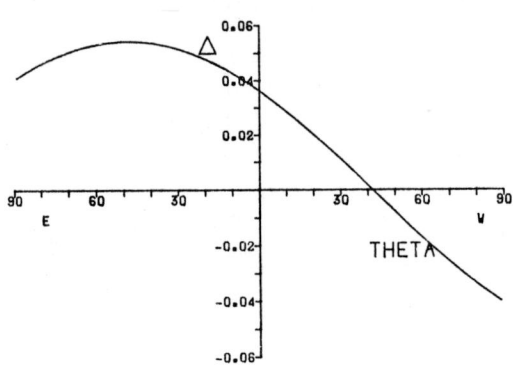

4 CONCLUSIONS

The analysis of the Wilson effect has led to the following conclusions: (a) it is maximum at 40°-50°E, (b) it is zero not at the centre of the disk, but at some 40°W, and (c) it is negative from 0° to 40°W. We believe that this asymmetry of the phenomenon must come out from the fact that the studied spots are the remainder of the leader part of sunspot groups.

Reference list

Collados, M. & Bonet, J.A. (1984). Two numerical processes for the calibration of photographic plates. App. Opt., 23, 2827.
Hejna, L. & Solov'ev, A.A. (1985). On the asymmetry of Wilson's sunspot effect. Bull. Astron. Inst. Czechosl., 36, 183
Obashev, S.O. et al. (1982). The angle of inclination of the sunspot symmetry axis to the solar surface. Solar Phys., 78, 59.
Prokakis, Th. (1974). The depth of sunspots. Solar Phys., 35, 105.
Wilson, P.R. & Cannon, C.J. (1968). The structure of a sunspot. III: Observations of the Wilson effect. Solar Phys., 4, 3.
Wilson. P.R. & McIntosh, P.S. (1969). THe structure of a sunspot. V: What is the Wilson effect. Solar Phys., 10, 370.

A NEW MECHANISM FOR THE EVERSHED EFFECT

F.H. Busse
Theoretical Physics IV, University of Bayreuth,
D-8580 Bayreuth, FRG

1. Introduction

It is commonly assumed that small scale fluid motions in the presence of a mean flow enhance the effects of the molecular viscosity and thereby are capable of balancing the forces driving the mean flow. There are many examples, however, in the geophysical and astrophysical contexts, where mean flows are generated or amplified by the action of fluctuating small scale motions. Usually the influence of rotation causes an anisotropy of the small scale motion such that finite Reynolds stresses are generated. The best known example is the differential rotation of the sun which originates from large convection cells which are aligned with the axis of rotation of the sun (Busse, 1970; Gilman, 1976). Another example is the mean shear generated by three-dimensional convection in a horizontal plane layer rotating about an axis of rotation inclined with respect to the vertical (Hathaway and Somerville, 1983; Busse, 1982). In the latter problem, thermal convection assumes two states. At the onset of convection no mean shear is generated because the two-dimensional convection rolls aligned with the horizontal component of the rotation vector are governed by the same dynamics as in an isotropic horizontal layer. Only after a second bifurcation in which convection assumes a three-dimensional form does the anisotropy of the layer manifest itself in the generation of a mean shear.

This mechanism is closely related to the new mechanism discussed in this paper. It is wellknown that rotation and a homogeneous magnetic field have rather similar effects on the motion of an electrically conducting fluid. The onset of oscillatory convection and the statement of the magnetic analogue of the Proudman-Taylor-Theorem (Chandrasekhar, 1961) are but two examples of common properties. In this paper we like to point out that convection in a horizontal layer which is intersected by an inclined magnetic field generates Reynolds stresses in the direction of the inclination as soon as a transition to three-dimensional forms of motion has occurred.

In order to outline the mechanism in its most simple form we shall restrict the attention to convection in a fluid layer of high magnetic diffusivity with stress-free boundaries. From the general nature of the argument it will become evident that the mechanism is likely to operate in a wide variety of circumstances. It seems plausible that one of its manifestations is the Evershed flow observed in the penumbrae of sunspots. In the discussion section at the end of the paper we shall comment on this possibility and point out the need for further work.

2. Mathematical Formulation of the Problem

We consider a horizontal fluid layer of thickness d with the fixed temperatures T_1 and T_2 ($T_2 > T_1$) at the upper and lower boundaries. The fluid layer is intersected by a homogeneous magnetic field with the flux density B_0 at an angle χ with the horizontal plane. Using d as length scale, d^2/ν as the time scale where ν is the kinematic viscosity, $(T_2-T_1)P/R$ as scale of the temperature and B_0 as scale of the magnetic field we write the equations of motion for the velocity field $\underline{u}$, the heat equation for the deviation θ of the temperature field from the static solution and the equation of induction in the following form,

$$(\tfrac{\partial}{\partial t}+\underline{u}\cdot\nabla)\underline{u} = -\nabla\pi+\underline{k}\theta+\nabla^2\underline{u}+Q\tfrac{\lambda}{\nu}\underline{B}\cdot\nabla\underline{B} \qquad (2.1a)$$

$$P(\tfrac{\partial}{\partial t}+\underline{u}\cdot\nabla)\theta = R\underline{u}\cdot\underline{k}+\nabla^2\theta \qquad (2.1b)$$

$$(\tfrac{\partial}{\partial t}+\underline{u}\cdot\nabla)\underline{B} = \underline{B}\cdot\nabla\underline{u}+\tfrac{\lambda}{\nu}\nabla^2\underline{B} \qquad (2.1c)$$

$$\nabla\cdot\underline{u} = 0, \qquad \nabla\cdot\underline{B} = 0 \qquad (2.1d,e)$$

The Boussinesq approximation has been assumed and the following dimensionless parameters have been introduced,

Chandrasekhar number $\quad Q \equiv B_0^2 d^2/\rho_0\mu\lambda\nu$

Rayleigh number $\quad R \equiv \gamma(T_2-T_1)gd^3/\nu\kappa$

Prandtl number $\quad P = \nu/\kappa$

The letters λ and κ denote the magnetic and thermal diffusivities, respectively; γ is the coefficient of thermal expansion, g is the force of gravity directed in the direction of the unit vector $-\underline{k}$ and ρ_0 is the average density. In order to eliminate equations (2.1d,e) we introduce the following general representations for the solenoidal fields $\underline{u}$ and $\underline{B}$,

$$\underline{u} = \nabla \times (\nabla \times \underline{k}\phi) + \nabla \times \underline{k}\psi + U\underline{j}; \quad \underline{B} = \underline{k}\sin\chi + \underline{j}\cos\chi + \left[B_y \underline{j} + \nabla \times (\nabla \times \underline{k}h) + \nabla \times \underline{k}g\right]\frac{\nu}{\lambda}$$

We have separated the mean and fluctuating parts such that the average over the functions ϕ, ψ, h and g over planes normal to $\underline{k}$ vanishes. In writing the mean quantities we have anticipated that a mean flow U and the corresponding mean distortion B_y of the magnetic field will occur only in the direction of the horizontal component of the imposed magnetic field given by the unit vector $\underline{j}$.

We shall use a Cartesian system of coordinates with the z- and y-coordinates in the directions of $\underline{k}$ and $\underline{j}$, respectively. By taking the z-components of the curl and of the curlcurl of equation (2.1a) and by taking the z-components of equations (2.1c) and of its curl we arrive at the following equations for ϕ, ψ, h and g,

$$\frac{\partial}{\partial t}\nabla^2 \Delta_2\phi + \underline{k}\cdot\nabla\times(\nabla\times(\underline{u}\cdot\nabla\underline{u})) = -\Delta_2\theta + \nabla^4\Delta_2\phi + Q\left[\underline{j}\cos\chi + \underline{k}\sin\chi\right]\cdot\nabla\nabla^2\Delta_2 h \quad (2.2a)$$

$$\frac{\partial}{\partial t}\Delta_2\psi - \underline{k}\cdot\nabla\times(\underline{u}\cdot\nabla\underline{u}) = \nabla^2\Delta_2\psi + Q\left[\underline{j}\cos\chi + \underline{k}\sin\chi\right]\cdot\nabla\Delta_2 g \quad (2.2b)$$

$$0 = \nabla^2\Delta_2 h + \left[\underline{j}\cos\chi + \underline{k}\sin\chi\right]\cdot\nabla\Delta_2\phi \quad (2.2c)$$

$$0 = \nabla^2\Delta_2 g + \left[\underline{j}\cos\chi + \underline{k}\sin\chi\right]\cdot\nabla\Delta_2\psi \quad (2.2d)$$

Assuming $\nu \ll \lambda$ we have neglected all terms multiplied by ν/λ. We shall return to this assumption in the discussion section. The operator Δ_2 denotes the horizontal Laplacian, $\Delta_2 = \nabla^2 - (\underline{k}\cdot\nabla)^2$.

It is wellknown that the solution of (2.1c), (2.2) corresponding to the minimum Rayleigh number has a vanishing dependence on the y-coordinate. Using stress-free boundary conditions

$$\phi = \frac{\partial^2}{\partial z^2}\phi = \frac{\partial}{\partial z}\psi = 0 \quad \text{at} \quad z = 0,1 \quad (2.3)$$

we write this solution in the form

$$\phi_0 = A\cos\alpha x \sin\pi z + O(A^3 p^2) \quad (2.4a)$$

$$\theta_0 = \{(\pi^2+\alpha^2)^2+Q\sin^2\chi\pi^2\}\left[\phi_0-PA^2\alpha^2\sin 2\pi z/8\pi\right]+O(A^3P^2) \quad (2.4b)$$

$$R = \{(\pi^2+\alpha^2)^2+Q\pi^2\sin^2\chi\}\left[(\pi^2+\alpha^2)/\alpha^2+P^2A^2\alpha^2/8\right]+\ldots \quad (2.4c)$$

The term independent of A in the expression for R gives the critical value of the Rayleigh number after it has been minimized with respect to the wavenumber α. For large values of Q the critical wavenumber α_c grows proportionally to $\pi(Q\sin^2\chi/2)^{1/4}$. Since the magnetic field enters equations (2.2) only through the term $\nabla^2\Delta_2 h$ in the case of y-independent solutions, there is no need to specify magnetic boundary conditions at this point.

3. Properties of the Oscillatory Instability of Convection Rolls

Among the instabilities that induce a transition to a three-dimensional form of convection the oscillatory instability predominates in low-Prandtl-number fluids. Following the analysis in the case $Q = 0$ (Busse, 1972) we superimpose disturbances of the form

$$\tilde{\phi} = (\tilde{\phi}_0+b\tilde{\phi}_1+b^2\tilde{\phi}_2+\ldots)\exp\{iby+\sigma t\}$$
$$\sigma = \sigma_0+b\sigma_1+b^2\sigma_2+\ldots \quad (3.1)$$

where the quantities $\tilde{\phi}_n$ are functions of x and z only. The expressions for $\tilde{\psi},\tilde{\phi},\tilde{g},\tilde{h}$ are analogous to that for $\tilde{\phi}$. The dependence on the wavenumber b has been assumed in the form of a power series because we anticipate that the onset of the oscillatory instability occurs for small b. In fact this instability can be regarded as a modification of the translational disturbances given by

$$\tilde{\phi}_0 = (-\alpha A)^{-1}\partial\phi/\partial x, \quad \tilde{\theta}_0 = (-\alpha A)^{-1}\partial\theta/\partial x, \quad \tilde{h}_0 = (-\alpha A)^{-1}\partial h/\partial x,$$
$$\sigma_0 = 0 \quad (3.2)$$

Another property that contributes to the instability is the presence of a horizontal motion independent of z described by $\tilde{\psi}_0 = $ const. which is barely damped by viscous friction. This property remains essentially unchanged as Q is increased from zero if highly conducting boundaries are assumed. In that case the Fermi boundary conditions,

$$h = \partial g/\partial z = 0 \quad \text{at} \quad z = 0,1, \quad (3.3)$$

permit a z-independent solution $\tilde{g}_0$ and the effect of the Lorentz force in equation (2.2b) is of the order b^4 if $\tilde{g}_0$ is independent of x as well. From this and similar arguments it becomes evident that the analysis of the oscillatory instability of Busse (1972) is changed only slightly in quantitative aspects if the extension to finite Q is considered. The expression for the frequency of oscillations,

$$\sigma_1 = \pm i\pi\alpha A/2 \tag{3.4}$$

is independent of Q and the critical value A_c for onset of the instability increases with increasing Q. This conclusion agrees with extensions to finite values of Q of the numerical analysis of the oscillatory instability in the presence of rigid boundaries (Busse and Clever, 1982, 1983).

Instead of pursuing the general analysis we focus the attention on some peripheral aspects of the instability which are important for the generation of a mean flow. The equations for $\tilde{\phi}_1, \tilde{\theta}_1$ and $\tilde{h}_1$ now include terms proportional to $Q\cos\chi$,

$$\nabla^4 \Delta_2 \tilde{\phi}_1 - \Delta_2 \tilde{\theta}_1 + Q\sin\chi \frac{\partial}{\partial z} \nabla^2 \Delta_2 \tilde{h}_1 = +iQ\cos\chi\sin\chi \frac{\partial}{\partial z} \Delta_2 \tilde{\phi}_0 \tag{3.5a}$$

$$\nabla^2 \tilde{\theta}_1 - R\Delta_2 \tilde{\phi}_1 = 0 \tag{3.5b}$$

$$\nabla^2 \Delta_2 \tilde{h}_1 + \sin\chi \frac{\partial}{\partial z} \Delta_2 \tilde{\phi}_1 = -i\cos\chi \Delta_2 \tilde{\phi}_0 \tag{3.5c}$$

To simplify the analysis we have neglected terms proportional to A on the left hand side of equations (3.5). Those terms will generate additional contributions to the solutions $\tilde{\phi}_1, \tilde{\theta}_1$ and $\tilde{h}_1$ of different symmetry, but will not change qualitatively the result obtained in the following. After eliminating $\tilde{\theta}_1$ and $\tilde{h}_1$ from equations (3.5) and using expressions (3.2) for $\tilde{\phi}_0$ we obtain the following equation for $\tilde{\phi}_1$,

$$\{\nabla^6 - Q\sin^2\chi \frac{\partial^2}{\partial z^2} \nabla^2 - R\Delta_2\} \tilde{\phi}_1 = -iQ\sin 2\chi (\pi^2+\alpha^2)\pi\sin\alpha x \cos\pi z \tag{3.6}$$

The solution of equation (3.6) satisfying the boundary conditions (2.3) can be written in the form

$$\tilde{\phi}_1 = i\sin\alpha x \sin 2\chi \sum_{n=1}^{\infty} D_n 8n(4n^2-1)^{-1} \sin 2n\pi z \tag{3.7a}$$

where the coefficients D_n are given by

$$D_n \equiv Q(\pi^2+\alpha^2)\left[(4n^2\pi^2+\alpha^2)^3+Q\sin^2\chi\, 4n^2\pi^2(4n^2\pi^2+\alpha^2)-R\alpha^2\right]^{-1}$$
(3.7b)

The main properties of solution (3.7) that will be needed in the following are the antisymmetric dependence with respect to the mid-plane of the layer, $z = 1/2$, and the positivity of the coefficients D_n.

4. The Reynolds Stress of Oscillatory Convection

Since the oscillatory disturbance of the convection rolls has a sinusoidal dependence on the y-coordinate, its interaction with the steady component of convection does not yield a mean flow. As the oscillations grow to finite amplitude, however, the terms quadratic in $\tilde\phi$ cause a mean vertical transport of y-momentum. The equations for the mean velocity U in the y-direction and the corresponding component of the magnetic field are given by

$$\frac{\partial^2}{\partial z^2}U + Q\sin\chi\frac{\partial}{\partial z}B_y = -\tilde{A}^2\frac{\partial}{\partial z}\overline{\Delta_2\tilde\phi\,\partial^2\tilde\phi/\partial y\partial z}$$
(4.1a)

$$\frac{\partial}{\partial z}B_y + \sin\chi U = 0$$
(4.1b)

where $\tilde A$ denotes the amplitude of the oscillatory component of convection and the bar indicates the average over the x,y-plane. We have used the integrated form of the equation of induction, since it can be presumed that there does not exist a mean electric field in the y-direction. In evaluating the Reynolds stress given by the overlined quantity in (4.1a) we must take into account that only the real parts of expressions (3.1) have physical meaning. Accordingly we obtain from (4.1)

$$\frac{\partial^2}{\partial z^2}U - Q\sin^2\chi U = \tilde A^2\sin 2\chi b^2\alpha^2\pi^2\sum_{n=1}^{\infty}nD_n\{\cos(2n-1)\pi z - \cos(2n+1)\pi z\}$$
(4.2)

and finally

(4.3)

$$U = \tilde A^2\sin 2\chi b^2\alpha^2\pi^2\sum_{n=1}^{\infty}nD_n\left\{\frac{-\cos(2n-1)\pi z}{(2n-1)^2\pi^2+Q\sin^2\chi}+\frac{\cos(2n+1)\pi z}{(2n+1)^2\pi^2+Q\sin^2\chi}\right\}$$

The fact that U is antisymmetric with respect to $z = 1/2$ guarantees that the condition $B_y = 0$ at $z = 0,1$ can be satisfied. In the summation over n the first term gives by far the largest contribution and clearly indicates that for a positive angle χ ($|\chi| \leq \pi/2$ is always assumed) the mean flow increases from negative values in the lower half of the layer to positive values in the upper half.

The physical interpretation of the Reynolds stress exhibiting the same sign as χ is straightforward. Because of the magnetic analogue of the Proudman-Taylor-theorem the y-dependent part of the motion tends to form a circulation elongated in the direction of the magnetic field as indicated in figure 1. Such an eddy tends to transport positive y-momentum in the upward direction and negative y-momentum in the downward direction if χ is positive as is assumed in the figure. The resulting mean shear flow results from the balance between the viscous and magnetic forces on the one

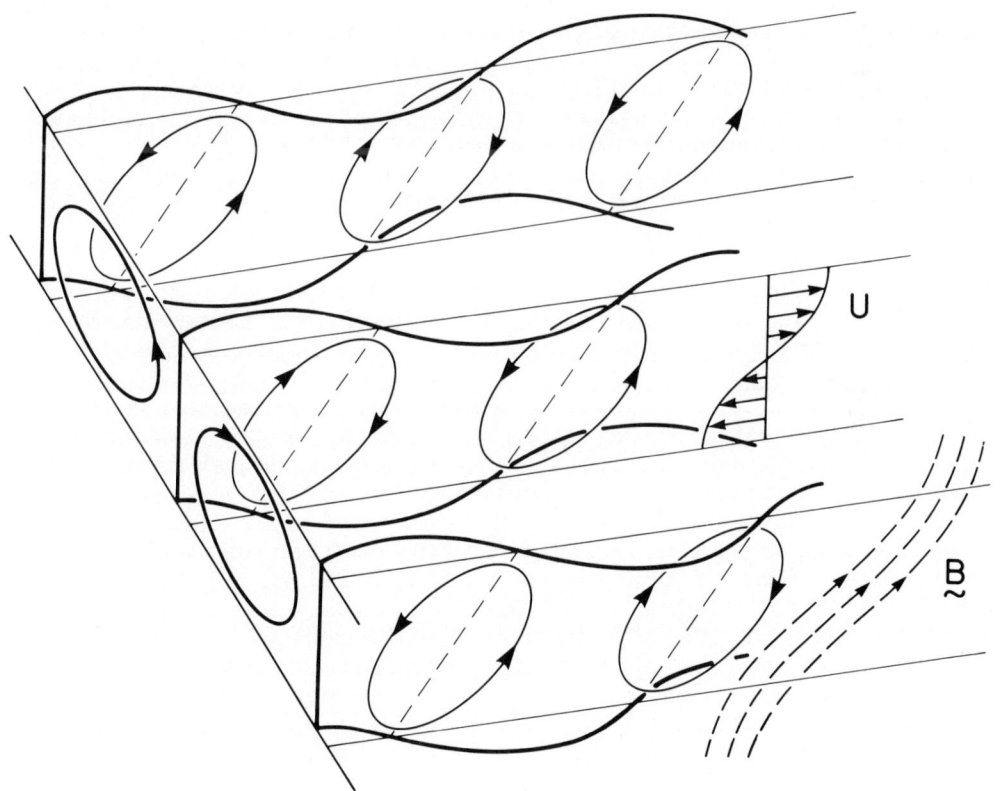

Figure 1: Sketch of oscillating convection rolls in a fluid layer intersected by an inclined magnetic field.

side and the divergence of the Reynolds stress on the other side. This process does not depend on the symmetry of the problem and the main consequence of the imposition of a no-slip condition at the lower boundary, for example, is likely to be an additional constant flow in expression (4.3) such that the new boundary condition becomes satisfied.

5. Discussion

The Evershed effect has been a wellknown phenomenon in solar physics for over 60 years; but few theoretical models have been constructed for its explanation. As Thomas (1982) observes in his recent review of sunspot theory there are two schools of thought. Meyer and Schmidt (1968) have created the "siphon"-model of the Evershed effect. A recent extension of the model which also deals with the nature of the bright and dark filaments of the penumbra has been given by Schmidt et al. (1986). The other school of thought goes back to Danielson (1961) who derived a penumbra model based on longitudinal convection rolls. Galloway (1975) used this model and outlined a theory of the Evershed effect in which the flow is driven by the Lorentz force originating from the curvature of the magnetic field lines in the dark filaments. A detailed analysis of this suggestion has not yet been carried out, however. The proposal of the present paper is similar to that of Galloway in that the process of the generation of the Evershed flow occurs locally. But instead of the Lorentz force due to curved field lines we invoke the Reynolds stress of three-dimensional convection in the presence of an inclined field.

A particularly unrealistic feature of our model is the limit of low magnetic Reynolds number. When the assumption $\nu \ll \lambda$ is dropped the influence of accumulation of magnetic flux in the region of flow convergence at the boundaries of the rolls can be described and the effect of the magnetic Reynolds stress can be included. But such an analysis goes beyond the scope of the present paper. We just like to mention that the magnetic Reynolds stress appears to have the same sign as the ordinary Reynolds stress derived in this paper. Since both types of Reynolds stress increase with the magnetic field strength it is reasonable to expect that the flow generated by this mechanism is strongest in the dark filaments where the magnetic field appears to be strongest. An additional effect that tends to strengthen the mean flow in the dark filaments is connected with the observation that the magnetic field is more horizontal in those areas. Since the magnetic "braking" of the mean flow grows approximately with $\sin^2 \chi$ for large Q according to expressions (4.3) while the Reynolds stress effect is proportional to $\sin 2\chi$, small angles of inclination are especially favorable for the mechanism.

The example of the oscillatory instability has been used in this paper for reasons of mathematical simplicity. It is clear from the analysis, however, that any other instability introducing three-dimensional motions will lead to a similar Reynolds stress. Unfortunately there is not much observational evidence availabe about the three-dimensional nature of motions in the penumbral filaments. The variation with depth of the Evershed flow is another feature for which observational data are desirable in order to stimulate the future development of the theory. It is hoped that the new instruments established in the Canary Islands will eventually provide some new insights into the origin of the Evershed effect.

Acknowledgement: The author is indebted to Dr. David Galloway for introducing him to the Evershed Effect. The research has been supported under NSF-Grant EAR-82 05668.

References

Busse, F.H. (1970b). Astrophys. J. 159, 629-639.
Busse, F.H. (1982). Zeitschrift f. Naturforschung 37a, 753-758.
Busse, F.H, and Clever R.M. (1983). J. de Méchanique Théor. Appl. 2, 495-502.
Busse, F.H., and Clever, R.M. (1982). Physics of Fluids 25, 931-935.
Chandrasekhar, S. (1961). Hydrodynamic and Hydromagnetic Stability, Oxford: Clarendon Press.
Danielson, R.R. (1961). Astrophys. J. 134, 275- .
Galloway, D.J. (1975). Solar Phys. 44, 409-415.
Gilman, P.A. (1976). Proc. IAU Symposium No. 71, Basic Mechanisms of Solar Activity, ed. V. Bumba and J. Kleczek (Dordrecht: Reidel), pp. 207-228.
Hathaway, D.H., and Somerville, R.C.J. (1983). J. Fluid Mech. 126, 75-89.
Meyer, F., and Schmidt, H.U. (1968). ZAMM 48, T218-T221.
Schmidt, H.U., Spruit, H.C., and Weiss, N.O. (1986). Astron. Astrophys. 158, 351-360.
Thomas, J.H. (1981). pp. 345-358 in "The Physics of Sunspots", L.E. Cram & J.H. Thomas, eds., Sacramento Peak Observatory Report.

DECAY RATES OF SUNSPOT GROUPS FROM 1874 TO 1939

F. Moreno-Insertis, M. Vazquez
Instituto de Astrofisica de Canarias
38200 La Laguna, Tenerife, Spain

Abstract

GPR material from 1874 to 1939 is used to study sunspot area decay rates. Various possible decay laws (exponential, linear, parabolic) are tried to fit the observational material. The dependence of the decay rates on the spot's maximum area as well as their variation along the activity cycle are studied.

1 INTRODUCTION

One of the simplest observational inputs for sunspot theory is the decay rate of the sunspot area, which also provides a measure for the loss of magnetic flux from the spot in its decay phase. In this contribution we present the first results of a study of the sunspot area decay using the information contained in the digitized version of the Greenwich Photoheliographical Results (GPR) from 1874 to 1939. All areas are given in millionths of visible hemisphere (MVH).

Three simple possible decay laws have been considered in this study:

(a) Decay rate proportional to the instanteneous sunspot area: $dA/dt \propto -A$. This results in an exponential law,

$$A = A_{max} \exp[-c(t-t_o)], \quad c>0 \qquad (1)$$

(b) Decay rate proportional to the length of the spot boundary, $dA/dt \propto -A^{1/2}$, so that a second-order polynom

$$A(t) = q(t-t_o)^2 + l(t-t_o) + c \qquad (2)$$

should provide the best fit to the data. This law would apply, for instance, if the decay were caused by convective "gnawing" at the spot boundary.

(c) Decay rate constant in time,

$$dA/dt = \text{const}, \qquad A = bt + a, \qquad (3)$$

but, in principle at least, different for the different spots.

The exponential law did not provide the best fit to the data except for a small part (6%) of the spots studied. For the remaining 94% a test of the quadratic as compared to linear character of the decay was carried out. To that end the quadratic law (2) was expressed in dimensionless form and the resulting (dimensionless) coefficients q and l were compared. For that a unit of time has to be chosen. Here we have used the time elapsed from t_o, the time of maximum recorded area, A_{max}, to the time when the spot had reached an area $A_{max}/10$. Finally, (2) was rendered fully dimensionless by dividing it through A_{max}; this last operation, of course, does not change the ratio q/l.

2 RESULTS

(a) The linear decay term is significantly greater than the quadratic term. In fact the averages for the different types of spots give q values which are smaller by a factor from 5 to 100 than the corresponding l values.

(b) The isolated spots and the recurrent groups decay more slowly than the non-recurrent groups, in agreement with the results of Bumba (1963).

(c) From the results it is apparent that the linear decay rate is not the same for all spots but depends rather on the spot's maximum area, A_{max}. In Figs 1a and 1b we have represented the linear (dimensional) decay rate coefficient in abscissas versus A_{max} in ordinates. Two different families are clearly distinguishable for recurrent and isolated spots. The non-recurrent groups can be found only in the lower (steepest) branch.

(d) The decay rates corresponding to different phases of the cycle (ascending phase, maximum, descending phase) have been compared. Differences are found in the descending phase, but only for the total area decay.

Full details of the analysis carried out and of the results obtained will be given in a forthcoming paper.

REFERENCES

Bumba, V. (1963). Development of sunspot group areas in dependence of the local magnetic field. Bull, Astron. Inst. Czech. 14, 91.

Fig 1 a: Area decay rate of the whole sunspot group (U+P) versus maximum area for recurrent and non-recurrent groups.

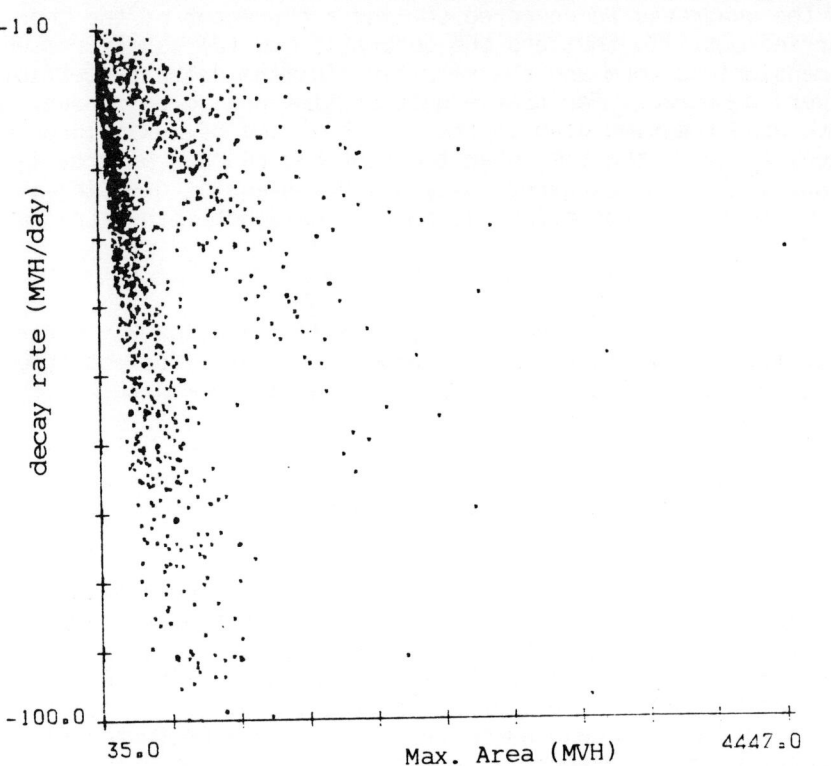

Fig 1 b: Area decay rate of the umbra versus maximum umbra area for recurrent and non-recurrent groups.

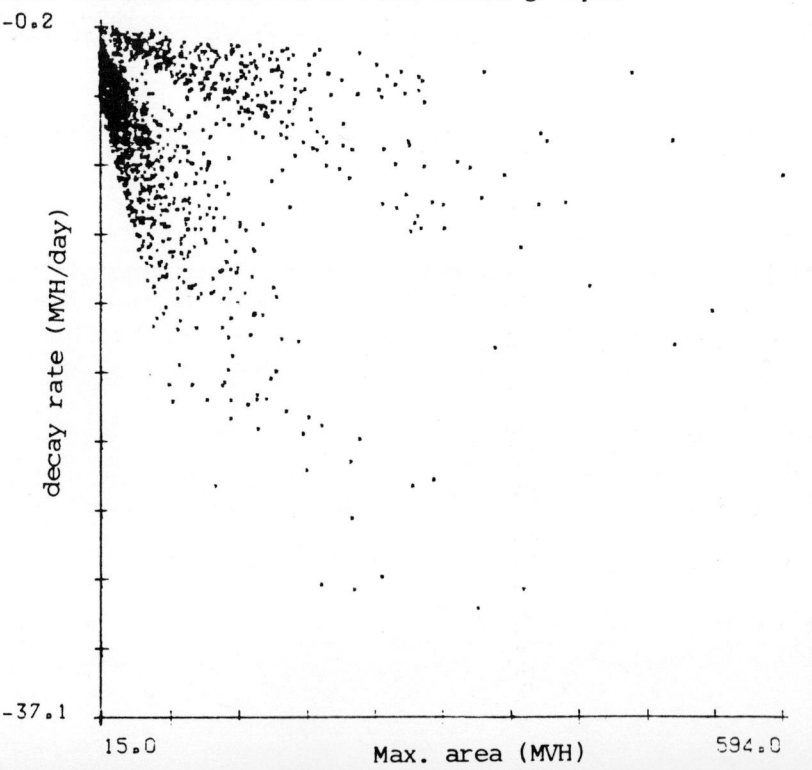

HOW IS THE PENUMBRA FORMED?

H.C. Spruit
National Solar Observatory, Sunspot, New Mexico 88349, USA
and
Max Planck Institut für Physik und Astrophysik, Karl-Schwarzschildstr. 1, 8046 Garching, West Germany

Abstract. Instead of answering the question posed by the title, simple theoretical arguments are reviewed that restrict the directions in which the answer should be sought. They make plausible that the field of the penumbra is in quasistatic equilibrium except with respect to motions along the field, that penumbral filaments are unlikely to be elevated with respect to the photosphere or their bright neighbors, that the boundary between the field and the underlying convection zone is at a shallow depth of the order of a pressure scale height and that convective processes in the penumbra take the form of flows nearly parallel to the field rather than Danielson's (1961) rolls. Why there should be a penumbra at all, i.e. why a spot does not look like a giant pore, seems to be a more difficult question.

1. Observational indications

Applied to observational facts, a number of very simple theoretical arguments yield robust but moderately strong conclusions. This is done in the next section. I believe that these conclusions should be convincing to a significant majority. Somewhat more involved and therefore less leakproof arguments yield more spectacular conclusions. These are given in section 3. Though I have only included deductions that I believe to be compelling, they inevitably contain a certain amount of personal bias.

The observational facts that I will appeal to are a selection from the many facts known about the penumbra (see the review by Garcia de la Rosa, this volume):
- The nonsteady nature of the penumbra, with individual features living 20 to 60 min.
- The ratio of penumbral to umbral field strength, roughly 1/3 at the edge of the spot.
- The fact that the motions in the penumbra are primarily parallel to the field, with only small sideways and vertical components.
- The motion of bright grains towards the umbra (Muller, 1973a).
- The appearance of dark filaments near the spot boundary (Moore, 1981; Title et. al. 1986)
- The sudden drop in field strength and (Evershed) flow speed at the spot boundary (see Figure 1).

From these (and other) facts several meaningful theoretical inferences can be drawn, but they do not add up to a complete picture. This reflects, in my opinion, the fact that the deeper reason for the existence of penumbrae is not well understood at present.

2. Simple theoretical deductions

A few things which theoreticians will be very reluctant to give up are:
- The Lorentz force has no component along the field.
- The magnetic field strength is continuous along a field line. In the absence of shock waves the flow speed is also continuous along the flow.
- The high conductivity of the solar plasma implies that at the observed time and length scales the field is essentially 'frozen in'. As a consequence, any stationary flow is exactly parallel to the field. Proposed theories must be valid at high magnetic Reynolds numbers (as opposed to low R_m calculations such as that by Busse in these Proceedings).
- Structures which change slowly compared with the time to reestablish equilibrium are close to equilibrium

2.1. How much can we do with magnetic forces?

Though strong magnetic fields limit the freedom of solar plasma to move about, the first of the above considerations means that one cannot fix mass at arbitrary points in space using magnetic forces, since there is always one direction in which the mass is free to move under the influence of the other (nonmagnetic) forces. For example if a field line is inclined with respect to the surface of the Sun, gas is free to drop along it under the influence of gravity. It is like making a house mouse-proof: it is not sufficient to close *most* entrances.

2.2. Sharpness of the spot boundary

The second of the premises stated above has implications for the observed sharpness of the spot boundary. An example of this is shown in Figure 1 which reproduces observations of the Evershed flow by L.J November. In most places around the spot boundary the drop in flow speed on passing from the spot into the surroundings is not resolved (at scale better than 1 arcsec). Where do the field and the Evershed flow go at the edge of the spot? At the edge, the boundary between field and field free gas (as seen in a vertical plane) passes through the $\tau = 1$ level, and the observed sharpness provides a limit on the inclination of this boundary with respect to the horizontal plane. Due to the spatial resolution of the observations, 'sharp' means on a scale of 1000 km (1.3 arcsec) or less. The continuum contrast, magnetic and velocity signals are all formed in a layer close to $\tau_{5000} = 1$. Since the magnetic fields and velocities are measured in the wings of spectral lines, they refer to a layer not more than about 100km thick above $\tau_{5000} = 1$. Once the boundary between field and field free gas is higher than this, the signals are drastically reduced. Thus the inclination with respect to the horizontal is around $1/10 = 6°$, or larger. Evidently, the observed sharpness does not put strong limits

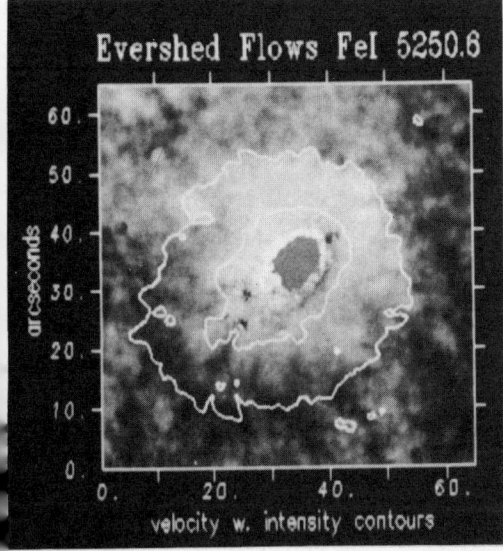 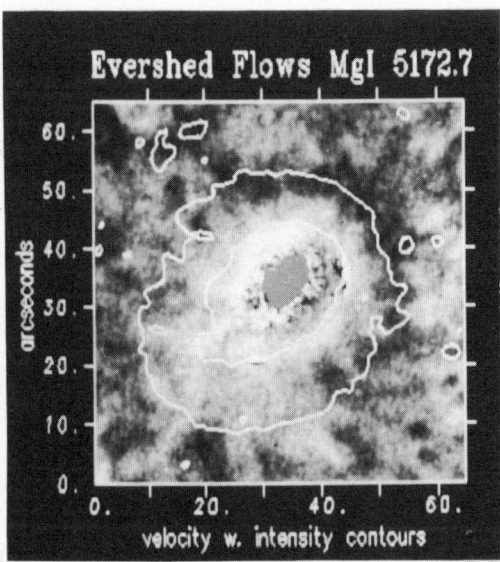

Fig. 1. The Evershed flow in a photospheric (left) and a chromospheric line (right) in the spot at 11 S, 12 E on March 1 1983. North is approximately in the direction of the bottom of the image, East to the right. The white contour shows the outline of the spot and the umbra in intensity. Positive velocity white, negative dark, resolution of the original data better than 1 arcsec. Note the reversal of the flow direction in the chromosphere, and the abrupt vanishing of the photospheric Evershed flow at the spot boundary. Observations (with the UBF at the Sacramento Peak Vacuum Tower Telescope) and reduction by L.J. November (see November, 1984). Copyright the National Optical Astronomy Observatories.

on the field configuration at the spot edge, and this is mostly because 'small scale' in observational terms is still 'large scale' with respect to the intrinsic vertical scale of the atmosphere. An entirely satisfactory answer to the question posed is therefore: Both the flow and the field just continue along a weakly inclined path.

2.3. How close to equilibrium are penumbral structures?

A comparison of the observed life times of individual structures with theoretically determined timescales for establishing equilibrium can tell us to what extent they are in equilibrium, and this limits strongly what properties can be reasonably be ascribed to them. Since the influence of a magnetic field is quite different parallel to the field lines and perpendicular to them, and the vertical direction behaves different from the horizontal due to gravity, time scales for equilibrium are different in the three spatial directions. Consider first the equilibrium *perpendicular* to the field lines, in a *horizontal* plane. Let us call this for the time being the *transversal* direction. Gravity has no effect here, and neglecting the effects of field line curvature, the structure is in magnetostatic equilibrium in this direction if

$$P + B^2/8\pi = const. \qquad \qquad 1$$

in this direction. Disturbances in transversal equilibrium propagate with the speed of the fast mode, $v_f = (c^2+v_A^2)^{1/2}$, where c is the sound speed and v_A the Alfven speed. At $\tau = 1$ we have $c \approx v_A \approx 7\ kms^{-1}$. If the structure is not in equilibrium, it will change on a time scale $t_{trans} = d/v_f$, where d is its transversal size. Taking for d the width of a penumbral filament, 500 km say, we have $t_{trans} \approx 1$ min. Since changes in filament structure typically take place on much longer time scales, we conclude that the field is approximately in magnetostatic equilibrium in the transversal direction.

In the *vertical* direction, the adjustment time depends on whether the disturbance has a large or a small scale in the transversal direction. If this scale is of the same order as that in the vertical direction, the time scale is of the order of the period of a buoyancy oscillation (if the stratification is stable) or a convective turnover time (if it is unstable). The presence of the magnetic field also influences this time scale (if the field decreases with depth, it stabilizes the stratification, see also section 4). Unless the stratification is close to convective instability (including the magnetic field) the resulting time scale is of the order $t_{vert} \approx (H/g)^{1/2}$, where g is the acceleration of gravity and H the pressure scale height. For the penumbra this yields $t_{vert} \approx 30\ sec$. This is short compared with the observed life time of penumbral structures. If on the other hand the transversal length scale is large, the vertical adjustment time is just the sound travel time over a pressure scale height, since buoyancy effects are absent in that case. This time scale is of the same order as the previously determined t_{vert} however since $c^2 = \gamma g H$. We conclude that an approximate magnetostatic equilibrium also holds in the vertical direction. This means:

$$\frac{d}{dz}(P + B^2/8\pi) \approx g\rho, \qquad \qquad 2$$

where the curvature of the field lines has again been neglected.

In the *longitudinal* direction (parallel to the field lines) the readjustment time t_{long} will depend on whether the field is taken to be exactly horizontal, or slightly inclined as the observations indicate. If the field were horizontal, disturbances in longitudinal equilibrium which have a small transversal length scale propagate with the *cusp* or *tube speed*, $c_t = [c^2 v_A^2/(c^2+v_A^2)]^{1/2}$ (and with the sound speed if the transversal length scale is large). With an assumed longitudinal length scale of 3000 km (4 arcsec) this yields $t_{long} \approx 10$ min. If the field is inclined, there is also a buoyancy-influenced time scale, equal to the period of those buoyancy waves which have their fluid motions parallel to the field. If the inclination with respect to the horizontal is ϕ then this time scale is $(H/g)^{1/2}/\sin\phi$. For $\phi = 3°$ this is also on the order of 10 min. We conclude that for the longer and nearly horizontal structures the adjustment times are not much smaller than the observed life times of penumbral structures. These *need not be in longitudinal equilibrium*. The absence of equilibrium would manifest itself in the form of slow motions parallel to the field, and it is gratifying that the observed motions seem to be mostly of this kind.

Shorter or more inclined structures would have shorter adjustment times, and they must be closer to longitudinal equilibrium if their life time (see Muller, 1973a) is as long as that of the longer structures.

Summarizing, we have deduced from the observed life times and length scales of penumbral structures that they are roughly in magnetostatic equilibrium perpendicular to the field lines, satisfying Equations 1 and 2. Along their length they are not very close to equilibrium, that is there are flows along the field lines that fill or empty field lines, while the structure remains in balance in the other two directions. This can be compared to what happens to the barges that are used in dredging operations in rivers and harbours. As mud (penumbral gas) from the river bottom is pumped into an empty barge (a field line), it sinks increasingly deeper into the water (the surrounding penumbral material). The flow of the sludge through the filling pipe cannot be described by hydrostatic equilibrium, but the vertical motion of the barge is a series of quasistatic equilibria.

Making the obvious assumption that the flows along the field lines are closely related to what causes the inhomogeneity of the penumbra, it follows that the inhomogeneities in temperature, density and field strength are also intrinsically *field aligned*.

3. More involved or less conclusive deductions

3.1. Are dark filaments elevated?

I will argue that this is unlikely because dark filaments are too heavy, but before doing so let us formulate the question more precisely. This can be done in various ways, two of which are:

A. Is a dark structure as a whole elevated above the level $\tau_{5000} = 1$ in the normal photosphere? This is what is claimed by Moore(1981) for filaments near the edge of the spot.

B. Is the level $\tau = 1$ in the dark structure, i.e. its visible surface, higher than the surrounding photosphere (as defined by the mean $\tau_{5000} = 1$ level)?

Using Equation 2 we can show that for dark structures near the spot edge the answer to the more stringent form (A) is negative. It is harder to make definite statements about the less stringent question (B).

In order for a structure to be dark in the continuum, it must have a sufficient optical depth, or mass column density. For a reduction to 60% of the photospheric brightness (the value quoted by Muller, 1973b) a column density m of about $10\ g\ cm^{-2}$ ($\tau \approx 3$) is needed. The pressure exerted by this mass on the underlying gas is $P = mg = 2.7\,10^5$. With a penumbral field strength of 1500 G the total pressure exerted is $3.6\,10^5\,dy.cm^{-2}$. Since equilibrium must hold according to Eq. 2, this pressure must be balanced by something below. The photospheric pressure at $\tau = 1$ is only $1.3\,10^5$ however. Hence the structure as a whole cannot overlie the normal photosphere. It will press down until it meets a pressure of $3.6\,10^5$; this happens at a depth of 150 km. This argument applies near the spot boundary, where the structure is in direct contact with the normal photosphere. Elsewhere in the penumbra, structures can in principle be supported at greater heights by a strong magnetic field below. A pressure of $3.6 10^5$ can be balanced by a field strength of 3000 G. It can not be excluded that such field strengths exist deeper in the penumbra. Nevertheless, this yields a somewhat awkward construction because the gas density in the underlying field can not be very high. If it were higher, of the order of the photospheric density at similar depths for example, it would be too heavy and would itself have to be supported by even higher fields deeper down. At some depth however the field strength vanishes, so that the densities that can be accomodated this way are limited. We then have the situation that a heavy ($10\ g\ cm^{-2}$) structure is supported by a light mainly magnetic medium below. Such configurations are unstable by the Rayleigh-Taylor mechanism (the magnetic field does not suppress these instabilities, see, e.g., Acheson, 1979) and are destroyed on a time scale on the order of the free fall time over the depth of the density inversion.

In its less restrictive form (B) a similar reasoning can be applied. The column density at $\tau_{5000} = 1$ is about $4\ g\ cm^{-2}$ and to support this a pressure of $1.3 10^5$ is needed. This is the pressure at $\tau = 1$ in the photosphere, so that the observed surface of the structure cannot be much higher than this level. In principle some support can be provided by field strengths of the order of 1800 G, but as before stability considerations limit what can be done with magnetic support.

Note that these considerations apply only to structures seen as dark *in the continuum*. Dark structures in H_α for example correspond to far smaller mass column densities, so that equilibrium considerations do not yield significant constraints for the relative heights of bright and dark structures observed in chromospheric lines.

The subject of elevated filaments can be illustrated using an analogy with the earth's atmosphere and oceans; this is done in the Appendix.

3.2. Inclination of field lines

From the observed lengths of filaments one can estimate the inclination of field lines. We have already deduced that the intrinsic inhomogeneity is aligned with field lines. This implies that the outline of a bright or dark filament corresponds to the *cross section of a field aligned structure* with the $\tau = 1$ level. Assume now that the vertical length scale of the structures seen at the surface is not much greater than their observed lateral dimensions (one may argue whether this is reasonable or not). Then the ratio of width to length of the filament is just the inclination with respect to the level $\tau = 1$. If the width of the filaments is 0.5 arcsec (Muller, 1973a) then filaments of 4 arcsec length would be inclined by 7°. To this one has to add the inclination of the $\tau = 1$ level with respect to the horizontal. The shorter structures that are seen closer to the umbra would be correspondingly more vertical.

3.3. Magnetostatic spot models

Some more can be learned by comparing models for the equilibrium of the spot as a whole with observations. Such models can be constructed for example as follows (Schmidt and Wegmann, 1983). From the observed time scales of changes in spots one deduces in the same way as above that a spot is close to a mechanical equilibrium. This means that

$$\frac{dP}{dz}\bigg|_\Phi = g\rho, \qquad 3$$

where the derivative is with respect to depth, but taken along a field line. Assume in addition that the gas pressure is constant on horizontal planes inside the spot. The homogeneity of the umbra suggests that this could be a fair approximation in the umbra. With these approximations, the field must be *force free* except at the boundary between the spot and the surrounding convection zone. The vacuum above the photosphere into which the field fans out does not support magnetic torques. It follows from this that the field must be untwisted, i.e. it is not just a force free but in fact a potential field. The structure of the field is then uniquely determined when the pressure difference between the spot and its surroundings is specified as a function of depth, and the total magnetic flux of the spot is given. There are no analytic methods to solve this problem. A numerical method has been developed by Wegmann (1981). If the assumption of constant pressure on horizontal planes is not made, the field is no longer a potential field, and different numerical methods are needed, such as that of Pizzo (1986). With such models some easily understood properties of the field configuration can be verified:

1. Above the photosphere and not too close to the umbra the field is nearly that of a monopole with its origin slightly below the photosphere. From this it follows that the magnetic flux Φ of the spot is related to the penumbral field strength by

$$\Phi = 2\pi R^2 B_p, \qquad 4$$

where B_p is the field strength at the outer boundary of the penumbra, and R the spot radius. This value can be calculated for observed spots and compared with the measured magnetic flux of the umbra. One finds (Schmidt, see his contribution in the Discussion) that *most of the magnetic flux of the spot crosses the surface not through the umbra but through the penumbra.*

2. The inclination of the field lines with respect to the horizontal increases with height. The inclination at $\tau = 1$ at some point in the penumbra is therefore greater than at the boundary with the convection zone below this point. The level $\tau = 1$ is less inclined than this boundary, at least in some average sense, since the optical depth of the penumbra is evidently greater near the umbra than at the spot boundary. Taken together this shows that the field lines are more vertical than the $\tau = 1$ level.

This fact can be used (Spruit, 1981) to interpret the inward motion of bright grains found by Muller (1973a). Assuming that the bright-and-dark structure represents some form of convective motion, the bright elements must be moving upward and the dark ones downward (otherwise the convective energy flux would not be upward). The inclination of the field lines with respect to $\tau = 1$ then makes the intersection of the bright elements move towards the umbra. From the observed maximum speed of $0.5 km^{-1}$ and an assumed inclination of 7° one derives an upward speed of $60 ms^{-1}$, which is consistent with the small vertical motions in the penumbra (Beckers and Schröter, 1969), believed to be of the order $200 ms^{-1}$ or less (Schröter, personal communication).

3.4. Why is there a penumbra?

In the context of the magnetostatic models discussed above we can give some meaning to this question. If we adhere strictly to the simplifying assumption that the pressure within the spot is constant on horizontal planes, and equal to the umbral pressure (taken from an empirical umbral stratification), one finds that at the photospheric level the gas pressure is small compared to the magnetic pressure. From this one can calculate the depth of the boundary between the penumbral field and the convection zone simply by balancing the magnetic pressure of a 1500 G field against the external gas pressure. This yields a depth of 50 km *above* the photosphere; the penumbral field would be transparent, and hence the spot would not show a penumbra at all. It would look like a giant pore. Obviously the assumption of constant pressure on horizontal planes is wrong. The gas pressure on the field lines that form the penumbra is apparently much higher than on those that form the umbra. This can be understood qualitatively as a result of the thermal contact between the boundary of the spot and the surrounding convection zone. This will keep the temperature on the outermost field lines higher than on the umbral ones. Hence the pressure scale height is also larger on the penumbral field lines, so that the gas pressure decreases with height less rapidly. This ties the question of the origin of the penumbra to the difficult question how energy is transported, horizontally and vertically, in a spot. It seems reasonable to say that there is energy transport by 'some' form of convection, but a much better answer than this is needed before we can even answer the question why the spots have penumbrae at all.

3.5. Mode of energy transport in the penumbra

The small values for the vertical velocities in the penumbra suggest strongly that, though some convective process seems be taking place, it does not contibute strongly to the energy transport. I assume here that the vertical velocities are of the order $200 ms^{-1}$ (Schröter, personal communication) or less. This is an order of magnitude smaller than in the photosphere. The convective energy flux at any level in the atmosphere is roughly

$$F_c = \rho |v| c_p \Delta T, \qquad 5$$

where $|v|$ is (the absolute value of) the vertical velocity, c_p the specific heat, ΔT the temperature difference between upward and downward moving elements. Since ρ, c_p, and ΔT have quite similar values in the photosphere and in penumbral filaments, the convective flux would be an order of magnitude smaller in penumbral filaments. The energy flux in the penumbra however is about 85% of the normal photospheric flux. The implication is that *in the penumbra the energy is transported mainly by radiation.* This applies not just to the observed layers (which would not be surprising) but to deeper optically thick layers as well: for reasons of mass continuity the vertical velocity varies with depth much more slowly than the optical depth so that the observed velocities are representative for the first scale height (150 km) below $\tau = 1$. If most of the flux were carried by convection, we would expect vertical velocities similar to those in the granulation. Though this has not been observed, the implications of this issue are substantial, and new observations (at high resolution) to corroborate the previous results are called for.

4. A shallow penumbra model

If the energy in the penumbra is indeed carried mostly by radiation, as suggested above, one can calculate the depth of the penumbra from the observed intensities, using a simple radiative transport calculation. One integrates a radiative equilibrium atmosphere, starting at $\tau = 1$ using the energy flux inferred from the observed continuum intensities. When the temperature in the model reaches that of the photosphere at the same depth (which it does very soon), one has the depth of the penumbra. For dark filaments we have already done this in a slightly different way in section 3.1. In these structures the depth of the penumbra would be thus about 150 km. In the bright filaments, which are almost as bright as the photosphere, the depth must be much smaller.

By more indirect means the same conclusion was reached by Schmidt, Spruit and Weiss (1986), using theoretical arguments about the energy transport in a penumbra-like field. I briefly describe here the basic idea of this calculation, and the conclusions.

The starting point is the observation that a magnetic field overlying the convection zone is quite stable to convective motions, at least near the boundary with the convection zone. Equilibrium condition (2), applied at this boundary (which we call the *magnetopause*) shows that the gas pressure above it is less than below, the difference is made up by the magnetic pressure of the penumbra. Since the temperature must be continuous across the boundary (due to radiative exchange) the gas density is reduced above the boundary by roughly the same factor. The relative difference is small if the boundary is at a large depth, where the magnetic pressure (for a given field strength) is small compared to the gas pressure, but near the surface, where the two are comparable, the density difference is substantial. This difference in density, due to the presence of the field, makes the interface stable to overturning motions, in the same way a water surface on earth is stable. There may be convective motions in the penumbra and below it, but no motions that cross the interface, as long as the density difference exceeds a certain minimum. Taking into account the stabilizing effect of the magnetopause, a mixing length calculation of convective energy transport near and above the magnetopause was made. It was assumed that the magnetic field itself (assumed roughly constant with depth above the magnetopause) has little effect on the efficiency of convection and that the stabilization is effective mainly near the magnetopause. As the depth of the magnetopause was increased in these models, the energy flux through it varied as follows. At zero depth, the flux is nearly the same as the normal photospheric flux. With increasing depth the flux decreases until a minimum is reached at a depth on the order of a pressure scale height. Thereafter the flux increases again. Thus for a given energy flux (namely the observed flux) the calculation gives in general *two* possible models. The first is shallow, on the order of a scale height. In this model the magnetopause is so close to the surface that it effectively suppresses convection, and the energy transport is mostly by radiation. In the second model, the magnetopause is much deeper, so that there is room above it for efficient convection to develop. The depth of this second model depends on the details of how the convective efficiency in the magnetic field (away from the magnetopause) is modeled. For the models reported it was on the order of 800 km. In the first kind of model the convective (vertical) velocities are small, in the second kind they are comparable to granulation velocities. As the energy flux is increased, corresponding to filaments closer to the boundary of the spot, the depth of the first model decreases, that of the second model *increases*. The second kind was then rejected as unrealistic, because the depth of the penumbra must vanish at the boundary.

What the calculation shows is that due to the stabilizing effect of the magnetosphere, one is not free to choose the depth of the penumbra arbitrarily if one wants to reproduce the observed penumbral brightness. This restriction turns out to be strong enough to limit the choice to shallow models in which the penumbra (defined as everything above the magnetopause) is almost in radiative equilibrium. It is gratifying that this agrees with the observational argument given above. Some predictions from this model are:
- in bright filaments the magnetopause lies very close to the local $\tau = 1$ level; they are almost transparent.
- the depth of the $\tau = 1$ level is about the same in bright and dark structures (the penumbra is thus predicted to be a rather smooth surface).
- the difference in field strength between bright and dark structures is small (less than 100 G).

Acknowledgement

It is a pleasure to thank Dr. L.J. November for permission to use his sunspot observations, and for stimulating discussions on sunspot physics.

References

Acheson, D.J.: 1979, *Solar Phys.* **62**, 23

Beckers, J.M. and Schröter, E.H.: 1969, *Solar Phys.* **10**, 384

Danielson, R.R.: 1961, *Astrophys. J.* **134**, 275

Moore, R.L.: 1981, *Astrophys. J.* **249**, 390

Muller, R.: 1973a, *Solar Phys.* **29**, 55

Muller, R.: 1973b, *Solar Phys.* **32**, 409

November, L.J.: 1984, in *Small Scale Dynamical Processes in quiet stellar Atmospheres*, ed. S.L. Keil, National Solar Observatory, Sunspot, N.M. 88349, USA, p.74

Pizzo, V.I.: 1986, strophys. J.**302**, 785

Schmidt, H.U., Spruit, H.C. and Weiss, N.O.: 1986, *Astron. Astrophys.*, **158**, 351

Schmidt, H.U. and Wegmann, R.: 1983, in *Dynamical Problems in mathematical Physics*, eds. B. Brosowski, E. Martensen, Lang, Frankfurt, p.137

Spruit, H.C.: 1981 in *The Physics of Sunspots*, eds. L.E.Cram and J.H. Thomas, Sacramento Peak Observatory, Sunspot, N.M. 88349, USA, p.359

Title, A.M., Tarbell, T.D., Simon, G.W., and the SOUP Team, to be published in *Proceedings of the COSPAR 26th Plenary Meeting*, Toulouse, June 1986

Wegmann, R.: 1981, in *Numerical Treatment of free Boundary Value Problems*, ed. J. Albrecht, Birkhäuser, Stuttgart, p.335

Appendix: a terrestrial analogy for elevated filaments.

Suppose there were an extraterrestrial civilization observing the earth from a distance. I assume that it can do this with a resolution of 40 km. It so turns out that among their scientists there is a school maintaining that the oceans on earth are elevated above the continents. Evidence for this would be their observations of the Malay archipelago, where the color of the water surrounding the islands clearly shows that land extends below the oceans. From this they conclude that the oceans are absorbing structures elevated above the mean level of the land masses, which they call earthosphere. Another school would use its knowledge of physics (which I assume would be shared by both schools) to reach the opposite conclusion: The oceans lie lower than the mean level defined by the land masses. From infrared spectroscopy the composition and temperature of the atmosphere and the oceans would be known. The time scales for motions in the oceans would be known (for example from observations of tides). The second school would point to the very strong surface gravity of the planet, reflected in the well known fact that the scale height of the atmosphere is considerably smaller than the observational resolution, and show that hydrostatic adjustment times for structures of the observed length scales in the oceans and atmosphere are much shorter than the observed time scales. They would then use arguments quite similar to those I have used above to show that the oceans and atmosphere are close to hydrostatic equilibrium, and that the atmosphere will not support oceans floating above land masses.

Nonlinear Compressible Convection in Regions of Intense Magnetic Fields

N. E. Hurlburt

Department of Applied Mathematics and Theoretical Physics, University of Cambridge, Cambridge, CB3 9EW, England

Abstract. Two-dimensional numerical simulations are used to explore the behaviour of nonlinear compressible convection extending over multiple scale heights within regions of intense magnetic fields, e.g. sunspots and pores. The large magnetic pressures possessed by such magnetic fields give rise to buoyancy forces which can dramatically alter the behaviour of both oscillatory and steady types of convection.

There are two models for the structure of sunspots below the surface: the plug and the jellyfish. In the plug model the field is more or less uniform throughout the spot and the convection must coexist with it. In the jellyfish model the spot consists of tubes which break apart beneath the surface, so there are well defined regions of strong field interspersed with field-free, convecting regions. The latter model explains phenomena such as umbral dots as fingers of field-free flows pentrating to the surface. Part of the motivation for our research is to see if such phenomena can also be explained in the plug model. Some noteworthy elements of sunspots which must be considered are magnetic pressures with magnitudes comparable to local gas pressures and motions which can span a wide range of densities, pressures and material properties in depth. The effects of compressibility are thus fundamental to the dynamics.

One material property of particular importance to magnetoconvection is the ratio of the magnetic to thermal diffusivity ς, which controls the coupling between the magnetic field and the convection. For the first 10000km below the photosphere, ς increases by several orders of magnitude. Near the surface it is very small, so the field is nearly frozen to the fluid. About 2000km below the photosphere it passes through unity, and is larger than one from there to roughly 20000km. In Boussinesq simulations (Weiss 1981a and b), where ς is invariant in depth and where the fluid is essentially incompressible, oscillatory convection is preferred in regions of strong magnetic field when $\varsigma < 1$. When $\varsigma > 1$ only steady motions are predicted by Boussinesq theory. However this theory is incapable of predicting how convection should behave in regions where ς varies with depth, particularly for convection on supergranular scales which span both the steady and oscillatory regimes.

We have conducted a series of simulations to examine how magnetoconvection in the outer layers of the solar convection zone might behave by taking ς to be proportional to the density of the fluid, thus increasing with depth. We solve the equations for fully-compressible, time-dependent magnetoconvection in two-dimensional, plane-parallel layers. The upper and lower impenetrable boundaries are stress free and at fixed temperatures. Details on the numerical techniques and model equations can be found in Hurlburt and Toomre (1987). We take the density ρ (and hence ς) to vary by a factor of 11 across our computational domain, and impose an initially uniform vertical magnetic field across the layer. We can measure the strength of the field by the Chandrasekhar number Q and the strength of the thermal driving by the Rayleigh number R.

In Figure 1 we display the velocity and magnetic field of a typical simulation where $\varsigma = 1.2$, $R = 50000$ and $Q = 1000$. Although the values of the Rayleigh and Chandrasekhar numbers are far below what those of sunspots, the value of ς and its variation are representative of that experienced by supergranular scales of con-

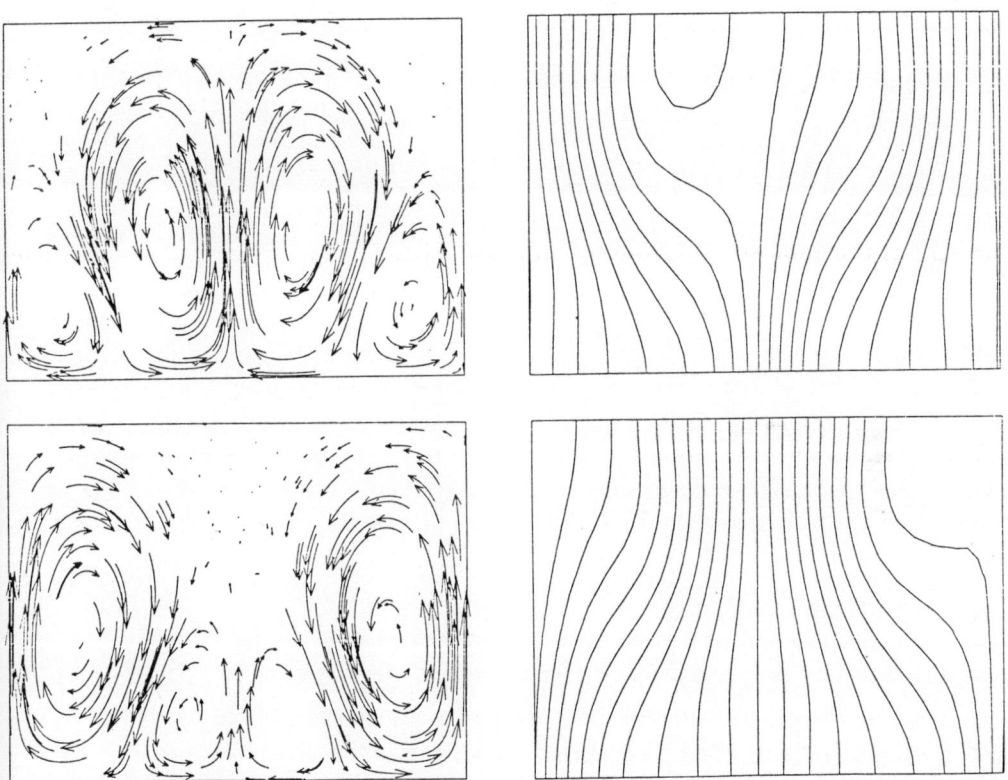

Figure 1. A typical simulation with moderate magnetic Prandtl number ς displayed at two instants of its oscillation. The velocity is displayed as streaklines in the left panels while the magnetic field lines are displayed in the right panels. Here $\varsigma = 1.2$, the Rayleigh number $R = 50000$ and the Chandrasekhar number $Q = 1000$.

vection. Two cells exist within out periodic computational domain. They are dominated by upward-directed plumes which alternately ascend, sweep the field aside and decay. Motions in the lower half of the layer, where $\varsigma > 1$ continually overturn as Boussinesq analysis predicts. Similarly, motions in the upper half of the layer oscillate as that simpler theory also predicts. The mixture of the two behaviours leads to the novel, modulated convection evident in the figure, which has a wide range of possible timescales. If the simulation in Figure 1 were viewed from above, one would see a narrow region of bright, rising fluid surrounded by cooler, broader regions of slowly descending fluid. Both the observed timescale and the observed spatial stucture of umbral dots might arise from such modulated convection.

Linear analysis of compressible magnetoconvection predicts that the solution presented in Figure 1 and all other solutions should be steady. For lower values of R such steady solutions exists, suggesting that the time-dependence of Figure 1 is due to nonlinear effects. Nonlinearity also introduces a multiplicity of possible solutions. Figure 2 displays two typical steady solutions which have identical parameters but

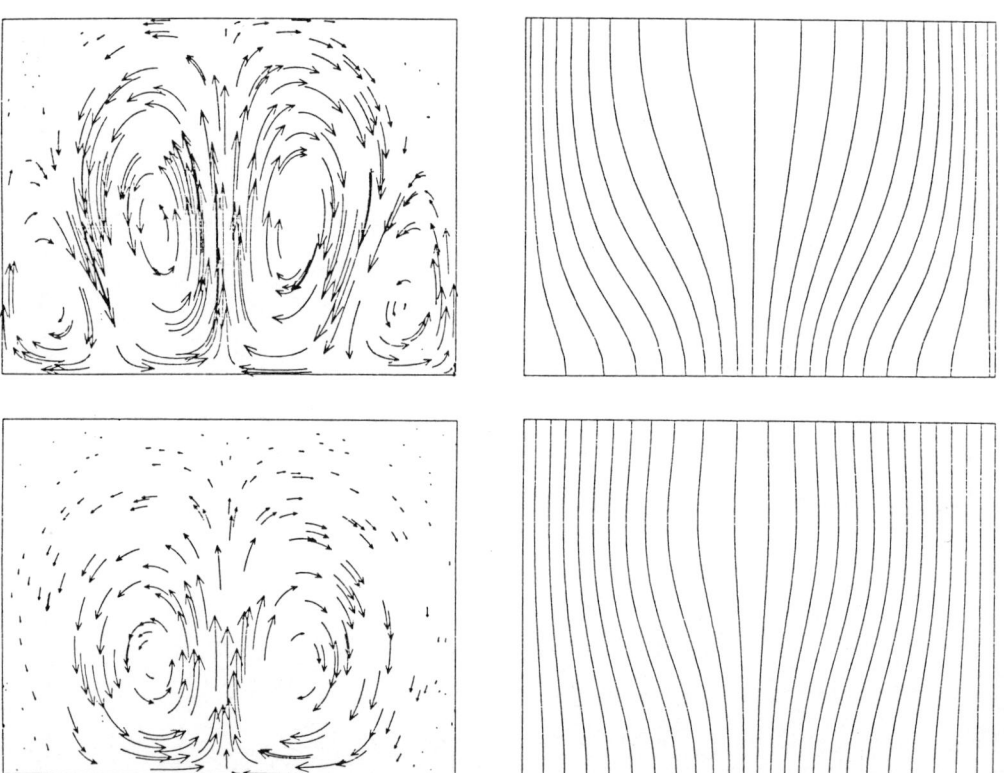

Figure 2. Two different steady simulations of compressible magnetoconvection both with $\varsigma = 1.2$, $R = 33000$ and $Q = 1000$ obtained from different initial conditions.

differ in their initial conditions. The solution shown in Figure 2a used that shown in Figure 1 for initial conditions while the solution displayed in Figure 2b was started from a polytropic stratification with small-amplitude, random velocity perturbations. While they are quite different in their strength, both solutions display a preference for upward motions.

In the two-dimensional, fully-compressible simulations of Hurlburt, Toomre and Massaguer (1984; 1986), stronger downward flows were preferred rather than upward ones as found here. The pressure fluctuations in their two-dimensional simulations possess horizontal structures which satisfy nearly a Bernoulli-like relation with the horizontal velocities. As a result, pressure fluctuations accentuate buoyancy driving in regions of descending motions by working in concert with temperature to enhance the density fluctuations. However pressure works against temperature within the ascending motions to weaken the density fluctuation, and thus the flows are broader and weaker there. The presence of a magnetic field introduces a new kind of pressure balance. Rather than the Bernoulli balance between pressure and velocity found in field-free compressible convection, the gas pressure can now serve to balance variations in the magnetic pressure. The resulting reduction in density within regions of concentrated field drive ascending flows within them. Hence a change in behaviour can occur once the fluctuations in gas pressure due to the magnetic pressure overcome those due to the horizontal velocity. Such a change requires both a sufficiently strong magnetic field and a sufficiently small value of ζ so large variations in field strength exist.

Since magnetic buoyancy, generated by the magnetic pressures influence on density, plays an important role in modifying the symmetry of the vertical flows, one might expect a change in convective behaviour to occur between quiet and active regions on the solar surface. High-resolution observations of Doppler shifts and C profiles in solar convection might test such theoretical predictions on the dynamical behaviour of compressible convection.

I am grateful for advice from M.R.E. Proctor and N.O. Weiss. This research has been supported by grants from the UK Science and Engineering Research Council.

Hurlburt, N.E., Toomre, J. and Massaguer, J.M. (1984). Two-dimensional compressible convection extending over multiple scale heights. Astrophys. J. **282**, 557-573.

Hurlburt, N.E., Toomre, J. and Massaguer, J.M. (1986). Nonlinear compressible convection penetrating into stable layers and producing internal gravity waves. Astrophys. J., **311**, 563-577.

Hurlburt, N.E. and Toomre J. (1987). Magnetic fields interacting with nonlinear compressible convection. Astrophys. J., submitted.

Weiss, N.O. (1981a). Convection in an imposed magnetic field I. The development of nonlinear convection. J. Fluid Mech., **108**, 247-272.

Weiss, N.O. (1981a). Convection in an imposed magnetic field II. The dynamic regime. J. Fluid Mech., **108**, 273-289.

THE INTENSITY DISTRIBUTION IN SUNSPOT PENUMBRAS

Collados, M.; del Toro, J.C.; Vázquez, M.
Instituto de Astrofísica de Canarias, 38200 - La Laguna,
Tenerife, Spain.

ABSTRACT
The intensity distribution of the penumbra at different
stages of evolution have been analyzed. The results have
been different for both evolved and primitive penumbras.
While the former present almost symmetrical, single-peaked
histograms, the same does not occur for the latter, their
distributions being, preferentially, asymmetrical or
double-peaked. We have interpreted these results in terms
of bright and dark elements. Thus, an evolutionary process
has been proposed to explain the diverse characteristics
found at the different stages.

1 INTRODUCTION

As is known, the penumbra is the outer part of spots,
presenting a filamentary structure in which bright and dark filaments
alternate around the umbra, being elongated in the radial direction.
During its formation, in its very first stages, it does not appear
uniformly around the spot, but rather forms in patches (Collados et al.
(1985)). In addition, the structure is, already from the beginning,
filamentary (Collados et al. (1985)). Thus, the primitive penumbra
looks like the evolved one, from a morphological point of view.

Concerning its photometric characteristics, Maltby (1972) finds that
the mean penumbral intensity relative to that of the photosphere is
0.72 at 500 nm. Moreover, Muller (1973) indicates that the mean
relative intensity of the bright features is 0.95, and that of the dark
background is just 0.60, occupying, respectively, 0.40 and 0.60 of the
total penumbral surface. In contrast, Grossmann-Doerth and Schmidt
(1981), analysing the penumbral intensity distribution, conclude that
the observed histogram is not consistent with the duality bright
filament-dark background in those terms expressed by Muller.

Regarding the initial stages of the formation of the penumbra, just
morphological descriptions appear in the literature (Bray and Loughhead
(1964), Bumba and Suda (1984)). No photometric studies have been
carried out on the subject up to now.

In the present study, we have analysed the intensity distribution of
the penumbra at different stages of evolution, separating those
well-formed and stable penumbras (hereinafter called evolved) from

those being born and rapidly changing penumbras (hereinafter referred as to primitive). As it will be seen, there is no contradiction between a single-peaked histogram and Muller's statements. But, on the contrary, the comparison done of both kinds of penumbras does lead to a new view of the penumbra of sunspots.

2 OBSERVATIONS

A total of 106 spot images have been analysed. The pictures were taken at the Observatorio del Teide at Izaña (Tenerife), with the 40 cm vacuum telescope, property of the Kiepenheuer Institute with AHU Kodak film ($\gamma \sim 3$) at a wavelength of 500 nm. The scale on the images is 5."5/mm, giving all the images a resolution of 0."5-0."6. The photographs were digitized with the IAC's microdensitometer, using a square slit 20 μm long and a step of 20 μm in both directions.

Each image was converted from densities into intensities, by means of its corresponding characteristic curve, which was calculated following the PFD process, based on the half-filter method (Collados and Bonet (1984)).

Restoration was then applied to each one of the pictures, applying the following filter (Collados (1986)):

$$F(u,v) = \frac{1}{H(u,v)} \quad \text{if } |\sqrt{u^2+v^2}| \leq f_o$$

$$F(u,v) = \frac{H(u,v)}{H^2(u,v) + \alpha [|\sqrt{u^2+v^2}| - f_o]} \quad \text{if } |\sqrt{u^2+v^2}| > f_o$$

with two free parameters: f_o, the frequency which determines where the noise must be eliminated from, and α, a constant related with the signal to noise ratio. H(u,v) is the transfer function of the system, and has been evaluated assuming an ideal 40 cm telescope, and an equivalent one of 30 cm for the atmosphere.

Penumbral zones were, later, selected by means of a high resolution graphic display, where their limits were determined, and extracted from the images to be analyzed.

3 DATA ANALYSIS AND RESULTS

For regular spots, mean penumbral histograms have been evaluated at 10° intervals from the center of the disk to 70° (see fig. 1a. for an example). The obtained distributions have proved to be independent of the heliocentric angle: all of them are single-peaked, more or less symmetrical, and have a width which is, approximately, the same in all the cases.

The interpretation of these results requires a previous analysis. As noted, Grossmann-Doerth and Schmidt (1981) state that a single-peaked

intensity distribution is incompatible with Muller's (1973) results
that the penumbra is formed by bright structures, which intensity is

Fig. 1(a): Mean intensity distribution obtained at a
heliocentric angle between 10° and 20°. It is single-peaked
and, more or less, symmetrical (agreeing with
Grossmann-Doerth and Schmidt (1981)). These characteristics
have proved to be independent of the position of spots on
the solar disk. Fig. 1(b): Numerical histogram obtained by
assumming the mean intensity of bright and dark elements to
be, respectively, $0.85 I_{ph}$ and $0.65 I_{ph}$. The widths of
the individual distributions are $0.10 I_{ph}$. The similarity
with the observed ones is apparent.

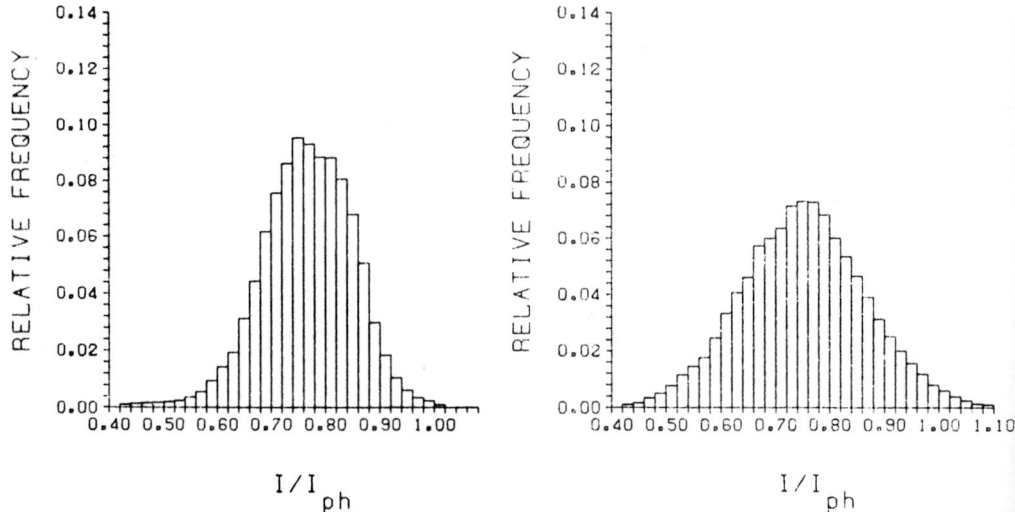

Fig. 2: Intensity distribution of a penumbra during its
first stages of formation.

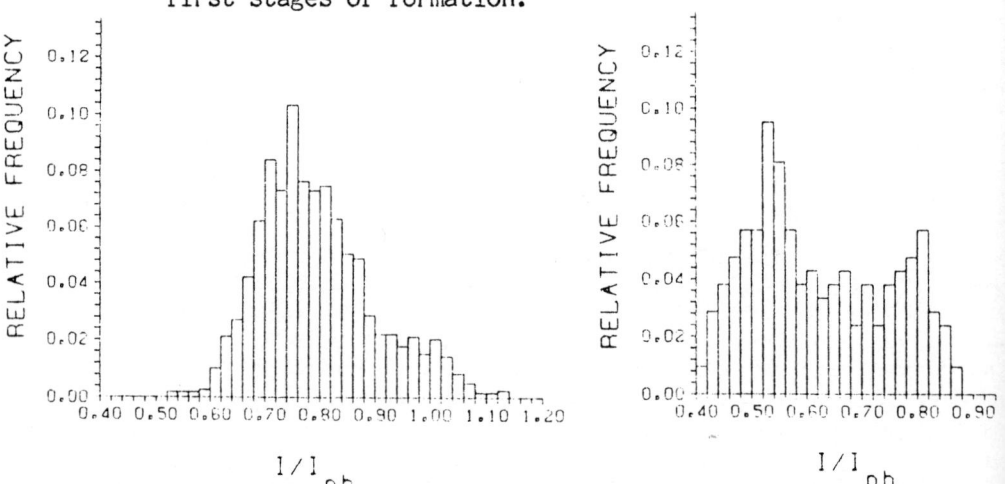

0.95 times that of the photosphere, superimposed to a dark background,
with relative intensity 0.60.

We have tried to verify this statement, and, with that aim, we have generated random gaussian numbers to represent the extreme values of bright and dark features. It has also been assumed that the photometric profiles of these structures are sinusoidal. The mean intensity of each kind of filament, the width of the individual distributions and the surface fraction occupied by each one of them have been left as free parameters.

The results show, on the one hand, that if the widths of the individual distributions are small, a double peak always appears, in a clear contradiction with the observational results. The partial surface fraction influence is very small. On the other hand, if the individual distributions are wide enough, the resulting histogram has one maximum and is rather symmetric, agreeing with observations. Intermediate situations give rise to double-peaked or asymmetric distributions. The main difference between our numerical analysis and the one carried out by Grossmann-Doerth and Schmidt (1981) lies in the fact that they did not take into account the influence of the intermediate points in the profiles of the dark and bright elements. When these are included into the numerical distribution, a single-peaked histogram is obtained (see fig. 1b for an example). Thus, it can be concluded that the observed intensity histograms agree with Muller's results if the bright filaments have a relative intensity of 0.85-0.90, and the dark ones 0.60-0.65, and their individual distributions have a standard deviation greater than 0.10. This implies that the latter do overlap and the definitions of a bright or a dark penumbral filament has only a local sense, as Wiehr et al. (1984) state, but only within a certain range of intensities.

The same analysis has been carried out on the early penumbra. Among the 54 studied cases, several different kinds of intensity distributions have been distinguished: (i) those presenting a single peak and, which are, approximately, symmetrical; (ii) those which are asymmetrical, with a mean intensity greater than that at which the maximum appears; (iii) those which are asymmetrical, with a mean intensity smaller than that of the maximum; (iv) those with a clear double-peaked structure; and (v) those which are, more or less, constant (see fig. 2). The relative frequencies of appearance are, respectively, 11%, 34%, 28%, 19%, and 8%, allowing to deduce that the just born penumbra intensity distribution is, preferentially, asymmetrical or double-peaked.

These results suggest, nevertheless, that the characteristics of the penumbra are different at the various stages of evolution. It seems as though the penumbra is composed, in these first stages, by two kinds of clearly different structures, the opposite as is the case in the evolved one. The only explanation which can account for this apparent contradiction is that the magnetic field structures can not be the same at different heights. It can be imagined that the penumbra is formed as the magnetic field lines become more and more inclined outwards. As this process goes on, the penumbra is more and more horizontal (i.e., longer) and thick. If the horizontal distribution of bright and dark elements is not the same in the different layers, then they should alternate vertically in a progressively more homogeneous proportion. The intensity distribution, thus, tends to be symmetrical around a single peak.

This simple evolution process is able to explain why Casanovas (1973) observed two dark filaments crossing each other. As the observations are made with wide band filters, radiation comes out from a diversity of layers, and, in a certain sense, all of them are "seen". A different horizontal inclination of the magnetic field lines at different layers, in complex groups, could give rise to this phenomenon.

4 CONCLUSIONS

It has been shown that the penumbra intensity distribution is not in contradiction to Muller's results. A different histogram for the penumbra during its early stages of evolution has been found, being, preferentially, asymmetrical or double-peaked. A possible evolutionary process to explain the observations is proposed.

Reference list

Bray, R.J.; Loughhead, R.E.: 1964, "Sunspots". Chapman and Hall. London
Bumba, V.; Suda, J.: 1984, Bull. Astron. Inst. Czechosl., 35, 28.
Casanovas, J.: 1973, Doctoral Thesis. University of La Laguna.
Collados, M.: 1986, Doctoral Thesis. University of La Laguna.
Collados, M.; Bonet, J.A.: 1984, App. Opt., 23, 2827.
Collados, M.; Garcia de la Rosa, J.I.; Moreno-Insertis, F.;Vázquez, M.:
    1985, in R. Muller (ed.), "High Resolution in Solar
    Physics". 8th Regional Meeting of the IAU. Toulouse.
    France.
Grossmann-Doerth, U.; Schmidt, W.; 1981, Astron. Astrophys., 95, 366.
Maltby, P.: 1972, Solar Physics, 26, 76.
Muller, R.: 1973, Solar Phys., 32, 409.
Wiehr, E.; Koch, A.; Knölker, M.; Küveler, G.; Stellmacher, G.: 1984,
    Astron. Astrophys., 140, 352.

JOINT DISCUSSION ON TOPICS OF SESSIONS 7 AND 8, SUMMARIZED BY THE CHAIRMAN OF SESSION 7

H.U. Schmidt
Max-Planck-Institut für Astrophysik
Garching, Germany

General discussion:
The observed magnetic profiles of sunspots seem to violate the conservation of flux: The whole flux leaving the spot through the halfsphere of the radius of the spot R should be $B_1 2\pi R^2$ when B_1 is the observed strength of the field at the edge where it is known to be horizontal. The strength should stay constant over the whole halfsphere, as it dominates the gas pressure in chromosphere and corona completely. This flux should balance the flux crossing the spot through the flat horizontal umbra with maximum fieldstrength B_0 in the center. Taking an outer edge of the umbra at $\frac{R}{2}$ we get a flux $<B_0 \pi (R/2)^2$ and consequently $B_0 > 8B_1$ which contradicts many if not all observed profiles, e.g. Beckers and Schröter (Solar Physics 10, 384, 1969) with $B_0 = 2B_1$. The factor 8 could be reduced if the radius of the flux at the umbra level extends beyond R/2 but even an extension to R would need a fieldstrength $2B_1$ throughout this cross-section not just in the center. Before theory of sunspots can proceed we need more exact measurements of fieldstrength and inclination throughout the spot (H.U. Schmidt).

I ask the observers about turbulence in spot umbrae. Beckers claims 1.5 kms^{-1} r.m.s. unresolved velocity, horizontal and vertical. It is unconfirmed. Such measurement bears directly on energy transport mechanisms in umbrae. Further it seems incompatible with a convective blueshift <60ms^{-1} observed by Beckers. New observations are needed (H.C. Spruit). One must distinguish macro- and microturbulence (W. Mattig). Both give rise to an I-v correlation, only pure horizontal motion would not (H.C. Spruit).

Discussion relating to invited review of F. Moreno-Insertis:
Sunspots rotate differentially, magnetic patterns of coronal holes rotate rigidly. If sunspots are rooted at the bottom of the convection zone, where then are coronal holes rooted? (S.O. Stenflo). Apparent rigid rotation of a coronal hole pattern in interplanetary field may be due to latitudinal averaging in the low harmonics that dominate the interplanetary field (A. Title). But they rotate with the equator not with an average speed (Stenflo). Sunspots may rotate with the plasma independent of the coronal holes which may rotate with a rigid convective pattern (N.O. Weiss). Buoyant flux tubes producing sunspots and rooted at the bottom of the convection zone follow a weighted average of rotation throughout the convection zone (F. Moreno-Insertis).

How to interprete the observed formation and decay of sunspots from and into certain fragments as reviewed earlier by J.I. Garcia de la Rosa? (U. Anzer). First both pictures of a monolithic flux column or a cluster of many tubes are too extreme. The fragments may evolve from flute or other instability along the rise or exist already before the rise (F. Moreno-Insertis).

Is the heat flow necessarily blocked within the flux tubes of the spot or could the turbulence be kept from entering the tubes altogether? (C. Zwaan). Meyer et al. (1979) argue that below 2000 km the magnetic diffusivity exceeds the thermal one and so allows overturning convection within the field. In a cluster model overturning convection may be inhibited in the thin tubes but may be possible in the intertubular plasma (F. Moreno-Insertis).

Why should the tubes at the convection zone base match their internal and external density? (R. Rosner). We do not know. We need to study the process of flux concentration there and the MHD of the underlying overshoot region. Adiabatic evolution of a kink-unstable tube causes the lower loop to enter the underlying subadiabatic region but not deeper than a fraction of a scale height due to the enormous increase of buoyancy. Thus the tube can only stay in a narrow regime near the interface between the convective and radiative zones (F. Moreno-Insertis).

Can downward magnetic field transport be effective in 3D? (R. Rosner). Arter, Galloway and Proctor studied the effect of 3D cellular convection on horizontal field. There is a net downward pumping but it is associated with strong fields linking adjacent cells. This downward pumping by tesselated cells seems less effective than presumed (N.O. Weiss).

Discussion relating to contribution of M. Collados et al. on the Wilson effect:
Similar analysis at Oslo (Solar Physics $\underline{9}$, 497, 1969) resulted in a Wilson depression of 400 km (E. Jensen). If smaller west penumbrae are caused by braking of the higher rotation speed of spots the effect should vanish for old spots (H. Wöhl). The dependence on spot age has not yet been studied (M. Collados). In Zs. f. Astrophys. $\underline{62}$, 228 (1965) the inclination of the Evershed flow to the surface was found systematically larger in the western penumbra. Perhaps the west penumbrae are not so much smaller but steeper due to forces coming from rotation (E.H. Schröter). Is the effect different for leading and following spots as they may differ in inclination? (E. Priest). Following spots lack stability and could not be analyzed (M. Collados). Asymmetry may exist because the leading parts are usually more regular and spots do not often cross the full disk in circular shape (P. Maltby). Asymmetry may be intrinsic to leading part, however, evolution of spots does not seem to be the cause of the anomalies (M. Collados). The Wilson depression reduces the rotation velocity determined from single passage positions as compared to recurrent rotations. This may in the future allow an independent check (H. Wöhl).

Discussion relating to contribution of F. Busse:
You demonstrated that nonlinear effects can cause a mean flow in a model of penumbral convection rolls provided the penumbra is deep enough for such convection rolls. You still have to show that the mean flow is confined to regions of downward motion corresponding to the dark filaments (N.O. Weiss). The rolls tend to concentrate the flux in these regions. Since the mean flow in our model is proportional to the Chandrasekhar number squared for small values of it, the flow is expected to be strong in the dark regions (F. Busse). Your calculation assumes small magnetic Reynolds number; what is the relevance to sunspots? (H.C. Spruit). Realistic calculations of dynamic penumbra models are still impossible. My simple model points out the relevance of a new mechanism of mean flow generation by fluctuating motions in an inclined magnetic field for the Evershed effect. Because the mechanism rests on the magnetic analogue of the Proudman-Taylor theorem, it probably persists even at large Reynolds numbers (F. Busse).

Discussion relating to invited review of H.C. Spruit:
Why do you prefer Wittman's (1974) more horizontal field influenced by stray light from outer penumbra? (J.I. Garcia de la Rosa). Why should stray light affect the measured inclination? (H.C. Spruit). What are current feelings about implication of the SMM observation of 0.1% luminosity variations with spot occurence for the heat blockage? (E. Priest). Very confused. My own is a minority view but very strong. In A&A 1982 I calculated the thermal response of the convection zone to a sudden heat blocking by an object near its surface. For a depth large compared to the pressure scale height all the blocked flux stays within the convection zone, heating it as a whole. This takes $2 \cdot 10^5$ years. On all shorter time scales the zone acts as a reservoir of infinite capacity due to the convection communicating thermal disturbances through the star on a turbulent diffusion time scale (H.C. Spruit).

You rejected elevated penumbral filaments because their magnetic and gas pressure can only be balanced by photospheric gas pressure. But diamagnetic flow in those filaments may push aside the magnetic field (J.I. Garcia de la Rosa). The structures cannot be too far from vertical equilibrium, the magnetic field itself exerts similar forces also when expelled (H.C. Spruit). Does the SOUP Spacelab 2 white light movie support a shallow penumbra? (P. Maltby). Yes, just as any good spot pictures (H.C. Spruit). But it shows rapid motions (P. Maltby).

Can your model explain the absence of penumbra in pores? (B. Roberts). No, I don't think we understand what makes the penumbra appear (H.C. Spruit).

Discussion relating to contribution of M. Collados on penumbra Intensity:

Your density distribution is affected by the correction (telescope MTF) applied in reconstruction. Did you check this correction e.g. by reconstruction of granulation for which there exist corrected power spectra? (O. v. d. Lühe). The intensity distribution is more affected by the restoration method than by the MTF. A good restoration should not influence the intensity distribution but only the contrast, i.e. the histogram must widen but not change in form, as is the case here. We have not done the comparison you suggest. Surely we underestimate the degradation of the images by atmospheric turbulence but that should not affect our results (M. Collados).

STRUCTURE AND DYNAMICS OF SMALL MAGNETIC FLUX CONCENTRATIONS: OBSERVATION VERSUS THEORY*

M. Schüssler
Kiepenheuer-Institut für Sonnenphysik
Schöneckstr. 6, D-7800 Freiburg,
Federal Republic of Germany

Abstract. Observational results and theoretical model calculations for small magnetic flux concentrations in the solar photosphere are compared. We consider the formation of flux concentrations and their magnetic, thermal and velocity structure. Flux tube geometry, interaction with the environment and, finally, their destruction are discussed. In the conclusion, the necessity of further development and support of MHD model calculations in connection with new observational projects is emphasized.

1. INTRODUCTION

Since the discovery of the concentrated nature of non-spot magnetic fields in the solar photosphere (Beckers and Schröter, 1968; Stenflo, 1973), small scale magnetic flux concentrations have become increasingly important for our understanding of the evolution of active regions, the interaction of convection and magnetic fields, the heating of chromosphere and corona and, last not least, the basic dynamo mechanism itself (Spruit and Roberts, 1983). Progress in theory and observation was considerable in the last 15 years with breakthroughs on both sides: The concept of the "thin flux tube" made it possible to treat dynamical problems analytically (see review by Spruit, 1983), 2D and 3D numerical simulations have been carried out successfully (see reviews by Nordlund, 1985; 1986); the application of Fourier transform spectroscopy and polarimetry (Stenflo et al., 1984) and the development of innovative techniques to make efficient use of these data (see review by Solanki, these proceedings), led the way for a detailed analysis of the physical properties of concentrated magnetic structures in the solar photosphere. We should not forget to mention the progress in the field of (multidimensional) radiative transfer of polarized radiation (Landi Degl´Innocenti, 1983; van Ballegooijen, 1985a; Jones, 1986): These results will be crucial for the comparison of results from model calculations with observations.

In this situation and in view of the new solar facilities on Tenerife it seems useful to have a comparison between existing observations and the results of theoretical calculations. Limited space and time (as well as limited capability of the author) force us to select some problem areas which are of current interest. We shall concentrate on the photospheric layers, exclude wave propagation (see Hammer, these

* Mitteilungen aus dem Kiepenheuer-Institut No. 267

proceedings) and only consider unresolved flux concentrations, i.e. magnetic elements with diameters ≤ 0.3 arcsec. We shall as well not attempt to account for all observational and theoretical results concerning a specific problem but rather try to select the most important ones (admittedly introducing a fair amount of personal bias). More comprehensive reviews of observations have been given by Stenflo (1984) and Harvey (1986) while the theoretical results have been reviewed by Schüssler (1986).

2. BIRTH OF MAGNETIC ELEMENTS

The "convective collapse" mechanism (Parker, 1978; Spruit and Zweibel, 1979) for a flux concentration in a strongly superadiabatic environment has been recognized in 3D simulations (Nordlund, 1983; 1986) and investigated in the nonlinear 1D case by Hasan (1984, 1985) and Venkatakrishnan (1983, 1985). These calculations essentially confirmed the results of linear stability analysis carried out earlier: Perturbations associated with a downflow along the fieldlines lead to a concentration of an initially weak magnetic field until a new equilibrium with $\beta \equiv 8\pi p/B^2 \simeq 1$ is reached (p: gas pressure, B: magnetic flux density in the flux tube). The time scale for this process is 5...15 minutes. The final state, however, is not static: The radiative heat exchange with the surrounding gas leads to overstability and, in the nonlinear case, stationary oscillations with a typical period of 1000 sec are found.

What can be said from the observational side about the formation of magnetic elements? First of all, there is strong evidence that more than 90% of the magnetic flux observed in the photosphere at any given time is in the form of concentrated fields (Howard and Stenflo, 1972). If a lifetime of the magnetic structure could be determined, this would give an indication for the formation and destruction times that could be compared with the 5...15 minutes timescale for the convective collapse. The lifetime of ≃ 20 min for the intensity enhancement in bright points (Muller, 1983), taken as lifetime of the magnetic structure, would require a formation time ≤ 2 min, not compatible with convective collapse. Although there are some indications (Brants, 1985a,b; Wiehr, 1985), the collapse process itself has not been observed without doubt. Besides the big observational difficulties (high spatial and temporal resolution required), there are ambiguities in the interpretation: The downflow phase of an overstable oscillation in an already formed flux concentration leads to apparently the same signature as the convective collapse, i.e. a downflow, associated with cooling and magnetic field enhancement. Only if the initial field is weak enough, the convective collapse could possibly be distinguished from an oscillation of already established concentrations.

We can conclude that, as far as the birth of flux concentration is concerned, theory is ahead of observations: A process has been investigated in great detail, specific predictions have been made which at present have neither been confirmed nor refuted by observations.

3. LIFE OF MAGNETIC ELEMENTS

After the formation process, the magnetic element stays in the concentrated, strong-field state for an unknown period of time. Most of the existing observational results are concerned with this phase which may be one of static, stationary (steady flows) or oscillatory equilibrium.

The simultaneous measurement of the Stokes I- and V-profiles for many spectral lines with the Fourier transform spectrometer (Stenflo et al., 1984) marks a turning point in our understanding of this phase of the evolution of magnetic elements. We shall see, however, that these observations have to be complimented by measurements with the highest possible spatial and temporal resolution in order to derive definite conclusions.

3.1 Magnetic field

As already mentioned, more than 90% of the magnetic flux in the solar photosphere is in the form of strong fields. The flux density can be derived using the line ratio technique (Stenflo, 1973) which gives values of 1200...1700 Gs, depending on the atmospheric structure within the flux tube (Solanki and Stenflo, 1984). These values are consistent with the convective collapse mechanism (Spruit, 1979). The magnetic field signal (e.g. line ratio), on the other hand, is nearly identical for regions with different total magnetic flux (see Fig. 2 in Stenflo and Harvey, 1985) which indicates a <u>nearly unique flux tube structure</u> in regions of different filling factor. An exception are flux tubes in strong plages (large filling factor) which have a lower temperature and a higher field strength (by ≃ 20%) than network tubes outside active regions (Solanki and Stenflo, 1984). We will discuss the difference in temperature structure in Section 3.2 below. The difference in magnetic field signal is not unrelated to this because a variation of temperature introduces a change in the optical depth scale. Consequently, we cannot distinguish at present between the following effects on the magnetic field signal:

- "real" change of B (at constant geometrical depth z),
- change of optical depth scale (due to the temperature sensitivity of the opacity, correlation effects of overstable oscillations),
- 2D effects: A variation of the intensity distribution I(x,y) over horizontal planes in the tube (brighter regions contribute more to the magnetic field signal).

As will be discussed further in the next section, all these variations could be introduced by

- a variation of the typical flux tube size,
- a change of the properties of the non-magnetic gas surrounding the tubes (or both).

Observations of the center-to-limb variation of Stokes I, V and Q have shown that the field strength decreases with height and that the tube flares out (Stenflo et al., 1987a). This is predicted, of course, by any model calculation which takes the stratification of the external medium into account.

3.2 Thermodynamics

3.2.1 Temperature profile $T(\tau)$

A variety of semi-empirical model atmospheres for faculae, plages, network etc. have been presented in the past. We take those of Solanki (1986) as a reference since they seem to have to broadest foundation: Hundreds of spectral lines have been used and line profiles <u>from the interior of the flux tubes</u> derived by integration of the Stokes V-profiles. Fig. 1 gives the temperature as a function of continuum optical depth derived by Solanki (1986) for network and plage tubes:

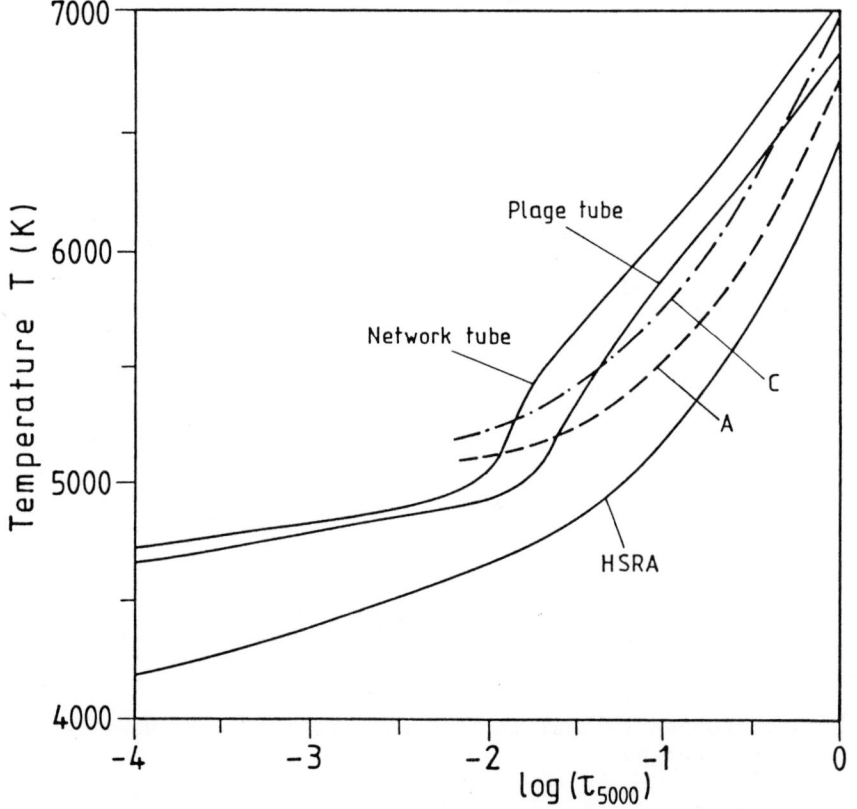

Fig. 1.: Temperature as a function of continuum optical depth for the semi-empirical models of Solanki (1986) and the tube models A and C of Deinzer et al. (1984b). The HSRA model is shown for comparison.

Solanki's models are compared with results from the MHD model calculations of Deinzer et al. (1984a,b): Curves A and C give the temperature profiles along the axis of two of their flux tube models having different degree of evacuation of the tube interior. Although the temperature on the axis underestimates the temperature averaged over the cross section of constant optical depth (the bright ring effect of a tube flaring out with height), the models seem to be too cool by a few hundred degrees. Several possibilities to remove this discrepancy can be considered:

- stronger evacuation of the tube,
- incorporation of an additional heating process (flows, waves),
- improved treatment of radiative transfer.

While the last point is obvious, the amount of evacuation is limited: An increasing "Wilson depression" means that we are looking deeper and deeper into the throat of the tube where the field strength increases rapidly. Since the observations seem to indicate an upper bound of about 2000 Gs, this limits the amount of evacuation and the value of the Wilson depression to ≤ 200 km below $\tau = 1$ of the non-magnetic atmosphere. Heating by flows has been considered in some detail (Unno and Ribes, 1979; Hasan and Schüssler, 1985), but the stationary downflows required run into conflict with the observed V-profiles (see Section 3.3 below). Wave heating in the photospheric layers of flux tubes has not been investigated in detail up to the present time (see Ulmschneider and Muchmore, 1986).

We should not forget to mention the results from differential photometry of faculae (Foukal and Duvall, 1985). They indicate that the temperature <u>gradient</u> $dT/d\tau$ around $\tau = 1$ is reduced in comparison to the normal photosphere (see also Elste, 1985). This is in agreement with flux tube models which are thermally influenced by lateral radiative exchange between tube and surroundings around $\tau = 1$.

Unfortunately, other authors (e.g. Spruit, 1976; Unno and Ribes, 1979; Osherovich et al., 1983) have not presented their temperature profiles as a function of optical depth. Thus they could not be included in Fig. 1. The temperature given by Ferrari et al. (1985) for optically thin flux tubes could not be included either because it is out of the scale: For $-1 \leq \log \tau \leq 0$ they have $7000 \text{ K} \leq T \leq 9000 \text{ K}$, i.e. these tubes are much too hot. Since the assumption of optically thin tubes effectively leads to $T_i(z) = T_e(z)$, i.e. internal temperature T_i equal to external temperature T_e <u>at the same geometrical level</u>, the internal density must be drastically reduced with respect to the external gas in order to achieve pressure equilibrium. Consequently, the Wilson depression is large and the tube is much hotter than the surroundings at equal optical depth. This is in strong disagreement with Solanki's and all other two-component semi-empirical models. Consequently, flux concentrations in the solar photosphere are not optically thin in the sense of Ferrari et al. (1985): They must be cooler than the surroundings at the same geometrical height. This is only possible if

- convection is suppressed in the tube to a sufficient degree and if
- the tube is wide enough that the lateral optical thickness is larger than unity in the region around $\tau \simeq 1$ vertically within the tube.

The latter condition is realized if the tube diameter is larger than 100 km at the relevant levels around $\tau \simeq 1$. Consequently, the photospheric flux cannot reside in magnetic concentrations which are much narrower than 100 km. This limit is considerably larger than the thermal boundary layer width of a few km which is the minimum size of the flux tubes formed by convective collapse (Schüssler, 1986).

3.2.2 Continuum intensity

It is well known that, at low spatial resolution, network faculae and plages exhibit negligible contrast in the continuum near the center of the disk. Foukal and Fowler (1984) showed that the mean continuum contrast in the network is about 0.1%. Furthermore, Hirayama et al. (1985) determined the energy flux of photospheric faculae by wideband photometry and integration over the angle of inclination. They found a flux increase of less than 0.5%, equivalent to a change in effective temperature of ≤ 7 K with respect to the quiet photosphere. Hence, there is no significant excess heat flux in facular, plage or network regions.

On the other hand, the bright points found in high-resolution observations as the basic elements of network and faculae near the center of the solar disk (Dunn and Zirker, 1973; Mehltretter, 1974; Muller, 1977) show a high continuum intensity contrast with respect to the normal photosphere: Values between 1.3...1.5 (Muller and Keil, 1983) and 2.0 (Koutchmy, 1978, for one single point) have been reported.

Consider a simple two-component model: A bright-point (magnetic) component with continuum intensity I_{cm} filling an area fraction α and an "environment component" with continuum intensity I_{ce} and area fraction $1-\alpha$. All intensities are normalized to the quiet mean photospheric intensity, I_{co}. The mean continuum intensity $\langle I_c \rangle$ measured with low spatial resolution is then given by

$$\langle I_c \rangle = \alpha \, I_{cm} + (1-\alpha) \, I_{ce} \tag{1}$$

Taking very moderate values of $\alpha = 0.02$, $I_{cm} = 1.3$ and assuming $I_{ce} = 1$, i.e. undisturbed environment, we find $\langle I_c \rangle = 1.006$, already significantly larger than the measurements of Foukal and Fowler (1984). To be consistent with these, we have to take $I_{ce} < 1$, i.e. the environment of bright points has to be cooler than the normal photosphere, a conclusion also drawn by Hirayama et al. (1985). Consequently, the excess flux in bright points is nearly exactly compensated by a flux deficit in the environment. This is a strong support for the view that bright points are magnetic flux tubes heated by lateral radiative influx of heat (Zwaan, 1967): The low excess flux indicates that the heat flux is mainly redistributed and channelled locally by the magnetic field

concentration and only a small fraction of the heat flux disturbance
propagates into the deep convection zone (cf. Spruit, 1977). The flux
tube models of Deinzer et al. (1984b) show just this effect: The bright
flux tube is surrounded by a dark ring of material cooler than average
which nearly compensates the excess flux of the tube.

As far as I_{cm}, the continuum intensity of the magnetic component is
concerned, we can make inferences from observations. First of all, the
semi-empirical models of Solanki (1986) give I_{cm} = 1.4 for network flux
tubes and I_{cm} = 1.3 for plage tubes. Furthermore, from the amount of
line weakening of the FeI λ5250.2 spectrum line a value of $I_{cm} \geqslant 1.6$
can be deduced (Schüssler and Solanki, in preparation). Optically thick
flux tube models with about 200 km Wilson depression (e.g. Model 8 of
Spruit, 1976; Model C of Deinzer et al., 1984b; see also Fig. 1 above)
reproduce these values, while the optically thin models of Ferrari et
al. (1985) predict $I_{cm} \simeq 3. ...5.$, much higher than any value derived
from observations so far. Steady flow models with a downflow rapidly
<u>increasing</u> with depth (Ribes et al., 1985) yield Icm ≃ 1.4 .. 1.9.
Anyway, the high continuum intensities predicted by models and derived
by observations strongly support the identification of magnetic flux
concentrations with photospheric bright points.

3.2.3 The difference between plage and network

Solanki and Stenflo (1984, 1985) showed that plage flux
tubes are cooler than network tubes (cf. Fig. 1). At the same time the
magnetic field in plage tubes appears to be somewhat stronger (if
"measured" with the same spectral lines). There are at least two
possibilities to interpret this result:

a) plage tubes are larger than network tubes,
b) the environment is cooler in plages than in the network.

Of course, both interpretations do not contradict each other and a
combination is possible.

Larger tubes are less efficiently heated by lateral influx of
radiation, they will be cooler and have a larger Wilson depression due
to the temperature sensitivitiy of opacitiy: The field is measured
deeper in the tube and therefore larger (Knölker et al., 1985).
Interpretation b) is related to the increased area fraction α ("filling
factor") of magnetic structures in plages. A value of α = 0.25, for
example, leads to a mean distance between magnetic structures of the
order of their own diameter: The "cool rings" around flux tubes will
begin to overlap and the mean environment will be cooler than that of
network tubes. Lateral heat influx will be less efficient because more
flux tubes have to share the heat flux reservoir and the individual
tube will be cooler than a network tube which can be viewed as an
isolated structure.

An indication for a variation of the non-magnetic component with
increasing α is the change of the spectrum line bisectors (Brandt and
Solanki, these proceedings) and the decrease of the convective

blueshift (Solanki, 1986).

However, other effects may be important:
Overstable oscillation or flows (see Section 3.3) could influence the thermal structure for plage and network in a different way and also the horizontal distribution of intensity, magnetic field, flow velocity, opacity and their correlations effect the measured quantities. An iteration between observation and 2D/3D MHD models including radiative transfer is a possibility to take these effects into account. They may be crucial for a reliable interpretation of the measurements.

3.2.4 Spurious magnetic flux variation

In view of some reports of mysterious magnetic flux losses in decaying active regions (e.g. Wallenhorst and Howard, 1982; Wallenhorst and Topka, 1982) as well as dramatic short-term variations of magnetic flux in high-resolution observations (e.g. Topka et al., 1986) it seems appropriate to draw the attention of the reader to the fact that all determinations of magnetic fields (except the direct measurement of a fully resolved Zeeman pattern) depend sensitively on the thermal structure of the magnetic component of the atmosphere. In particular, if the magnetic structure is spatially unresolved, the signal from a Babcock-type magnetograph depends strongly on

- the continuum intensity I_{cm}
- the profile φ_m of the spectrum line considered

within the magnetic structure. These quantities are determined by the temperature, density, etc. (depth dependence and horizontal distribution) within the flux concentration. They can vary in time due to fragmentation processes during the decay of active regions (spots → pores → magnetic elements) as well as through dynamical processes within flux tubes (e.g. oscillations). The spurious flux loss due to fragmention of a spot into magnetic elements (without real flux change) can be up to factor 10 (Grossmann-Doerth et al., 1987). A reliable calibration of any magnetographic device in terms of magnetic flux can only be performed if the thermal structure (and its time dependence) of the magnetic regions under consideration is known. Until that has been realized, any quantitative conclusions from magnetographic results should be viewed with suspicion.

3.3 Flows, Internal Dynamics

The relation to and interaction with external velocity fields (granulation, supergranulation, vortex flows) will be discussed in Section 3.5. In this chapter we shall address the problem of flows within the magnetic structure, i.e. quasisteady flows along the field, oscillations and disordered flows ("turbulence").

3.3.1 Diagnostics

The main problem is to distinguish between the magnetic structure and its surroundings. This can be done directly if the magnetic structure is spatially resolved (provided a proper consideration of scattered light is performed). If the structure is unresolved, one can use the Stokes V-profile which originates exclusively in the magnetic regions. In principle, ordered flows could then be detected by measuring the wavelength shift of the V-profile zero-crossing, while disordered motions are reflected in the width of the components of the profile. Potentially there is an additional (and unexpected!) diagnostics of internal motions: The asymmetry of the observed V-profiles, i.e. the difference in area and amplitude between the red and the blue wings of the profile. Such an asymmetry has been found for many spectrum lines in observations of network, facular and plage areas (Stenflo et al., 1984; Scholiers and Wiehr, 1985). Auer and Heasly (1978) have shown that in LTE a combination of velocity and magnetic field gradients along the line of sight is required to produce asymmetries. With data for many spectral lines available, the asymmetry could indeed be used as a diagnostic tool for velocities inside flux tubes. However, it has been proposed by Kemp et al. (1984) and Landi Degl'Innocenti (1985) that also a NLTE effect (atomic polarization in a magnetic field) might contribute to the observed asymmetries.

On the other hand, there are additional hints that velocities inside flux tubes are present and possibly involved in the generation of the asymmetries. From 1D/LTE model calculations Solanki (1986) found that he could only reproduce the spectrum line profiles formed inside the flux tubes (wavelength integrated V-profiles) if a significant turbulent velocity broadening is included (3 ... 3.5 km/s). The value for the macroturbulent velocity required for any given line and the amplitude asymmetry of the V-profile for the same line are found to be correlated. The observed decrease of the asymmetry towards the solar limb, if interpreted in terms of velocities, indicates that the relevant flows are mainly vertical, along the magnetic field lines (Stenflo et al., 1987a).

3.3.2 Steady flows

Flows of a quasisteady nature along the fieldlines of a magnetic flux concentration can be detected by the shift in wavelength of the zero-crossing of the V-profile. Early attempts with low spectral resolution (e.g. Giovanelli and Slaughter, 1978) have been compromised, however, by the asymmetries which can lead to spurious shifts (Solanki and Stenflo, 1986). From the FTS data Stenflo and Harvey (1985) and Solanki (1986) did not find steady flows in the photospheric layers of flux tubes larger than an upper limit of 250 m/s for the absolute flow velocity. Also in the high resolution spectrum of Koutchmy and Stellmacher (1978) no downflow in the magnetic structure was detected.

It must be borne in mind, however, that the large aperture and the long integration time used for the FTS measurements imply a severe averaging. It has been argued, therefore, that oscillations in the tubes could be averaged out in time and space (many tubes in the aperture oscillating with random phases). However, overstable oscillations driven by lateral heat exchange should show _intensity-velocity-correlations_ (in analogy to the convective blueshift due to granulation) which themselves would lead to a wavelength shift. This is not observed.

Pahlke and Solanki (1986) calculated V-profiles for a variety of models
for the flux tube atmosphere and parametrized the magnetic field and
velocity gradients. They found that the observed asymmetries for the
FeI λ5250.2 and λ6302.5 lines can be modelled by a suitable choice of
the parameters, but that it is impossible to reproduce simultaneously
the vanishing wavelength shift of the zero-crossing.

The correct sense of the asymmetry (blue wing with larger area and amplitude) results for combinations of velocity and magnetic fields with

$$\frac{d|B|}{d\tau} \cdot \frac{dv}{d\tau} < 0 \qquad (2)$$

along the line of sight (v: velocity, B: magnetic flux density).

If the magnetic field decreases with height, we see from Eq. (2) that
either a

- downflow decreasing with depth or an
- upflow increasing with depth

is required. This can also be seen in the results of Ribes et al.
(1985) who calculate spectrum line and V-profiles for a variety of
steady flow models. It turns out that those models which yield the
correct sign of the asymmetry predict large zero-crossing shifts
uncompatible with the observations: The FTS results exclude all models
containing significant steady flows.

3.3.3 Overstable oscillations

In a magnetized plasma with $\mathcal{K} > \eta$ ($\mathcal{K}$: thermal diffusivity, η: magnetic diffusivity), an overstable oscillation with motion
preferentially along magnetic fieldlines is a possible mode of magnetoconvection. In the case of a thin flux tube, a convectively stable
(strong field) tube exhibits buoyancy oscillations (e.g. Hasan, 1984)
which can turn overstable (growing) if heat exchange with the
surrounding (superadiabatically stratified) gas is included (Hasan,
1985; Venkatakrishnan, 1985). Hasan (1986) has carried out a linear
stability analysis for a thin flux tube with constant $\beta = 8\pi p/B^2$
describing the heat exchange by way of Newton's law of cooling and the
radiative relaxation time given by Spiegel (1957). He found that all
tubes are unstable:

- for $\beta > \beta_c$: monotonic instability (convective collapse)

- for $\beta < \beta_c$: overstable oscillations with periods of 500 .. 1500 sec

β_c depends on the radius of the flux tube and varies between 4 and 8
for flux tube radii between 200 and 50 km. One could imagine, therefore, that a weak field tube first collapses monotonically until $\beta < \beta_c$
and then reaches a stationary state with oscillations of an amplitude
fixed by nonlinear effects.

Hasan (1986) also shows the phase relations between the perturbations of temperature and velocity which indicate significant correlations depending on the stratification (sub- or superadiabatic) of the surrounding gas. Such correlations are exhibited as well by the nonlinear calculations (Hasan, 1985; Venkatakrishnan, 1985): Upflows are hot, downflows are cool. Unfortunately, no calculations of the emergent intensity and its correlation with velocity and magnetic field have been carried out. It would be most surprising, however, if such correlations would not appear. The consequence of a correlation of intensity and velocity would be a systematic blue- or redshift of the spectral lines: Values larger than 250 m s^{-1} are excluded by observations (see above). An intensity-magnetic field correlation would introduce a bias in all magnetic field measurements which would be even more disturbing if it depends on other properties of the structure (size, environment, ..) which may change in time or during the evolution of an active region.

At present, no clear observational indication for the existence of overstable oscillations has been found. The results rather point to the contrary. However, a more sophisticated treatment of radiative energy transport is necessary before quantitative predictions can be derived from the MHD models. For example, the inclusion of radiative transport in the vertical direction seems to reduce the periods and amplitudes of the oscillations (Venkatakrishnan, 1985).

Whether overstable oscillations could be the origin of the observed V-profile asymmetries remains to be clarified in the future. The vanishing of the asymmetry in the infrared (deep layers of the photosphere) shown by Stenflo et al. (1987b) might be crucial for this interpretation.

3.4 Geometry

Obviously, magnetic flux concentrations have a lateral extension and they probably flare our with height. It follows that thin flux tube models, two-component models and 1.5D radiative transfer can only be applied for observations of structures at the center of the disk. The large diagnostic potential provided by observations of the center-to-limb variation of the Stokes parameters (e.g. Stellmacher and Wiehr, 1973; Stenflo et al., 1987a) can only be exploited if 2D or 3D MHD model calculations including radiative transfer are used. Otherwise it would be a hopeless task trying to disentangle the combined effects of geometry and height variation.

3.4.1 Intensity contrasts

The largest amount of observational material has been gathered for the facular intensity contrasts and their center-to-limb variation (CLV) in white light or selected spectrum bands. We have seen in Section 3.2.2 that it is very important to distinguish between observations resolving the individual structures and those averaging structure and environment which is not necessarily identical to the "quiet Sun". At the center of the disk, high resolution observations resolve faculae into facular points with continuum intensities of

1.3...1.5 (Muller and Keil, 1983). On the other hand, the (low resolution) mean continuum intensity is ≃1. due to the cool environment of bright points. Most low or medium resolution measurements agree in finding that the continuum intensity contrast $\langle \Delta I \rangle / I_{co} = (\langle I_c \rangle - I_{co})/I_{co}$ nearly vanishes at disk center ($\mu = \cos\theta = 1$) and increases toward the limb up to a value ≃0.2 at $\mu = 0.2$ (cf. Muller, 1985). Whether it continues to grow by a further approach to the limb is presently under debate (see Chapman and Klabunde, 1982; Libbrecht and Kuhn, 1984). "Facular granules", the structure elements of limb faculae ($0.6 \geqslant \mu \geqslant 0.1$), have a typical size of 0.6 arcsec (Muller, 1977).

Facular points or bright network points with a size of ≃ 0.2 arcsec observed at disk centre behave quite differently. Their visibility dereases rapidly for $\mu < 1$ (Muller and Roudier, 1984), much faster than would be expected due to foreshortening effects. Does this mean that facular points (and the associated magnetic elements) are not responsible for the limb brightening of faculae? At first sight, model calculations seem to support this interpretation: <u>Small</u> flux tubes appear bright at disk center, but their CLV variation of intensity contrast is not in agreement with the measurements (Spruit, 1976; Chapman and Gingell, 1984; Deinzer et al., 1984b). <u>Large</u> tubes (d > 600 km) appear dark at disk center, but show a positive intensity contrast near the limb in qualitative agreement with the observations (Spruit, 1976; Knölker et al., 1985). The origin of this limb brightening is the "hot wall effect": viewed near the limb, the hot surrounding material becomes visible through the optically thin flux tube interior. Fig. 2 shows schematically this situation:

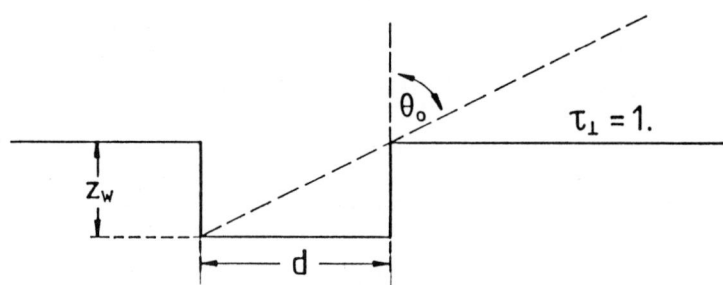

Fig. 2: Schematic sketch of the "hot wall effect" (cf. Spruit, 1976)

The maximum contrast will arise for the angle of inclination θ_o: For $\theta > \theta_o$ more and more of the hot wall is obscured by the near side photosphere, for $\theta < \theta_o$ a part of the dark bottom is visible and the wall is viewed under large inclination. It follows that the angle of maximum contrast is simply related to the ratio $\gamma = d/z_w$ of diameter to Wilson depression of the tube:

$$\mu_o = \cos\theta_o = (1 + \gamma^2)^{-1/2} \qquad (3)$$

The model calculations reproduce this simple relation quite well in spite of the much more involved thermodynamical description. We see from Eq. (3) that in order to reproduce $\mu_o < 0.2$ as indicated by observation we need $\gamma > 5$, i.e. a diameter >500 km for $z_w = 100$ km or >1000 km for $z_w = 200$ km.

However, there are problems with this interpretation: First, it is nearly impossible to reproduce the high contrast values measured during the Stratoscope flight (Rogerson, 1961). Second, a mixture of structures from $d < 150$ km up to $d \simeq 1000$ km is very difficult to reconcile with the observed small variation of the line ratios indicating a very similar structure for all flux concentrations (Stenflo and Harvey, 1985). Do we observe pore-like structures (> 1 arsec, dark) in evolved faculae?

Alternatively, we could attribute both kinds of observations, disk center bright points and limb faculae to the same structure, i.e. small magnetic elements with $d \simeq z_w \simeq 150$ km, $\gamma \simeq 1$. A hot bottom of the tube together with the hot wall effect leads to a contrast curve which peaks at $\mu =1$ (disk center) and steeply declines for decreasing μ: The bottom (as main contributor to excess brightness) is obscured for $\mu < 0.7$ and the wall reaches its maximum contrast just there. A similar obscuration effect has been discussed by van Ballegooijen (1985b) in the context of observations of polarized light. In this way we can understand the "visibility function" derived by Muller and Roudier (1984; see their Fig. 2a): The probability of detection decreases for diminishing contrast; less and less structures can be observed because only the high contrast tail of the distribution of bright points is visible by approaching the limb.

But how can we understand then the limb brightening of faculae and network? The coarse structure of limb faculae ("facular granules"; Müller, 1977) and their slowly decreasing width as a function of μ indicate an overlapping of single structures, the "hot cloud model" (cf. Rogerson, 1961). Fig. 3 shows schematically this effect:

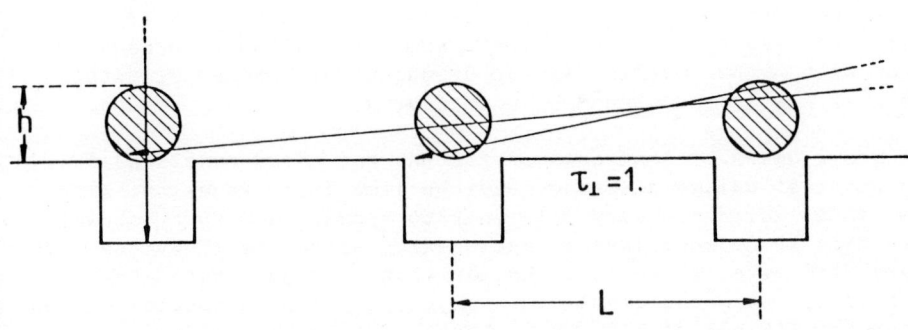

Fig. 3: Schematic sketch of the "overlapping hot cloud effect".

Assume that a cloud of hot gas is situated above the Wilson depression of each flux concentration. The individual cloud is optically thin for continuum radiation and therefore invisible for vertical incidence of the line of sight. However, if several clouds overlap along the line of sight they can become optically thick and a limb brightening ensues. It is clear from Fig. 3 that the condition for overlapping is given by

$$\frac{h}{L} \geqslant \frac{\mu}{(1-\mu^2)^{1/2}} \approx \mu \quad \text{for } \mu \ll 1 \qquad (4)$$

assuming $h \ll L$, i.e. the vertical extension of the hot cloud to be small compared to the mean distance L of flux tubes. L is related to α, the magnetic filling factor. For a rectangular grid of flux tubes, the relation between L and α is given by:

$$L = \left(\frac{\pi}{4\alpha}\right)^{1/2} \cdot d \qquad (5)$$

Consequently, we can relate μ_0, the value of μ for which the equality sign in Eq. (4) is valid, to d. If use $d = h$, the result is:

$$\mu_0 = \left(\frac{4\alpha}{\pi + 4\alpha}\right)^{1/2} \qquad (6)$$

The following table gives some values:

$\alpha [\%]$	1	2	5	10	15	25
μ_0	.11	.16	.24	.34	.40	.49

We see that an observational test is possible: The CLV due to the hot wall effect does not depend on α provided that the properties of the magnetic structures do not change drastically. The hot cloud effect, on the other hand, leads to a CLV of continuum contrast which depends on α: Regions with higher filling factor brighten for smaller distances from the disk center than those with smaller α.

Is there other observational support for the hot cloud model? Besides the high contrast values observed near the limb (e.g. Rogerson, 1961; Hirayama, 1978) which are very difficult to reconcile with a hot wall (remember that it is more like a "tepid wall" since the flux tubes are surrounded with material which is cooler than average) there are reports of a direct measurement of an <u>elevation</u> of limb faculae during solar eclipses (Akinov et al., 1982; 1984). The results of Chapman and Klabunde (1982) of a facular contrast increasing out to the extreme limb (leading to almost constant facular intensity) are strongly in favour of overlapping hot clouds. However, Libbrecht and Kuhn (1984) could not confirm this result. Further measurements are necessary to clarify the point.

3.4.2 Spectrum lines

The same problems and ambiguities as discussed for the continuum come into play if we consider the CLV of spectrum lines (e.g. Stellmacher and Wiehr, 1973). It has been very clearly demonstrated by van Ballegooijen (1985a) that the observed line profiles from spatially unresolved structures are the sum of very different profiles which originate from different parts of the structure. These are weighted with the continuum intensity which varies as well. Thus there is no hope of ever deriving reliable information from the CLV variation of spectrum lines (and the Stokes vector) if not detailed 2D/3D MHD model calculations are utilized as a diagnostic tool.

3.5 Relation to the environment

Magnetic flux concentrations are not isolated; they interact hydro-and thermodynamically with the surrounding gas. The relation of magnetic elements to photospheric velocity fields is well known: The magnetic field is organized preferentially in the super-granular network pattern and, on the granular scale, contained in the intergranular lanes (e.g. Muller, 1983; Title et al., 1987). The reaction of thin magnetic flux tubes on supergranular and granular flows has been simulated by Meyer et al. (1979) and Schmidt et al. (1985).

The general association of magnetic structures with "downdrafts" had been recognized for a long time, but only since the FTS/polarimeter was in operation it became clear that the downdrafts were outside the flux concentrations. Besides the association with downdrafts of convective flow patterns, the calculations of Deinzer et al. (1984b) predict the existence of an additional thermal circulation due to the cooling of the flux tube surroundings by lateral radiative heat transport. In the vicinity of the tube, this circulation leads to downflow. Possible consequences of such an additional downflow are:

- enhancement of the intergranular downflow adjacent to flux concentrations,
- deformation of the granular pattern (confirmed by Roudier, 1986),
- stabilization of the surrounding flow pattern (see Title, these proceedings),
- stabilization of the tube itself by establishment of a vortex flow (Schüssler, 1984).

With increasing filling factor, the signatures of granular convection are more and more disturbed: The convective blueshift decreases (Solanki, 1986) and the spectrum line asymmetries ("C-shape") decrease as well (e.g. Brandt and Solanki, these proceedings). In the wings of the lines, however, the redshift increases and this might indicate the predicted enhanced downflow around the flux concentrations (Immerschitt and Schröter, these proceedings). The significant changes of the line asymmetries with respect to the undisturbed photosphere are easily misinterpreted in terms of "downdrafts" if the full line profile is not available at sufficient spectral resolution, as pointed out by Miller et al. (1984).

We conclude that the models predict a strong backreaction of the magnetic structures on the surrounding medium. This seems to be confirmed by a variety of observations. The full diagnostic content of the measurements will be revealed with the aid of more sophisticated model calculations.

4. DEATH OF MAGNETIC FLUX CONCENTRATIONS

The lifetime of a magnetic flux concentration is presently unknown. Bright points typically exist for about 20 min (Muller, 1983) but it is not clear whether their decay signals the destruction of the magnetic structure as well. This seems rather implausible, however: If at least 90% of the magnetic flux at any given time is in strong field form and if the lifetime of this phase would be indeed only 20 min, flux dispersion and reformation by convective collapse together should take only 2 min for each flux concentration. However, the convective collapse alone takes 5-15 min to produce a flux concentration. Hence, the decay of bright points cannot be taken as evidence for the dissolution of the related flux concentrations. Bright points rather seem to be occasional brightenings (overstable oscillations?) of magnetic structures existing much longer. This is not implausible because the stabilizing whirl flow around a flux concentration is partly due to the existence of the flux tube itself: It creates a thermal circulation which will last as long as the flux tube exists.

Consequently, a flux concentration, once formed, is not necessarily subject to the timescale of granular flows and may exist much longer, but obviously it cannot live forever: Magnetic flux emerges during the solar cycle and apparently does not accumulate indefinitely. From the above considerations and from the Big Bear observations with the video-magnetograph (e.g. Martin et al., 1985) it seems plausible that the entire concentrated structures are involved in the reconnection /retraction/expulsion processes that clean the solar photosphere from magnetic flux (see Priest, these proceedings, for details of these mechanisms). It would be very interesting to see whether an interaction of the vortices surrounding the flux concentrations could be observed during their encounter. The relative motion of two flux tubes would depend on the signs of the vorticity of their surrounding whirls.

Furthermore, it should be investigated by observation whether magnetic flux concentrations stay in strong field form or go through a formation - dispersion - reformation cycle. Since formation and dispersion presumably take about half an hour, the lifetime of the strong field phase should at least be of the order of 5 hours.

5. CONCLUSION

During the last years the improvements of instrumentation, observational diagnostics and theoretical modelling have led to substantial progress in our understanding of the physics of small magnetic flux concentrations. The field is in rapid development on both the observational and the theoretical side, but further progress demands considerable efforts. The spatial and temporal resolution of our instruments has to be pushed to its limits: We need as much information as possible, i.e. the full Stokes vector resolved in space and time. This means new concepts for post-focus instrumentation and strong support for active/adaptive optics and image reconstruction methods (see von der Lühe, these proceedings) in connection with ground-based telescopes. This means the urgent need of a large solar telescope in space.

On the other hand, all investments of intellect, spirit and money will only pay off if theory is supported in a similar way: We need a new generation of 2D/3D MHD model calculations including sophisticated radiative transfer of polarized light. Without such tools it is impossible to disentangle the information about the structure and dynamics of photospheric flux concentrations contained in the profiles of the Stokes parameters. Simulations are urgently needed for the interpretation of ground-based observations and they have to be available when the data from space telescopes such as HRSO will arrive.

It is not at all an easy task to manufacture these tools, i.e. develop 2D/3D codes up to a stage where they can be used as standard diagnostic instruments. It will take intelligence, (wo)manpower, money and time; good access to supercomputers is essential and the groups must exceed some "critical size": MHD and LTE/NLTE radiative transfer have to be combined. I would consider it an act of remarkable foresightedness, if authorities would decide to spend a small percentage of the money allocated to large instrumental projects for the development of the tools necessary to make full use of the data gathered.

REFERENCES

Akinov, L.A., Belkina, I.L., Dyatel, N.P.: 1982, Sov. Astron. 26, 334
Akinov, L.A., Belkina, I.L., Dyatel, N.P.: 1984, Sov. Astron. 28, 310
Auer, H.L., Heasley, J.M.: 1978, Astron. Astrophys. 64, 67
van Ballegooijen, A.A:: 1985a, in M.J. Hagyard: "Measurements of Solar Vector Magnetic Fields", NASA CP-2374, p. 322
van Ballegooijen, A.A.: 1985b, in H.U. Schmidt (ed.), "Theoretical Problems in High Resolution Solar Physics", Proc. of the MPA/LPARL Workshop, Max-Planck-Institut für Astrophysik, Munich, p. 167
Beckers, J.M., Schröter, E:H.: 1968, Solar Phys. 4, 142
Brants, J.J.: 1985a, Solar Phys, 95, 15
Brants, J.J.: 1985b, Solar Phys. 98, 197
Chapman, G.A., Klabunde, D.P.: 1982, Astrophys. J. 261, 387

Chapman, G.A., Gingell, T.W.: 1984, Solar Phys. **91**, 243
Deinzer, W., Hensler, G., Schüssler, M., Weisshaar, E.: 1984a, Astron. Astrophys. **139**, 426
Deinzer, W., Hensler, G., Schüssler, M., Weisshaar, E.: 1984b, Astron. Astrophys. **139**, 435
Dunn, R.B., Zirker, J.B.: 1973, Solar Phys. **33**, 281
Elste, G.: 1985, in H.U. Schmidt (ed.), "Theoretical Problems in High Resolution Solar Physics", Proc. of the MPA/LPARL Workshop, Max-Planck-Institut für Astrophysik, Munich, p. 185
Ferrari, A., Massaglia, S., Kalkofen, W., Rosner, R., Bodo, G.: 1985, Astrophys. J. **298**, 181
Foukal, P., Duvall, T.: 1985, Astrophys. J. **296**, 739
Foukal, P., Fowler, L.: 1984, Astrophys. J. **281**, 442
Giovanelli, R.G., Slaughter, C.: 1978, Solar Phys. **57**, 255
Grossmann-Doerth, U., Pahlke, K.-P., Schüssler, M.: 1987, Astron. Astrophys., in press
Harvey, J.W: 1986, in W. Deinzer et al. (eds.): "Small Scale Magnetic Flux Concentrations in the Solar Photosphere", Vandenhoeck und Ruprecht, Göttingen, p. 25
Hasan, S.S.: 1984, Astrophys. J. **285**, 851
Hasan, S.S.: 1985, Astron. Astrophys. **143**, 39
Hasan, S.S.: 1986, Mon. Not. Roy. Astr. Soc. **219**, 357
Hasan, S.S., Schüssler, M.: 1985, Astron. Astrophys. **151**, 69
Hirayama, T.: 1978, Publ. Astron. Soc. Japan **30**, 337
Hirayama, T., Hamana, S., Mizugaki, K.: 1985, Solar Phys. **99**, 43
Howard, R., Stenflo, J.O.: 1972, Solar Phys. **22**, 402
Jones, H.P.: 1986, in W. Deinzer et al. (eds.): "Small Scale Magnetic Flux Concentrations in the Solar Photosphere", Vandenhoeck und Ruprecht, Göttingen, p. 127
Kemp, J.C., Macek, J.H., Nehring, F.W.: 1984, Astrophys. J. **278**, 863
Kneer, F.: 1986, in W. Deinzer et al. (eds.): "Small Scale Magnetic Flux Concentrations in the Solar Photosphere", Vandenhoeck und Ruprecht, Göttingen, p. 147
Knölker, M., Schüssler, M., Weisshaar, E.: 1985, in H.U. Schmidt (ed.), "Theoretical Problems in High Resolution Solar Physics", Proc. of the MPA/LPARL Workshop, Max-Planck-Institut für Astrophysik, Munich, p. 195
Koutchmy, S.: 1978, Astron. Astrophys. **61**, 397
Koutchmy, S., Stellmacher, G.: 1978, Astron. Astrophys. **67**, 93
Landi Degl'Innocenti, E.: 1983, Solar Phys. **85**, 3
Libbrecht, K.G., Kuhn, J.R.: 1984, Astrophys. J. **277**, 889
Martin, S.F., Livi, S.H.B., Wang, J.: 1985, in H.U. Schmidt (ed.), "Theoretical Problems in High Resolution Solar Physics", Proc. of the MPA/LPARL Workshop, Max-Planck-Institut für Astrophysik, Munich, p. 179
Mehltretter, J.P.: 1974, Solar Phys. **38**, 43
Meyer, F., Schmidt, H.U., Simon, G.W., Weiss, N.O.: 1979, Astron. Astrophys. **76**, 35
Miller, P., Foukal, P., Keil, S.: 1984, Solar Phys. **92**, 33
Muller, R.: 1975, Solar Phys. **45**, 105
Muller, R.: 1977, Solar Phys. **52**, 249
Muller, R.: 1983, Solar Phys. **85**, 113

Muller, R.: 1985, Solar Phys. 100, 237
Muller, R., Keil. S.L.: 1983, Solar Phys. 87, 243
Muller, R., Roudier, Th.: 1984, Solar Phys. 94, 33
Nordlund, Å: 1983, in J.O. Stenflo (ed.), "Solar and Stellar Magnetic Fields", IAU-Symposium 102, Reidel, Dordrecht, p. 79
Nordlund, Å.: 1985, in H.U. Schmidt (ed.), "Theoretical Problems in High Resolution Solar Physics", Proc. of the MPA/LPARL Workshop, Max-Planck-Institut für Astrophysik, Munich, p. 101
Nordlund, Å.: 1986, in W. Deinzer et al. (eds.): "Small Scale Magnetic Flux Concentrations in the Solar Photosphere", Vandenhoeck und Ruprecht, Göttingen, p. 83
Osherovich, V.A., Fla, T., Chapman, G.A.: 1983, Astrophys. J. 268, 412
Pahlke, K.D:, Solanki, S.K.: 1986, Mitt. Astron. Ges. 65, 162
Parker, E.N.: 1978, Astrophys. J. 221, 368
Ribes, E., Rees, D.E., Fang, Ch.: 1985, Astrophys. J. 296, 268
Rogerson, J.B.: 1961, Astrophys. J. 134, 331
Roudier, Th.: 1986, thesis, University of Toulouse
Schmidt, H.U., Simon, G.W., Weiss, N.O.: 1985, Astron. Astrophys. 148, 191
Schüssler, M.: 1984, Astron. Astrophys. 140, 453
Schüssler, M.: 1986, in Deinzer et al. (eds.): "Small Scale Magnetic Flux Concentrations in the Solar Photosphere", Vandenhoeck und Ruprecht, Göttingen, p. 103
Scholiers, W., Wiehr, E.: 1985, Solar Phys. 99, 349
Solanki, S.K.: 1986, Astron. Astrophys. 168, 311
Solanki, S.K., Stenflo, J.O.: 1984, Astron. Astrophys. 140, 185
Solanki, S.K., Stenflo, J.O.: 1985, Astron. Astrophys. 148, 123
Solanki, S.K., Stenflo, J.O.: 1986, Astron. Astrophys. 170, 120
Spiegel, E.A.: 1957, Astrophys. J. 126, 202
Spruit, H.C:: 1976, Solar Phys. 50, 269
Spruit, H.C.: 1977, Solar Phys. 55, 3
Spruit, H.C., Zweibel, E.G.: 1979, Solar Phys. 62, 15
Spruit, H.C.: 1979, Solar Phys. 61, 363
Spruit, H.C:: 1983, in J.O. Stenflo (ed.), "Solar and Stellar Magnetic Fields", IAU-Symposium 102, Reidel, Dordrecht, p. 41
Spruit, H.C., Roberts, B.: 1983, Nature 304, 401
Stellmacher, G., Wiehr, E.: 1973, Astron. Astrophys. 29, 13
Stenflo, J.O.: 1973, Solar Phys. 32, 41
Stenflo, J.O.: 1984, Adv. Space, Res. 4, 5
Stenflo, J.O., Harvey, J.W., Brault, J.W., Solanki, S.: 1984, Astron. Astrophys. 131, 333
Stenflo, J.O. Harvey, J.W.: 1985, Solar Phys. 95, 99
Stenflo, J.O., Solanki, S.K., Harvey, J.W.: 1987a,b , Astron. Astrophys., in press
Title, A.M., Tarbell, T.D., Topka, K.P.: 1987, preprint
Topka, K.P., Tarbell, T.D., Title, A.M.: 1986, Astrophys. J. 306, 304
Ulmschneider, P., Muchmore, D.: 1986, in W. Deinzer et al. (eds.): "Small Scale Magnetic Flux Concentrations in the Solar Photosphere", Vandenhoeck und Ruprecht, Göttingen, p. 191
Unno, W., Ribes, E.: 1979, Astron. Astrophys. 73, 314
Venkatakrishnan, P.: 1983, J. Astrophys. Astr. 4, 135
Venkatakrishnan, P.: 1985, J. Astrophys. Astr. 6, 21

Wallenhorst, S.G., Howard, R.: 1982, Solar Phys. 76, 203
Wallenhorst, S.G., Topka, K.P:: 1982, Solar Phys. 81, 33
Wiehr, E.: 1985, Astron. Astrophys. 149, 217
Zwaan, C.: 1967, Solar Phys. 1, 478

ADIABATIC LONGITUDINAL-TRANSVERSE MAGNETOHYDRODYNAMIC TUBE WAVES

Kurt Zähringer and Peter Ulmschneider
Institut für theoretische Astrophysik,
Im Neuenheimer Feld 561
D-6900 Heidelberg, Federal Republic of Germany

Abstract. We compute the propagation of adiabatic magnetohydrodynamic waves along thin flux tubes in the solar atmosphere. The time-dependent development and amplitude growth of the waves, excited at the foot of the photosphere as pure transverse waves, was followed into the low chromosphere. Strong mode-coupling to longitudinal waves was found. The swaying of the tube which increases with height resulted in a lifting of the entire tube mass which we attribute to centrifugal forces. This lifting resulted in adiabatic cooling of the tube.

1. Introduction

As discussed by Spruit (1982) thin magnetic flux tubes allow three types of tube wave modes: longitudinal, transverse and torsional waves. Despite the fact that these waves are probably all excited very efficiently it is interesting to investigate their non-linear coupling especially as this may have great importance for both chromospheric and coronal heating (c.f. Ulmschneider 1987). Hollweg et al. (1982) have computed the time-development of coupled longitudinal and torsional waves. They showed that from a purely torsional excitation longitudinal waves of considerable energy developed. In the present paper we want to report similar time-dependent work on coupled longitudinal and transverse waves.

2. Method

Following Spruit (1981) we assume a vertically oriented thin magnetic flux tube (c.f. Fig. 1) in which a mass element of width da at time t=0 can be uniquely identified by the Lagrange height, a, measured in the outward direction from the level where the continuum optical depth outside the tube is one. At a=0 the tube has a diameter of 100 km and a magnetic field strength of B_0=1500 G (c.f. Tab. 1). The tube is assumed

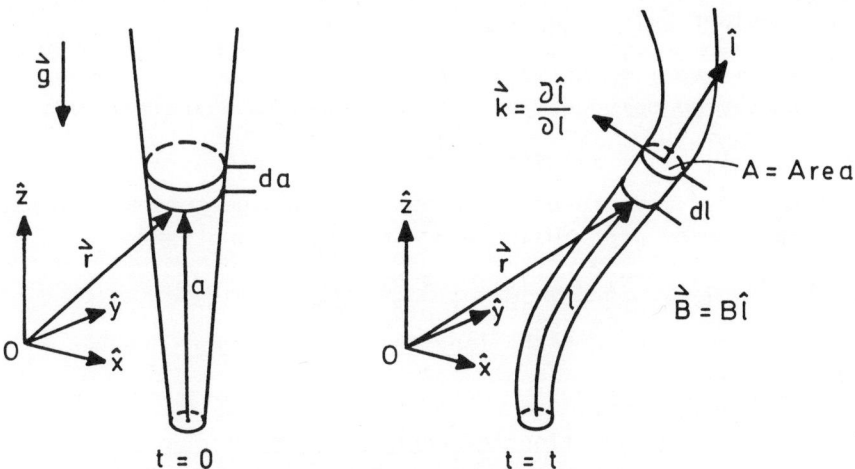

Fig. 1 Flux tube geometry

to spread with height almost exponentially in pressure balance in an external unmagnetized radiative equilibrium atmosphere similarly as discussed by Herbold et al. (1985). At some later time t (cf. Fig. 1) the tube is assumed to have kinks and the cross-sectional area A may be changed. The mass element has now the width dl and is displaced to a

a (km)	d_0 (km)	B_0 (G)	p_e $\left[\frac{dyn}{cm^2}\right]$	p_0 $\left[\frac{dyn}{cm^2}\right]$	T (K)	ρ_e $\left[\frac{g}{cm^3}\right]$	ρ_0 $\left[\frac{g}{cm^3}\right]$	c $\left[\frac{km}{s}\right]$	c_A $\left[\frac{km}{s}\right]$	c_T $\left[\frac{km}{s}\right]$	c_F $\left[\frac{km}{s}\right]$
0	100	1500	1.4(5)	4.7(4)	6049	3.6(-7)	1.2(-7)	8.0	12.2	6.7	6.1
832	637	37	8.3(1)	2.8(1)	4705	2.8(-10)	9.4(-11)	7.1	10.7	5.9	5.4

Tab. 1. Diameter d_0, magnetic field strength B_0, external p_e and internal gas pressure p_0, temperature T, external ρ_e and internal density ρ_0, soundspeed c, Alfvenspeed c_A, tube speed c_T, and speed c_F as function of the (Lagrange) height a in the initial tube model.

position described by the arc length l(a,t). The unit vector in direction l is defined by:

$$\hat{l} \equiv \left[\frac{\partial \vec{r}}{\partial l}\right]_t = \frac{1}{l_a}\left[\frac{\partial \vec{r}}{\partial a}\right]_t = (l_x, l_y, l_z), \quad \text{with} \quad l_x^2 + l_y^2 + l_z^2 = 1, \quad (1)$$

and l_a defined below. A curvature vector is defined by

$$\vec{K} \equiv \left[\frac{\partial \hat{l}}{\partial l}\right]_t = \frac{1}{l_a}\left[\frac{\partial \hat{l}}{\partial a}\right]_t \perp \hat{l} \quad . \quad (2)$$

The position of the mass element may also be described by the radius vector $\vec{r}$ from an Eulerian coordinate system (cf. Fig. 1). The magnetic field is assumed to be exclusively in the $\hat{l}$ direction. The conservation of mass requires

$$\rho(a,t) \, A(a,t) \, dl = \rho_0(a) \, A_0(a) \, da \quad , \quad (3)$$

where ρ is the density and the subscript o denotes values at time t=0. Eq. (3) defines a scale factor

$$l_a \equiv \left[\frac{\partial l}{\partial a}\right]_t = \frac{\rho_0 \, A_0}{\rho \, A} \quad . \quad (4)$$

We first consider <u>Eulers equation:</u>

$$\rho\left[\frac{\partial \vec{v}}{\partial t} + \vec{v}\cdot\nabla\vec{v}\right] = -\nabla p - \frac{1}{4\pi}\vec{B}\times(\nabla\times\vec{B}) + \rho\vec{g} \quad . \quad (5)$$

For the <u>transverse component</u> of this Eq. one has (Spruit 1981):

$$\rho\left[\left[\frac{\partial \vec{v}}{\partial t}\right]_a - \left[\hat{l}\cdot\left[\frac{\partial \vec{v}}{\partial t}\right]_a\right]\hat{l}\right] = -\left[\hat{l}\times\nabla(p+\frac{B^2}{8\pi})\right]\times\hat{l} + \frac{B^2}{4\pi}\vec{K} + \rho(\hat{l}\times\vec{g})\times\hat{l} \quad . \quad (6)$$

As the transverse wave moves also mass outside the tube we replace ρ in the inertia term by $\rho+\rho_e$. Using $p+B^2/8\pi = p_e(z)$ one finds

$$\left[\frac{\partial \vec{v}}{\partial t}\right]_a - \left[\hat{l}\cdot\left[\frac{\partial \vec{v}}{\partial t}\right]_a\right]\hat{l} = \frac{\rho c_A^2}{\rho+\rho_e}\frac{1}{l_a}\left[\frac{\partial \hat{l}}{\partial a}\right]_t + \frac{\rho-\rho_e}{\rho+\rho_e}(\vec{g}+gl_z\hat{l}) \quad , \quad (7)$$

with c_A = Alfvenspeed. For the <u>longitudinal component</u> one has:

$$\rho\hat{l}\cdot\left[\frac{\partial \vec{v}}{\partial t}\right]_a = -\frac{1}{l_a}\left[\frac{\partial p}{\partial a}\right]_t - \rho g l_z \quad . \quad (8)$$

We now consider the <u>induction and continuity equations:</u>

$$\frac{\partial \vec{B}}{\partial t} = \nabla\times(\vec{v}\times\vec{B}) = \vec{v}(\nabla\cdot\vec{B}) + (\vec{B}\cdot\nabla)\vec{v} - \vec{B}(\nabla\cdot\vec{v}) - (\vec{v}\cdot\nabla)\vec{B} \quad , \quad (9)$$

$$\frac{\partial \rho}{\partial t} + (\vec{v}\cdot\nabla)\rho + \rho(\nabla\cdot\vec{v}) = 0 \quad . \quad (10)$$

Using these Eqs. we find for the <u>transverse component:</u>

$$\left[\frac{\partial \hat{1}}{\partial t}\right]_a = -\frac{1}{I_a}\left[\left[\frac{\partial \vec{v}}{\partial a}\right]_t - \left[\hat{1}\cdot\left[\frac{\partial \vec{v}}{\partial a}\right]_t\right]\hat{1}\right] \quad , \tag{11}$$

and for the <u>longitudinal component</u>:

$$\left[\frac{\partial B}{\partial t}\right]_a = \frac{B}{\rho}\left[\frac{\partial \rho}{\partial t}\right]_a + \frac{B}{I_a}\hat{1}\cdot\left[\frac{\partial \vec{v}}{\partial a}\right]_t \quad . \tag{12}$$

From the time derivative of $B^2/8\pi = p_e - p$ we have

$$\frac{\rho c_A^2}{B}\left[\frac{\partial B}{\partial t}\right]_a = \left[\frac{\partial}{\partial t}(p_e - p)\right]_a = v_z \frac{dp_e}{dz} - \left[\frac{\partial p}{\partial t}\right]_a \quad , \text{ and with Eq. (12):}$$

$$\frac{\hat{1}}{I_a}\cdot\left[\frac{\partial \vec{v}}{\partial a}\right]_t = -\frac{1}{\rho c_A^2}\left[\frac{\partial p}{\partial t}\right]_a - \frac{1}{\rho}\left[\frac{\partial \rho}{\partial t}\right]_a + \frac{v_z}{\rho c_A^2}\frac{dp_e}{dz} \quad . \tag{13}$$

Eliminating ρ and p in favour of the sound speed c and the entropy S we now have the following 5 equations, <u>Eulers equation</u>:

$$\parallel \quad \hat{1}\cdot\left[\frac{\partial \vec{v}}{\partial t}\right]_a + \frac{1}{I_a}\left[\frac{2c}{\gamma - 1}\left[\frac{\partial c}{\partial a}\right]_t - \frac{c^2\mu}{\gamma R}\left[\frac{\partial S}{\partial a}\right]_t\right] + gl_z = 0 \quad , \tag{14}$$

$$\perp \quad \left[\frac{\partial \vec{v}}{\partial t}\right]_a - \hat{1}\cdot\left[\frac{\partial \vec{v}}{\partial t}\right]_a - \frac{\rho}{\rho + \rho_e}\frac{c_A^2}{I_a}\left[\frac{\partial \hat{1}}{\partial a}\right]_t - \frac{p - p_e}{\rho + \rho_e}(\vec{g} + gl_z\hat{1}) = 0. \tag{15}$$

<u>Induction and continuity equations</u>:

$$\parallel \quad \frac{\hat{1}}{I_a}\cdot\left[\frac{\partial \vec{v}}{\partial a}\right]_t + \frac{2c}{\gamma - 1}\frac{1}{c_T^2}\left[\frac{\partial c}{\partial t}\right]_a - \frac{\mu}{\gamma R}\left[\frac{c^2}{c_A^2} + \gamma\right]\left[\frac{\partial S}{\partial t}\right]_a - \frac{v_z}{\rho c_A^2}\frac{dp_e}{dz} = 0, \tag{16}$$

$$\perp \quad \left[\frac{\partial \vec{v}}{\partial a}\right]_t - \hat{1}\cdot\left[\frac{\partial \vec{v}}{\partial a}\right]_t - l_a\left[\frac{\partial \hat{1}}{\partial t}\right]_a = 0 \quad , \tag{17}$$

where $c_T = \left[c^2c_A^2 / (c^2 + c_A^2)\right]^{1/2}$ is the tube speed.

<u>Entropy conservation equation</u>:

$$\left[\frac{\partial S}{\partial t}\right]_a = \left.\frac{dS}{dt}\right|_{rad} \quad . \tag{18}$$

Going over to <u>characteristic form</u> these equations read:

$$\hat{1}\cdot d\vec{v} \pm \frac{2}{\gamma-1}\frac{c}{c_T}dc \mp \frac{\mu c^2}{\gamma R c_T}dS \mp \left[\frac{\mu c_T}{\gamma R}(\gamma-1)\frac{dS}{dt}\bigg|_{rad} + \frac{v_z c_T}{\rho c_A^2}\frac{dp_e}{dz} - gl_z\right]dt = 0, \tag{19}$$

along the $C_2^{\pm}$ characteristics given by $\frac{da}{dt} = \pm \frac{c_T}{I_a}$, (20)

where Eqs. (19), (20) are those of Herbold et al. (1985) for $l_z = 1$, and

$$(1-l_x^2)dv_x - l_xl_ydv_y - l_xl_zdv_z \mp c_Fdl_x - \frac{p - p_e}{p + p_e}gl_xl_zdt = 0 \quad , \tag{21}$$

$$(1-l_y^2)dv_y - l_xl_ydv_x - l_yl_zdv_z \mp c_Fdl_y - \frac{p - p_e}{p + p_e}gl_yl_zdt = 0 \quad , \tag{22}$$

along the $C_1^{\pm}$ characteristics given by $\frac{da}{dt} = \pm \frac{c_F}{I_a} \equiv \pm \sqrt{\rho/(\rho + \rho_e)}\frac{c_A}{I_a}$. (23)

3. Results

We have solved the timedependent equations described above for a case of an adiabatic wave and purely transverse shaking at the bottom of the tube assuming

$$v_{\perp x} = -v_o \sin(2\pi t/P), \quad v_{\perp y} = 0, \quad v_{\parallel} = 0, \tag{24}$$

with $v_o = 0.5$ km/s and transmitting boundary conditions at the top. Transmission was achieved assuming $v_{\parallel}$ = const along C_2^+ and $v_{\perp x}, v_{\perp y}$ =

const along C_1^+. The wave period was taken to be P=45s. As this shaking took place only in the x-direction the wave is confined to the x-z plane.

Fig. 2 shows a snapshot of the wave after 755 time steps. Note that the physical variables are shown here as function of the Lagrange height a. The curve labeled x shows the horizontal position of the center of

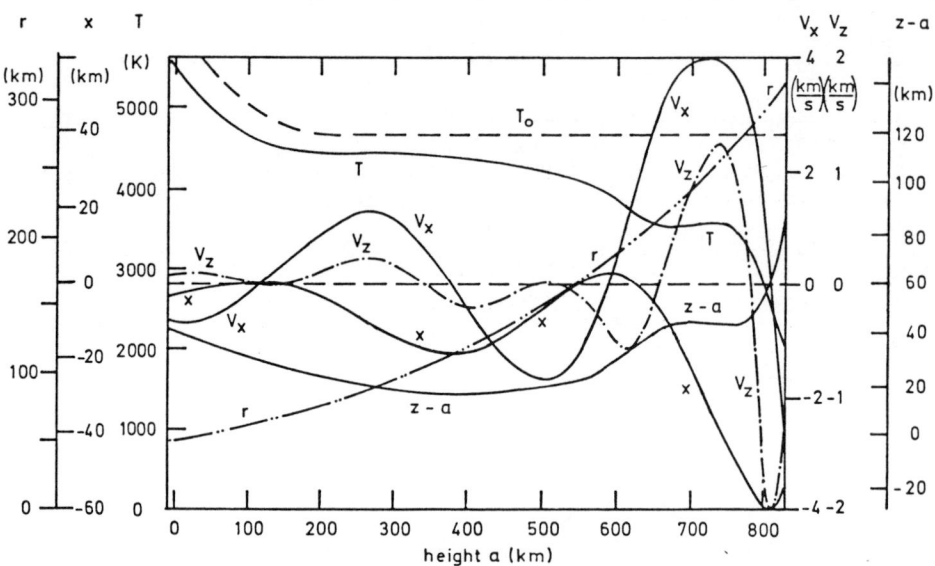

Fig. 2 Snapshot of the wave calculation at time t=1010.4 s, r=tube radius, other symbols are explained in the text.

the flux tube. At 800 km height the tube center is displaced by 60 km in the -x direction. This is about 10 percent of the tube diameter of 600 km at this height. The maxima and minima of the x curve corresponding to dx/da=0 coincide nicely with the nodes of the horizontal velocity v_x. The amplitude of v_x increases as a function of height due to flux conservation. With the propagation speed c_F roughly constant (Tab. 1) the amplitude of v_x grows roughly like $\rho^{-1/2}$. It is seen that v_x has a maximum of 4 km/s at 800 km height. At that height the x-curve has an inflection point where the gas elements in the tube, due to the motion of the wave in +z direction, are horizontally displaced with maximum speed in +x direction.

Fig. 2 shows that in addition to the transverse wave a longitudinal wave is generated where the vertical velocity v_z due to mode coupling shows a surprisingly large amplitude of 2 km/s, that is roughly one half the amplitude of v_x. Moreover it is seen in Fig. 2 that the longitudinal wave has twice the frequency of the transverse wave. This is understood if one decomposes the curvature force vector into vertical and horizontal components. In one wavelength of the x-curve the z-component of the curvature force vector changes sign four times while the x-component changes sign only two times.

Another surprising result of the present calculation is that the entire mass of the tube appears to be lifted relative to the initial position. Fig. 2 shows as function of height the distances z-a which are the height differences between the current (Eulerian) position of the gas elements and the (Lagrange) height (= the height at the start of the computation). It is seen that the bottom of the tube is lifted by about 40 km, the top by about 90 km, while everywhere else the tube is lifted by at least 20 km. The reason for this lifting appears to be the increase of the horizontal motion with height. As the bottom of the tube does not move much in the horizontal direction the large horizontal motion of the higher tube regions generates centrifugal forces, which, always outwardly directed, lead to the lifting of the tube. The result of this lifting is an adiabatic cooling of the entire tube as can be seen in Fig. 2 by comparing the current T and initial T_o temperatures.

4. Conclusions

We found that purely tranverse excitation of magnetic flux tubes will lead to transverse waves with amplitudes strongly growing with height. In addition due to strong mode coupling, longitudinal waves were generated with amplitudes only a factor of two smaller than those of the transverse waves. The wave period of the longitudinal wave was one half that of the transverse wave. The horizontal swaying of the tube increases strongly with height and resulted in centrifugal forces which lead to a lifting and adiabatic cooling of the entire gas column in the flux tube. This part of our calculation would certainly be modified if radiative damping were taken into account. In addition the analogy to other free-boundary organ pipe-type wave motions suggests strong resonance effects i.e. dependences on the chosen wave period. These effects will be considered in our forthcoming work elsewhere.

Acknowledgement The authors gratefully acknowledge generous support by the Sonderforschungsbereich 132 and by the Deutsche Forschungsgemeinschaft.

References

Herbold, G., Ulmschneider, P., Spruit, H.C., Rosner, R.: 1985, Astron. Astrophys. 145, 157
Spruit, H.C.: 1981, Astron. Astrophys. 98, 155
Spruit, H.C.: 1982, Solar Physics 75, 3
Hollweg, J.V., Jackson, S., Galloway, D.: 1982, Solar Physics 75, 35
Ulmschneider, P.: 1987, Proc. Symp. 11, Solar and Stellar Activity, COSPAR Conference, Toulouse

STABILITY OF MAGNETIC ARCADES

U. Anzer
Max-Planck-Institut für Physik und Astrophysik
Institut für Astrophysik
Karl-Schwarzschild-Str. 1, 8046 Garching b. München, FRG

and

A. W. Hood
Department of Applied Mathematics
University of St. Andrews
St. Andrews 9SS KY16, Scotland

Solar two-ribbon flares derive their energy from the shear of coronal magnetic arcades. Therefore the stability of such arcades is of primary interest. In this context Hood (1986) has investigated the stability of cylindrcal arcades against ballooning modes. He found that these arcades can become unstable only if the driving terms due to gas pressure are large enough. In the solar corona, however, the gas pressure is in general much smaller than the magnetic pressure and therefore the field has to be force-free, i.e. to a good approximation $\underline{B} \times \text{curl } \underline{B} = 0$ must hold. For this reason we have studied the stability of force-free cylindrical arcades (Hood and Anzer, 1986). We included in our investigations the effect of photospheric line-tying. Although we have not yet found a general proof of stability all our results point to a strong tendency towards stability of such fields. We selected one typical force-free field and sujected it to a thorough stability analysis. We took the following field

$$B_z = \begin{cases} B_o (1-r^2/a^2) & \text{for } r \leq a \\ 0 & \text{for } r > a \end{cases}$$

$$B_\vartheta = \begin{cases} \dfrac{B_o r}{a} \sqrt{1-2r^2/3a^2} & \text{for } r \leq a \\ \dfrac{B_o a}{\sqrt{3}\, r} & \text{for } r > a \end{cases}$$

This field turned out to be completely stable to all possible perturbations.

In addition to this investigation we could also prove that all force-free fields are stable against perturbations of the form

$$\xi = \xi(r) \, (e^{i(m\vartheta + kz)} - e^{i(-m\vartheta + kz)})$$

and also for

$$\xi = \xi(r) \, (2 e^{ikz} - e^{i(m\vartheta+kz)} - e^{i(-m\vartheta+kz)})$$

for all $m \geq 2$.

From these results we conjecture that all force-free cylindrical arcades are stable. The reason for this high degree of stability is that due to line-tying the critical $m = 1$ kink modes are suppressed.

This means that sheared force-free arcades cannot become unstable and produce solar flares. The arcades must either have magnetic islands in them – which would favour condensation – or they must be loaded with dense, heavy material. Only under such circumstances are instabilities possible. We therefore predict that two-ribbon flares must always be associated with extended solar prominences.

References:

A.W. Hood: 1986, Solar Phys. 103, 329
A.W. Hood, U. Anzer: 1986, submitted to Solar Phys.

THERMAL DISSIPATION IN SLENDER FLUX TUBES AND STRUCTURED MEDIA

P.M. Edwin
The Open University in Scotland, 60 Melville Street,
Edinburgh, Scotland

B. Roberts
Department of Applied Mathematics, University of
St Andrews, Fife, Scotland.

It is of interest to examine the behaviour of waves in flux tubes when dissipative effects are important. Such a theory has applications to isolated tubes and to sunspots (including umbral dots). Consider a magnetic flux tube of strength B_o and radius a within which is a gas with density ρ_o, pressure p_o and temperature T_o. In the absence of both stratification and dissipative effects, such a tube supports a longitudinal *surface* wave with phase speed c_T and a transverse kink mode with phase speed c_k, where

$$c_T^2 = \frac{c_o^2 v_A^2}{c_o^2 + v_A^2}, \qquad c_k^2 = \frac{\rho_o v_A^2}{\rho_o + \rho_e} \tag{1}$$

for sound speed $c_o = (\gamma p_o / \rho_o)^{1/2}$, Alfvén speed $v_A = B_o/(\mu \rho_o)^{1/2}$ and external gas density ρ_e. Additionally, the tube may guide slow *body* (both sausage and kink) waves with phase speed close to c_T.

Focussing attention on those waves with phase speed close to c_T, we may employ slender flux tube theory (see Thomas 1985, Roberts 1986 for recent reviews) to describe the effects of stratification. In the presence of gravity we obtain the local dispersion relation (Defouw 1976; Roberts & Webb 1978; Roberts 1983)

$$\omega^2 = k^2 c_T^2 + \omega_v^2 \tag{2}$$

for a wave of frequency ω and longitudinal wave number k. In Eqn.(2) the cut-off frequency ω_v is defined by

$$\omega_v^2 = N_o^2 + \frac{c_T^2}{\Lambda_o^2}\left[\frac{3}{4} - \frac{1}{\gamma}\right]\left[\frac{3}{4} - \frac{1}{\gamma} + \Lambda_o'\right] \tag{3}$$

for scale height $\Lambda_o(z) = p_o/\rho_o g$, and Brunt-Vaisala frequency N_o:

$$N_o^2 = -g\frac{\rho_o'}{\rho_o} - \frac{g^2}{c_o^2(z)}.$$

(A dash denotes differentiation with respect to height z.)
Non-adiabatic effects may also be included, most simply by adopting

Newton's Law of Cooling. Vertical motions v_z within the tube then satisfy

$$\frac{d^2v_z}{dz^2} + \left[\alpha_1 + \left[\frac{1-i\Omega}{1+\Omega^2}\right]\alpha_2\right]\frac{dv_z}{dz} + \left[\beta_1 + \left[\frac{1-i\Omega}{1+\Omega^2}\right]\beta_2\right]v_z = 0, \quad (4)$$

where

$$\Omega = \gamma\omega\tau_R, \quad \alpha_1 = -\frac{1}{2\Lambda_o}, \quad \alpha_2 = -\frac{\Lambda_o'}{\Lambda_o}, \quad \beta_1 = \frac{\omega^2 - N_o^2}{c_T^2} + \left[1 - \frac{\gamma}{2}\right]\frac{N_o^2}{c_o^2},$$

$$\beta_2 = \left[\frac{\gamma-1}{\gamma}\right]\left[\frac{\omega}{g\Lambda_o} + \frac{1}{\Lambda_o^2}\left[\frac{c_o^2}{\gamma v_A^2} + \frac{1}{2}\right]\right] + \frac{\Lambda_o'}{\Lambda_o^2}\left[\frac{c_o^2}{\gamma v_A^2} + 1\right] - \left[\frac{\Lambda_o'}{\Lambda_o}\right]'$$

for radiative decay time τ_R (assumed known from tables; e.g. Bray & Loughhead 1974; Spruit 1974; Giovanelli 1978, 1979). Eqn. (4) has been discussed by Webb & Roberts (1980) and Hasan (1986). Non-linear effects have also been considered by Herbold et al. (1985) and Venkatakrishnan (1985).

Consider now an energy equation of the form

$$\frac{\partial T}{\partial t} - \frac{1}{\rho_o c_p}\frac{\partial p}{\partial t} = \kappa \nabla^2 T, \quad (5)$$

where κ is a (uniform) thermal diffusivity and c_p is the specific heat at constant pressure. If Eqn.(5) is combined with the other (linearized) mhd equations and solutions inside and outside a finite tube are matched by requiring temperature, heat flux, total pressure and velocity to be continuous across the boundary of the flux tube, then there results a complicated determinantal equation involving the Bessel functions $I_o(\lambda r)$ and $K_o(\lambda r)$. The parameter λ satisfies the quartic (Edwin 1984):

$$\kappa\lambda^4 - (\kappa k^2 + i\omega + \text{factors involving } \omega, k)\lambda^2 + \text{terms involving } \omega, k = 0. \quad (6)$$

In principle this solves the finite tube model.

Adiabatic theory assumes that the thermal dissipation time scale $\tau_\kappa = \ell^2/\kappa$ is much greater than other time scales of the system. But what is the appropriate length scale, ℓ, of this system? Is it the reciprocal of the wave number, k, or the radius of the flux tube? Here we give explicit forms for τ_κ that are dependent on the *Peclet Number*, Pe = $a^2\omega/\kappa$.

Consider a slender, thermally dissipative, gravity-free flux tube. The heat equation for the temperature perturbation, T, and pressure perturbation, p, may be written in the dimensionless form:

$$iT = Pe^{-1}\left[\frac{\partial^2 T}{\partial r^2} + \frac{1}{r}\frac{\partial T}{\partial r} + \left[\frac{\omega^2 a^2}{c_o^2}\right]\frac{\partial^2 T}{\partial z^2}\right] + \frac{i(\gamma-1)}{\gamma}p. \quad (7)$$

The c_T body waves (those not affected by the environment of the tube) for this system are described by

$$\omega \simeq \pm k c_T + \frac{i}{2} \frac{(\gamma-1)}{\gamma} \frac{c_T^2}{c_o^2 \tau_\kappa},$$

where, for $Pe \le 1$ ($\kappa \ge a^2|\omega|$), the parameters λ of Eqn.(6) are given by $\lambda_n^2 = i j_n^{(o)^2} \kappa/a^2 \omega$ ($j_n^{(o)}$ are the zeros of the Bessel function J_o), and then

$$\tau_\kappa = a^2/\gamma\kappa j_n^{(o)^2}. \tag{8}$$

For $Pe > 1$ ($\kappa < a^2|\omega|$) we find

$$\tau_\kappa = 1/k^2 \gamma \kappa. \tag{9}$$

Now Webb (1980) has shown that it is possible to link the analyses for an unbounded medium (e.g. Antia & Chitre 1979; Zhugzhda 1979) with those for a system of close-packed slender flux tubes (Roberts & Webb 1978; Spruit & Zweibel 1979; Roberts 1983) by considering the limit of an infinitely large horizontal wave number in the infinite medium case. The slow body modes of close-packed tubes (i.e. ignoring the dispersive effects of each tube's exterior) in a gravitational and thermally dissipative medium are described by Eqn.(4) but with $\Omega = \omega a^2/j_n^{(o)^2} \kappa$. (The parameters λ of Eqn.(6) are given by $\lambda_n^2 = i j_n^{(o)^2} \omega\kappa/a^2(\omega^2-N_o^2)$.) It follows from this that the work of Webb & Roberts (1980), Hasan (1986), etc. may be exploited with the (radiative) decay time in their analysis set equal to $a^2/\gamma\kappa j_n^{(o)^2}$; that is, a *pseudo radiative time scale* is employed.

It is of interest to consider a Boussinesq fluid. Under adiabatic conditions the infinite set of body waves in each 'tube' of the closely-packed structure behaves as $\omega/k \simeq v_A(1-k_g^2/k^2)^{1/2}$, where $k_g^2 = \alpha g G_o/v_A^2 > 0$, $G_o = -dT_o/dz$ and α is the coefficient of volume (thermal) expansion. Including thermal dissipation for $0 < k_g < k$, we obtain *overstability* for $\kappa \ge a^2 |\omega|$ (Pe $\le$ 1):

$$\omega/kv_A \simeq 1 - iX^2 M/2j_n^{(o)^2} K. \tag{10}$$

Here $X = ka$, $K = \kappa k/v_A$ and $M = \alpha g G_o/k^2 v_A^2$. For the case $\kappa \ll a^2|\omega|$ (Pe $\gg$ 1), we obtain

$$\omega/kv_A \simeq (1 - k_g^2/k^2)^{1/2} - iMK/2(1-M). \tag{11}$$

The growth factor $\frac{1}{2}MK/(1-M)$ in Eqn.(11) is of the form suggested by Savage (1969), Moore (1973) and Parker (1979) but it appears to be appropriate only for *large* Peclet numbers. For *smaller* Peclet numbers the growth factor of Eqn.(10) is relevant.

A comparison of growth times and periods of oscillation as calculated above with umbral dot lifetimes (~25 min; Beckers 1981) and umbral oscillation periods (2-8 min; Moore 1981) is seen by taking M = 0.5, ka = 0.3 and κ = 400 km^{2}s^{-1} (Parker 1979) or κ = 10^{3}km^{2}s^{-1} (Roberts 1976) giving Pe > 1 or Pe < 1, respectively.

	Parker(1979)	Eqn.(11),Pe>1	Eqn.(10),Pe<1
Growth time	1-3 hr	37 min	42 min
Period of osclln. of gas beneath	100s	592s	428s

Parker argues that growth times should be comparable with the lifetimes of umbral dots. Those given here are shorter than those estimated by Parker but they are still rather long to explain overstable oscillations or umbral dot behaviour. For the most favourable case it would take three periods of oscillation to e-fold, which is not unreasonable. Recognising that compressibility effects will not be negligible and taking $\omega/k \sim c_T$ rather than phase speeds given in Eqns.(10) and (11) gives periods on the order of 300s (295s).

The removal of compressibility effects via the Boussinesq approximation must be treated with caution. If umbral dots are to be explained in terms of oscillatory behaviour then presumably they must relate to body waves that oscillate back and forth within the structure (as opposed to surface waves). However, the incompressible plasma supports only *surface* modes and so it is not a good guide as to the time scales of the compressible modes.

References

Antia, H.M. & Chitre, S.M. (1979). Solar Phys., 63, 67.
Beckers, J.M. (1981). *In* The Sun as a Star, ed. S. Jordan, NASA SP-450.
Bray, R.J. & Loughhead, R.E. (1974).The Solar Chromosphere. Chapman & Hall, London.
Defouw, R.J. (1976). Astrophys.J., 209, 266.
Edwin, P.M. (1984). Ph.D. Thesis, St Andrews University.
Giovanelli, R.G. (1978). Solar Phys., 59, 293.
Giovanelli, R.G. (1979). Solar Phys., 62, 253.
Hasan, S.S., (1986). Mon. Not. R. Astron. Soc., 219, 357.
Herbold, G., Ulmschneider, P., Spruit, H.C. & Rosner, R. (1985). Astron. & Astrophys., 145, 157.
Moore, R.L. (1973). Solar Phys., 30, 403.
Moore, R.L. (1981). Space Sci. Rev., 28, 387.
Parker, E.N. (1979). Astrophys.J., 234, 333.
Roberts, B. (1976). Astrophys.J., 204, 268.
Roberts, B. (1983). Solar Phys., 87, 77.
Roberts, B. (1986). *In* Small Scale Magnetic Flux Concentrations in the Solar Photosphere (Proc. Workshop, Gottingen, Oct.1985 eds. W. Deinzer, M. Knolker & H. Voigt), 169.
Roberts, B. & Webb, A.R. (1978). Solar Phys., 56, 5.
Savage, B.D. (1969). Astrophys.J., 156, 707.

Spruit, H.C. (1974). Solar Phys., 34, 277.
Spruit, H.C. & Zweibel, E.G. (1979). Solar Phys., 62, 15.
Thomas, J.H. (1985). *In* Theoretical Problems in High Resolution Solar
    Physics (Proc. Workshop, Munich, Sept.1985 ed. H.U.
    Schmidt), 126.
Venkatakrishnan, P. (1985). J. Astrophys. & Astron., 6, 21.
Webb, A.R. (1980). Ph.D. Thesis, St Andrews University.
Webb, A.R. & Roberts, B. (1980). Solar Phys., 68, 87.
Zhugzhda, Y.D. (1979). Sov. Astron., 23, 42.

WAVES AND THERMAL INSTABILITIES IN FLUX TUBES: THEIR ROLE
FOR THE STRUCTURE OF THE CHROMOSPHERE AND TRANSITION REGION

R. Hammer
Kiepenheuer-Institut für Sonnenphysik
Schöneckstr. 6
D-7800 Freiburg, FRG

1   INTRODUCTION

The outer Solar atmosphere is highly inhomogeneous and temporally variable. Many of the observed fluctuations in space and time are manifestations of magnetic field concentrations and of time dependent phenomena associated with them, such as waves and instabilities. The purpose of this review is to discuss some of the basic properties of these time dependent phenomena and their relevance to the chromosphere and transition region (hereafter TR) of the quiet Sun. Related reviews, which emphasize different aspects of the subject and can therefore serve as complementary sources of information, include those by Athay (1985a,b), Stein (1985), Thomas (1985), Mariska (1986), Roberts (1986) and Ulmschneider & Muchmore (1986).

2   ATMOSPHERIC STRUCTURE

2.1 Average Atmosphere

Before we discuss fluctuations, it is instructive to address ourselves to the "average" chromosphere. Fig. 1 shows the temperature and the net emission as functions of height for the global reference model (model C) of Vernazza et al. (1981). This static, plane-parallel model was constructed to fit the average emission that one would see at extremely low spatial and temporal resolution. The vertical extent of the model chromosphere is nearly 2000 km, corresponding to about 10 pressure scale heights. In the chromosphere, the pressure scale height $|dh/d\log p|$ is much smaller than the temperature scale height $|dh/d\log T|$, while in the TR the opposite is true. In the lower TR of this model, the temperature scale height becomes as small as a few km, whereas the pressure is roughly constant within the TR.

As a consequence of the steep temperature gradient in the model TR and of the high thermal conductivity in the overlying hot corona, thermal conduction is an important energy transport mechanism throughout the outer layers of the Solar atmosphere. It enables the Sun to collect energy from where the heating occurs and to transport it into regions where it is needed - in particular into the TR where the UV emission must be balanced. In this sense, the TR acts as an energy sink of the corona; its physics (and location: see below) can therefore not be investigated without considering the physics of the corona.

The chromospheric energy balance has a more local character. Since thermal conduction is inefficient at low temperatures, the net nonradiative heat input $|\nabla \cdot F_M|$ into a chromospheric volume element must be radiated out of the same volume element. The details of this emission, however, depend on the properties of the whole atmosphere because the chromospheric emission occurs mainly in optically thick spectral lines. Most of these lines, with the exception of the Lyman lines, are effectively thin. In this case, the net radiative energy loss $\nabla \cdot F_R$ can be approximated by

$$|\nabla \cdot F_M| = |\nabla \cdot F_R| = n_e \, n_H \, f(T), \qquad (1)$$

where n_e and n_H are the electron and neutral hydrogen densities, respectively. Athay (1981, 1985a) has emphasized that over most of the Solar chromosphere, where the electrons come mainly from hydrogen ionization, the combined T dependence of n_e and f is so strong that even large changes of the heating rate can be balanced by small changes of the temperature. This strong thermostatic effect associated with hydrogen ionization explains why the chromospheric temperature gradient is so small.

Once we reach a height where hydrogen is mostly ionized, this thermostatic control terminates, and the temperature gradient can become steeper. This is illustrated (after Athay 1985a) in Fig. 1 by a circle which indicates the level where hydrogen is half ionized. It has been

Fig. 1 Temperature (drawn) and net radiative energy loss (dashed) vs. height for the chromospheric reference model C of Vernazza et al. (1981).

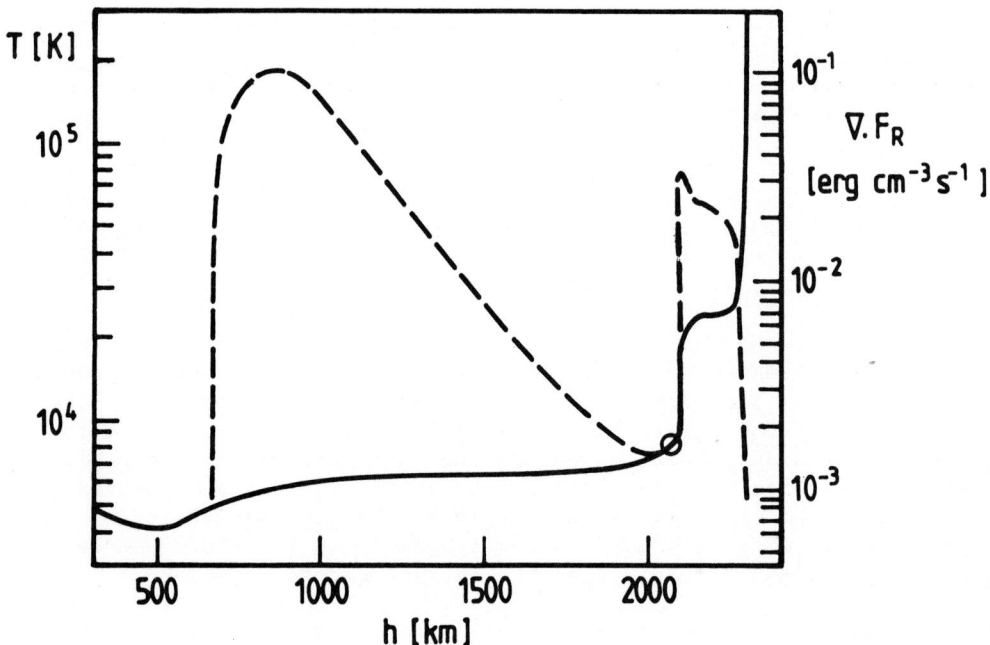

suggested that this <u>determines</u> the location of the TR. We should, however, be careful not to interchange cause and effect. Likely it is not the ionization of hydrogen that determines the location of the TR, but the opposite holds true: The very existence of the rapid temperature increase in the TR causes the fractional ionization to change from low to high; hence, there must be levels where hydrogen is half ionized and where the Lα line becomes optically thin; and these levels must obviously lie at fairly cool temperatures near the foot of the TR.

But what determines the location of the TR? I suggest it is ultimately the amount of energy that is deposited in the corona, in a way that is described elsewhere (Hammer et al. 1982, 1983, 1984). Theoretical models of magnetically closed (e.g., Rosner et al. 1978) and open coronal regions (Couturier et al. 1980; Hearn & Vardavas 1981a,b; Hammer 1982a,b) as well as semiempirical studies (Jordan 1980, Fig. 4) show that for a given type of corona (closed or open, with a specified heating law) there exists a unique positive relationship between the total energy losses and the base pressure of the corona. Physically, this relationship simply states that a density increase leads to enhanced radiation from the corona and its TR, and in open regions it leads also to a more massive wind. Since this coronal relationship must be fulfilled, the chromosphere alone cannot determine the location, and thus the pressure, of the TR.

The chromospheric net radiation loss shown in Fig. 1 must be balanced by some kind of nonradiative heating. This requirement provides an important test for any proposed heating mechanism. Fairly different mechanisms have been suggested to meet this requirement, like acoustic shock waves that develop out of compressional modes (e.g., Ulmschneider 1970) or the turbulent cascade of Alfvénic modes (Hollweg 1985; see, however, Ulmschneider 1986). Over most of the chromosphere, the characteristic damping length of the heat input, $|dh/d\log F_M|$, is slightly larger than the density scale height. The emission peak at 2200 km is due to Lα. Athay (1985a,b) argued that this peak cannot be reconciled, in such a one-dimensional model, with a heating mechanism that is powered from below. The Lα losses could be balanced, however, by energy that comes from above, for example the nonlocal heat flux associated with hot coronal electrons that leak through the thin TR. Over the past few years it became increasingly clear that this effect may be very important for the energetics of the lower TR (Roussel-Dupré 1980a,b; Shoub 1983; Dufton et al. 1984; Owocki & Canfield 1986).

2.2 Fluctuations

The Lα emission peak and the associated temperature plateau may also be an artifact of the simplifying assumptions underlying the reference model. The outer Solar atmosphere is neither plane-parallel, nor static, nor time independent. Lα spectroheliograms exhibit an impressive amount of fine structure (Bonnet et al. 1980; Foing et al. 1986). Most of the Lα emission appears to originate in small elongated structures, such as fibrils and loops, rather than a thin spherical shell. Similarly, the results from the HRTS experiments indicate a wealth of fine structure at TR temperatures (e.g., Brückner 1986). It has even been suggested that a considerable fraction of the emission that was traditionally ascribed to the lower TR (T < 1-2 10^5 K) comes from unresolved cool structures (Feldman 1983), such as magnetic loops (Antiochos & Noci 1986) or current sheets within small loops (Rabin & Moore 1984).

Like the TR, the chromosphere is also highly inhomogeneous. Outside of active regions, patches of bright emission in lines such as CaII K are mainly arranged along the boundaries of supergranulation cells. The resulting network pattern can be traced up into the TR. Above 3 10^5 K, its boundaries widen rapidly; and near 10^5 K it becomes unrecognizable. This behavior has often been interpreted in terms of the spreading with height of magnetic flux tubes, in particular since the CaII K emission was found to be linearly related to the magnetic flux density in the underlying photosphere (Skumanich et al. 1975). As to the chromosphere proper, it is extremely difficult to infer the magnetic field directly from observations (see the review by Jones 1985). We know, however, that in the photosphere the field is concentrated in thin flux tubes. Due to the rapid decrease of the ambient pressure, the field lines fan out with height until they finally merge with the field lines of neighbor flux tubes. The mergence, or "canopy", height has been debated in the literature between theoreticians (Spruit 1981; Anzer & Galloway 1983, 1985; Pneuman et al. 1986) and observers (Giovanelli 1980; Giovanelli & Jones 1982, 1983; Jones 1985). In the quiet Sun the mergence appears to occur somewhere in the mid chromosphere, at heights between 1000 km and 1400 km above $\tau(5000Å) = 1$. Obviously, this level is crucially important for the propagation of waves and steady flows both inside the flux tubes and in the external medium. Moreover, above the mergence level reconnection may become important (Parker 1982).

More or less steady downflows are observed in the chromosphere (usually at the sites of strong photospheric magnetic fields) and throughout the TR. The quiet TR shows persistent redshifts of about 5 km s^{-1} in CIV, but almost never blueshifts. At a first glance, this appears to contradict the principle of mass conservation - however, there exist at least two explanations of the phenomenon. First, it may be related to spicular material, which was shot up as a cool gas (1-2 10^4 K), then heated to a few times 10^5 K and distributed over a larger volume, and which finally cools and emits redshifted photons while it flows back along the network field lines. And second, for several types of global flows in coronal loops the redshifted emission dominates over the blueshifted emission. This applies to flows induced by asymmetric heating (Mariska & Boris 1983), by temporal fluctuations of the heating rate (Antiochos 1984), and during the nonlinear limit cycle oscillations proposed by Kuin & Martens (1982; see below, Sect. 4.2).

The observed downflows can severely affect the ionization balance. The downflowing matter traverses a temperature scale height so rapidly (in the lower TR within less than a second!) that some TR ions cannot establish ionization equilibrium (e.g., Joselyn et al. 1979). This may lead to errors in empirical models (Mariska 1984). Nonequilibrium ionization effects are also important in the upper chromosphere (Kneer 1980).

The quiet chromosphere and TR exhibit a large variety of temporal fluctuations of both impulsive and periodic nature. Well known examples of impulsive events are the spicules mentioned above, EUV bursts, and chromospheric and TR jets. Periodic wave motions have also been identified (e.g., Deubner 1984, 1985a,b; Staiger 1985; Kneer & von Uexküll 1985; and references therein). The analysis of waves is rapidly evolving into a powerful diagnostic tool for probing the Solar atmosphere. Various types of waves have been reported, including internal gravity waves, propagating as well as standing acoustic waves, transverse motions along Hα fibrils (Giovanelli 1975), and longitudinal waves along flux tubes (Giovanelli et al. 1978; Roberts 1983).

2.3 Time Scales

The dynamics of the atmosphere is closely coupled to the energetics. This coupling is best discussed in terms of characteristic relaxation time scales. In order to define them self-consistently, we begin with the energy equation,

$$\varrho T \frac{ds}{dt} = -\nabla \cdot F_R \quad -\nabla \cdot F_C \quad -\nabla \cdot F_M = -L. \quad (2)$$

The entropy s (per gram) can be changed by means of various energy sources and sinks, the sum of which is usually (e.g., Priest 1982) denoted by L, the net energy loss function. Here we are particularly interested in the effects of radiation, thermal conduction, and heating (both magnetic and nonmagnetic). Since simple thermodynamic relations and the continuity equation give

$$\varrho T \frac{ds}{dt} = \frac{1}{\gamma-1} \frac{dp}{dt} - \frac{\gamma}{\gamma-1} p \nabla \cdot v, \quad (3)$$

where γ is the ratio of specific heats, the energy equation reads

$$\frac{1}{p} \frac{dp}{dt} = \gamma \nabla \cdot v - \sum_i \frac{\gamma-1}{p} \nabla \cdot F_i . \quad (4)$$

The pressure p, and therefore also the internal energy density (~p), change by adiabatic expansion (first term) and by nonadiabatic heating or cooling (remaining terms). Since all terms have the dimension 1/time, the quantities

$$\tau_i = \frac{1}{\gamma-1} \frac{p}{\nabla \cdot F_i} \quad (5)$$

are the characteristic relaxation times associated with the individual processes. (Note that some authors prefer the related time scales $t_i = \gamma \tau_i$, which can be shown to be characteristic heating/cooling times for processes with constant pressure).

Radiative Time Scale. Radiative cooling can strongly affect the amplitudes, phases, and propagation characteristics of waves. In many photospheric flux tube calculations (e.g., those by Webb & Roberts 1980; Roberts 1983; Hasan & Schüßler 1985; Hasan & Kneer 1986), the radiative energy exchange between flux tube and external medium has been approximated by Newton's law of cooling with the classical Spiegel (1957) expression for the radiative relaxation time τ_R,

$$\nabla \cdot F_R = \rho c_V \frac{\delta T}{\tau_R} \quad \text{with} \quad \tau_R = \frac{c_V}{16 \kappa \sigma T^3} . \tag{6}$$

(A more general approach is discussed in Edwin & Roberts, these proceedings). When applied to the chromosphere, eq. (6) gives relaxation times as large as 10^3 s. On the other hand, Spiegel's expression cannot really be applied to the Solar chromosphere because it was derived for the damping of very small temperature fluctuations in a grey homogeneous atmosphere. In the chromosphere, the main emission occurs in lines, which after Giovanelli (1978) leads to a considerable reduction of τ_R to about 10^2 s. Recently Kneer (1986; see also Trujillo-Bueno and Kneer, these proceedings) gave an extensive discussion of the effects of stratification and lines in LTE and nLTE. Moreover, the geometry of flux tubes is complicated and height dependent. Finally, since the chromosphere extends over many density scale heights, any wave travelling upward must be expected to steepen and to become nonlinear, so that the associated temperature fluctuations are no longer small perturbations. Since line emission depends extremely nonlinearly on temperature, this may for example lead to fairly thin zones behind shocks in which the dissipated entropy is radiated away (Stein 1986). Consequently, for the calculation of chromospheric waves we must ultimately aim for a self-consistent treatment of nLTE radiative transfer in the few most important emission lines.

In the TR and corona, on the other hand, the emission is mainly optically thin, and the local radiative energy loss can be calculated from eq. (1). The emissivity function f(T) has been computed repeatedly; roughly speaking, it is an increasing function in the upper chromosphere and lower TR, but then it decreases with T up to about $2 \cdot 10^7$ K. The maximum is found near 10^5 K if optical thickness effects at lower temperatures, in particular in Lα, are taken into account in a crude way (McClymont & Canfield 1983). The corresponding radiative relaxation time,

$$\tau_R = \frac{1}{\gamma-1} \frac{p}{n_e n_H f(T)} \approx 10^{-31} \frac{T^2}{p \, f(T)} \tag{7}$$

(in cgs units), varies from about 10s below 10^5 K to 200s near $3 \cdot 10^5$ K and 10^4 s near 10^6 K. These figures refer to the reference model (Fig. 1); in active regions and high pressure coronal loops τ_R may be smaller by one or two orders of magnitude.

Conductive Time Scale. Under typical coronal conditions, the classical thermal conductivity along the field lines, $\kappa_\parallel \approx 10^{-6} T^{5/2}$, is five orders of magnitude larger than the cross field conductivity. Thus, when the transverse temperature gradient is not extreme (Rabin & Moore 1984) the conductive flux is channeled along the field lines. Conductive heating and cooling,

$$|\nabla \cdot F_c| \simeq |\nabla \cdot \kappa_\| \nabla_\| T| \simeq 10^{-6} \frac{T^{7/2}}{l^2}, \qquad (8)$$

can damp out coronal temperature fluctuations within a characteristic time τ_c that depends strongly on the temperature and on the characteristic length scale l associated with the fluctuation,

$$\tau_c \simeq \frac{1}{\gamma-1} \frac{p}{\nabla \cdot F_c} \simeq 1.5 \; 10^6 \frac{pl^2}{T^{7/2}}. \qquad (9)$$

3 WAVES

The simplest possible, and thus far best studied type of wave in the Solar atmosphere is a plane-parallel acoustic wave that propagates vertically upward from the convection zone. Below the photosphere, it suffers heavily from radiative damping. In the upper photosphere and lower chromosphere, however, damping is much less important, so that the energy flux of a propagating wave,

$$F_M \simeq \rho v^2 c_s, \qquad (10)$$

is approximately conserved. Since the sound speed $c_s \sim (T/\mu)^{1/2}$ varies only slowly in the chromosphere whereas the density ρ decreases rapidly outward, the velocity amplitude v of the wave must increase outward, on a scale height roughly twice as large as the density scale height. Large amplitude compressive waves rapidly form shocks, which can then dissipate the wave energy. We have now two counteracting effects: stratification alone would lead to an outward increase, and dissipation alone to a decrease, of the wave amplitude. Obviously, small amplitude shock waves continue to grow in amplitude (due to stratification) until shock dissipation becomes strong enough to counterbalance this effect. Above this level, the amplitude is nearly constant, so that according to eq. (10) the energy flux decreases roughly proportional to the density. Therefore, an acoustic shock wave automatically adjusts its damping length to the density scale height, consistent with the empirical requirements over most of the chromosphere (Fig. 1). This property is confirmed both by approximate (Ulmschneider 1970) and fully time dependent (Schmitz et al. 1985) shock wave calculations; but it need not necessarily be restricted to compressive waves - similar self-limiting nonlinear processes are conceivable for other types of wave modes.

Acoustic waves of different periods can interact nonlinearly with each other (Bohn & Stein 1985). Another very interesting problem is the interaction of a wave with the TR. Due to the rapidly changing propagation speed, part of the wave energy flux is reflected back into the chromosphere, while the rest is transmitted into the corona. The reflected part may lead to standing waves in the chromosphere (Mein 1977;

Schmieder 1978; Deubner 1984; Staiger 1985). In the corona, the radiative cooling time is much larger than the conductive cooling time associated with typical wavelengths. Therefore, compressive waves are damped predominantly by conduction (and viscosity, which I have not discussed here), rather than by radiation. Since the coronal temperature scale height is large compared with the wavelength, the wave can be thought of as a perturbation running along an average temperature profile. This is to be contrasted with the situation in the lower TR, where the temperature scale height is much smaller than the wavelength; hence, the TR is pushed up and down by the wave. This motion of the TR has been investigated numerically for various types of magnetic and nonmagnetic wave modes (Hollweg et al. 1982; Hollweg 1982; Mariska & Hollweg 1985). The region between the corona and the lower TR is also quite interesting. Tavakol & Tworkowski (1981) have analysed sinusoidal waves running along a model of the TR and corona. They found that these waves, even if they do not carry enough energy to heat the corona, nevertheless redistribute energy: they efficiently transport energy from the lower corona down into the upper TR.

This surprising result is ultimately caused by the nonlinearity of the conservation equations. The time averaged conductive flux, for example, cannot simply be calculated from the time averaged temperature alone,

$$\langle F_c \rangle \simeq \langle 10^{-6} T^{5/2} \frac{dT}{dh} \rangle \neq 10^{-6} \langle T \rangle^{5/2} \frac{d\langle T \rangle}{dh} , \qquad (11)$$

as is usually done in the derivation and interpretation of empirical models. The difference between the two terms depends on the temporal fluctuations and represents a net energy flux associated with the waves. Quite generally, waves cause nonlinear fluxes of mass, momentum ("turbulent pressure"), and energy. With respect to the energetics of the atmosphere this is particularly important for compressive waves, because they generate temperature fluctuations directly, and both conduction and radiation depend extremely sensitively on temperature.

A compressive wave can be confined to a magnetic flux tube. The energy flow is then channeled along the field lines, which diverge below the canopy. Thus with increasing height the energy is spread over an increasingly larger area. The wave amplitude cannot grow as rapidly as in the plane-parallel case, and nonlinear effects such as shock dissipation (Foukal & Smart 1981; Herbold et al. 1985; Ulmschneider 1985) and coupling with other wave modes are delayed to greater heights.

Magnetic flux tubes do not only play this passive role of channeling the flow. If the field is not infinitely strong, pressure fluctuations cause expansions and compressions in the shape of the flux tube, which propagate along the tube (cf. Fig. 2, left column). The associated magnetic tension represents an additional restoring force, which slows down the wave. The phase speed c_t is found to be smaller than both the Alfvén speed c_a and the sound speed c_{si} inside the tube,

$$\frac{1}{c_t^2} = \frac{1}{c_{si}^2} + \frac{1}{c_a^2} . \qquad (12)$$

The pulsation of the tube walls can lead to the emission of acoustic energy into the external medium and thus to a damping of the wave inside the tube. Spruit (1982) has shown that this effect vanishes for a linear monochromatic wave, provided that its phase speed c_t is smaller than the sound speed c_{se} in the external medium, because in that case the external acoustic wave is evanescent and does not transport energy away. After eq. (12) the requirement $c_t < c_{se}$ is clearly fulfilled if the temperatures, and thus sound speeds, inside and outside the flux tube are similar. However, it has been suggested recently that the external gas may be substantially cooler (cf. Sect. 4.1). Moreover, chromospheric waves cannot be expected to be linear. And finally, they are also not monochromatic. Wave packets of a finite length can feed energy into the ambient medium during the initial build up phase (Herbold et al. 1985). Even if that energy is not transported away, the energy transfer changes the propagation and dissipation properties of the wave inside the tube.

The magnetic tension restoring force does not only modify longitudinal compressive waves, but enables also other wave modes. Fig. 2 illustrates the most basic wave modes of a thin (compared with the wavelength; cf. Roberts & Webb 1978; Spruit 1981) flux tube. The kink mode is generated by shaking the tube, and the torsional mode by twisting it. Both types of waves appear to be produced efficiently on the Sun (Spruit 1984; Schüßler 1984). As long as they are linear they are non-compressive; thus they are not affected by radiative damping in the photosphere. The kink mode has a lower cutoff frequency than the longi-

<u>Fig. 2</u> Basic wave modes of a thin flux tube (Spruit 1981).

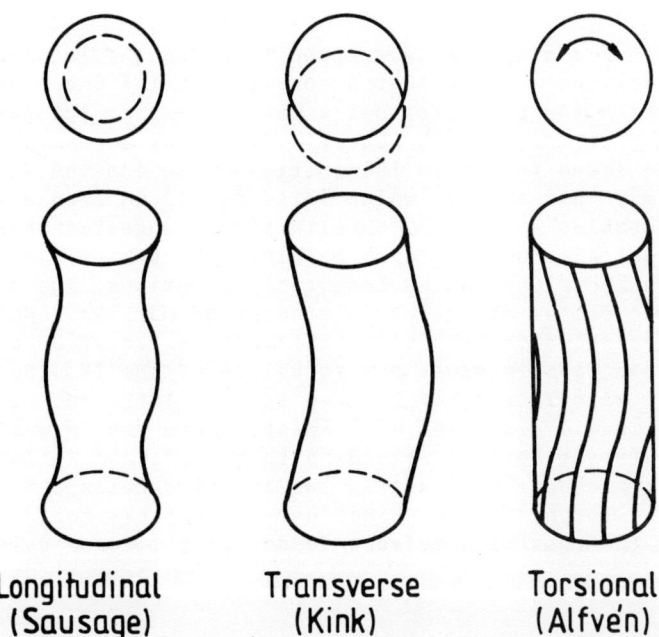

tudinal mode and can therefore pick up and transport energy from a larger portion of the granular power spectrum (Spruit 1984). The torsional mode is not handicapped by a cutoff. The magnetic wave modes can dissipate their energy either directly (phase mixing: see reviews by Heyvaerts 1984 and Stein 1985) or indirectly, after having been transformed into longitudinal wave motions which can then dissipate in shocks (Ferriz-Mas & Moreno-Insertis 1986). Such longitudinal motions are generated by both kink (cf. Zähringer & Ulmschneider, these proceedings) and torsional waves (Hollweg et al. 1982; Mariska & Hollweg 1985) when the wave amplitude approaches sound speed. However, it is presently still debated how much of the heating of the upper layers is due to waves of any kind and how much is due to reconnection in field lines that are rearranged by slow motions of the photospheric footpoints.

Roberts and collaborators (Roberts & Mangeney 1982; Roberts 1985; Edwin & Roberts 1986a) have shown that, at least in an unstratified medium, solitary waves are possible for the longitudinal mode, and likely also for the kink mode. A solitary wave is characterized by the balance between the effects of nonlinearity, which tends to steepen a wave front, and dispersion, which tends to flatten it. The net result is a wave front of permanent shape. This is analogous to the self-limiting behavior of acoustic shocks discussed above, where the effects of stratification and dissipation on the wave amplitude balance each other, again resulting in a wave of constant shape. The formation of a solitary wave, however, prevents a wave front from shock formation and thus from dissipating. It would be very important, therefore, to find out whether such waves are also possible in the stratified Solar atmosphere.

Both longitudinal and kink mode respond to a sudden, localized perturbation by generating a pulse that propagates at the phase speed of the wave mode. Behind the pulse follows a wake, in which the gas oscillates at the cutoff frequency (Rae & Roberts 1982). For acoustic waves Hollweg (1982) found that both the initial pulse and the following oscillations may form shocks, which shift the TR in such a way that the net result resembles a spicule. Spruit (1984) suggested that the same phenomenon might also be generated by kink wave pulses, after they first have nonlinearly produced longitudinal motions. The case of torsional Alfvénic pulses was recently studied by Mariska & Hollweg (1985) with the NRL flux tube code, which takes radiation and conduction into account and resolves the structure of the TR. (By comparison, in the earlier calculations of Hollweg et al. (1982) and Hollweg (1982), nonadiabatic processes were ignored, and the TR was represented by a contact discontinuity.) According to Mariska and Hollweg, a strong Alfvénic pulse produces nonlinearly two acoustic pulses, which shift the TR back and forth, but which neither form shocks nor generate wake oscillations. The resulting motions do not look like spicules.

4 THERMAL INSTABILITY

In a steady equilibrium situation, the energy losses of a fluid element balance the energy gains, so the net energy loss function (cf. eq. (2)) vanishes, $L = 0$. When we perturb such an equilibrium state, the perturbed energy loss and gain terms need no longer cancel. Suppose, for instance, that the net losses decrease ($\delta L < 0$) during a perturbation that causes a temperature enhancement ($\delta T > 0$). Then the perturbed plasma receives more energy than it loses, and it continues to heat up. Such a situation, in which $dL/dT < 0$ for any perturbation of the equilibrium state, is called thermally unstable. Obviously, thermal stability can only be investigated by taking <u>all</u> the terms in the energy equation into account, unlike what has often been done in the literature. One of these terms, namely the radiative term, has a strong potential for destabilizing the atmosphere, both at cool temperatures near the base of the chromosphere, and at high temperatures, in the TR. The discussion of possible thermal instabilities in these two atmospheric layers played an important role in the literature of the past few years.

4.1 Lower Chromosphere

Observations of the IR bands of CO indicate the existence of very cool material in the Solar atmosphere (Noyes & Hall 1972; Ayres & Testerman 1981), even in active regions (Ayres et al. 1986). To account for these observations, Ayres (1981, 1985) suggested that the atmosphere might consist of two vastly different types of regions: small scale regions (called "flux tubes") with intense chromospheric temperature inversions are embedded in a very cool (3000 K) ambient medium without temperature inversion. The temperature in the latter component is controlled by cooling in the strong CO lines. If this speculation is correct, the Solar chromosphere proper exists only inside the "flux tubes", at least up to a certain height. Clearly, more observations are needed to clarify this point. In any case, it has become obvious that CO and other molecules are extremely important for the energy balance of the outer atmosphere.

Kneer (1983, 1984, 1985) has emphasized that the autocatalytic nature of CO formation can drive a thermal instability: a negative temperature perturbation may lead to the formation of CO molecules, which further cool the atmosphere by line radiation, so that more CO molecules are formed, etc. The temperature minimum region of conventional Solar models, such as the reference model C of Vernazza et al. (1981), appears to lie in a temperature-density regime where this effect is so important that the radiation term in the energy equation acts destabilizing (cf. Fig. 2 of Muchmore & Ulmschneider 1985). The thermal stability of the atmosphere depends, however, also on the other terms, in particular on the heating term.

Pure radiative equilibrium models (i.e., models without heating) including CO have been constructed repeatedly (Kneer 1984, 1985; Muchmore & Ulmschneider 1985; Nordlund 1985; Muchmore 1986a,b). For a range of effective temperatures of the underlying photosphere the atmosphere can exist either in a hot or in a cool state, depending on its history (Kneer 1985, Muchmore 1986a). Muchmore & Ulmschneider (1985) investigated the hydrodynamics of a plane-parallel atmosphere that is forced from one state to the other by a change of the surface temperature. The effects of heating have also been investigated (e.g., Kneer 1985). Muchmore & Ulmschneider (1985) found that in the cool component acoustic waves can heat up the gas and destroy CO above the height of shock formation. For very large wave energy fluxes, this height drops below the level where CO cooling becomes important; then the cool gas disappears completely.

A very interesting point has been raised by Kneer (1984, 1985), who found at the base of the cool phase of his models a temperature gradient so steep that the atmosphere is convectively unstable. Hasan & Kneer (1986) investigated the linear stability of a thin flux tube embedded in such an atmosphere. They found instability for fields smaller than 20-40 G at 150 km above the temperature minimum, while for larger fields the radiative energy exchange between flux tube and environment drives overstable oscillations with periods between 5 and 10 min. The possible existence of a convectively unstable layer is still controversial, however. Nordlund (1985) obtained much shallower, and stable, temperature gradients from his radiative equilibrium models, and he argued this was due to a more realistic treatment of the CO lines. Muchmore (1986b), on the other hand, has recently recalculated his models with improved opacity distribution functions; his results confirm Kneer's finding of a convectively unstable layer. In any case, even the possibility of a convection zone in the atmosphere of the Sun, less than a scale height thick, is fascinating. It is well known that in very cool stars molecule formation causes convection zones (or an extension of existing ones). But the possibility that this might be relevant even for the Sun had not yet been appreciated.

4.2 Transition Region

Radiation acts destabilizing not only at low temperatures, due to the dissociation of molecules, but also at high temperatures, due to the ionization of the most important emitters. Above a certain temperature, the emissivity of an optically thin plasma decreases with T. Parker (1953) suggested that this would make the Solar corona unstable if thermal conduction were not important. Suppose a hot isothermal plasma is perturbed locally at constant pressure to a lower temperature. Then the radiative cooling (cf. eq. (1)) increases, and the perturbation tends to grow. Thermal conduction, however, can fill in the temperature dip provided that the spatial extent of the perturbed region is not too large. Therefore, radiation acts destabilizing at high temperatures, while conduction acts stabilizing. The relative importance of these two processes in the energy equation can be discussed in terms of the ratio of the associated time scales,

$$\frac{\nabla \cdot F_C}{\nabla \cdot F_R} = \frac{\tau_R}{\tau_C} \simeq 7 \; 10^{-38} \; \frac{T^{11/2}}{(\rho l)^2 \, f(T)} \; . \tag{13}$$

If this ratio is much larger than unity, conduction dominates; and radiation is so unimportant that we do not expect it to drive a thermal instability.

The thermal stability of an isothermal coronal plasma was studied in a classical paper by Field (1965). Compared with such an idealized situation, the real Solar corona presents a number of complications. First, the (unknown) heat input cannot be expected to remain constant, but will react somehow to a perturbation, and this may well affect the stability. Second, the TR and corona are permeated by magnetic fields. In the following I will restrict myself to review work in which they act only as ducts for mass and energy flows. (For recent reviews on more general MHD instabilities see Priest 1982; Einaudi 1984; Hood 1985). Third, the corona is not isothermal, but connected to the cool chromosphere via the TR. The TR can respond to coronal perturbations. For example, when the downward conductive flux is reduced during a perturbation, the radiative energy losses of the TR are no longer fully balanced. Thus some TR material must cool to chromospheric temperatures, which corresponds to an upward shift of the TR. Likewise, an excess conductive flux at the foot of the TR can heat up chromospheric material, again leading to a shift of the TR. These processes are commonly (mis)called coronal condensation and chromospheric evaporation, respectively. Many authors have stressed their importance for the dynamics of the outer Solar atmosphere (e.g., Krall & Antiochos 1980; Zweibel 1980; Hearn et al. 1982; Craig et al. 1982; Kuin & Martens 1982; Mariska & Boris 1983; Spruit & Roberts 1983; An et al. 1983; Antiochos 1984; McClymont & Craig 1985a,b; Craig & Schulkes 1985).

A problem that has been studied most extensively in the literature is the thermal stability of a simple static coronal loop. In such a loop, the product of pressure and loop length is related to the maximum temperature via a scaling law (e.g., Rosner et al. 1978). By using this relationship and a power law fit to the emissivity at high temperatures, of the form $f(T) \sim T^{-1/2}$, one can estimate from eq. (13) the ratio of the radiative and conductive time scales at the loop apex. The result is $\tau_R/\tau_C \simeq 7$, independent of the length and pressure of the loop. Consequently, near the top of the loop radiation is unimportant and thus unlikely to drive an instability. The energy balance is determined mainly by heating and conduction. The same applies to the whole high temperature portion of the loop, which occupies most of its volume. The coolest portion (i.e., the chromosphere and lowermost TR) can also not contribute to a radiative instability because at low temperatures the emissivity (even at constant pressure) is an increasing, rather than decreasing, function of T. Therefore, if a loop is unstable, one would expect that the instability is mainly driven by gas at temperatures just above the maximum of the emissivity function, where this function is decreasing, but still large enough that radiation is important, and where the temperature is not yet so high that conduction dominates. In fact, in those of the stability analyses discussed below which give instability (e.g., Antiochos et al. 1985), the eigenvectors of unstable modes tend to peak in this temperature regime. In the lower TR of standard loop models, heating is unimportant, and conduction balances radiation, $\tau_C \simeq \tau_R$. This suggests that a loop might be close to neutral stability and that the result of a stability analysis depends sensitively on how the physics of the lower TR is modeled.

These remarks explain some of the major findings of a large number of investigations of coronal loop stability. The first _linear_ analyses yielded intability (Antiochos 1979; Hood & Priest 1980). Antiochos derived a growth time of only 0.1 s. Chiuderi et al. (1981) and Craig & McClymont (1981), however, argued that by adding more cool, radiatively stable material to the base of the loop model one obtains stability, or at least growth times that are very large, of the order of days. (For reviews of this early work see Priest 1981 and Rosner 1981.) Of particular importance in this respect are radiative transfer effects in Lα, which reduce the emission and stabilize the gas at low temperatures (McClymont & Canfield 1983). Recently, Antiochos et al. (1985) presented a new discussion of the lower boundary conditions. They again obtained linear instability, with a growth time on the order of the coronal conductive cooling time, $\tau_C \simeq 10^3$ s. McClymont & Craig (1985a, b), who treated the loop footpoints differently, found neutral stability. They emphasized in particular the importance of a possible spatial dependence of the heating (see also Hammer et al. 1982) and of the chromospheric evaporation process.

Since 1982 it has been generally appreciated that this latter process, which provides a coupling between chromosphere and corona, is crucial for the stability of a coronal loop. Kuin & Martens (1982; see also Martens & Kuin 1983) modeled this coupling in a very approximate way in a study of the _nonlinear_ behavior of an unstable loop. They suggested that Solar coronal loops undergo limit cycle oscillations, with periods on the order of a day. During these oscillations, the TR periodically moves far down into the chromosphere and back again, and the coronal temperature and density vary by large amounts. This picture was challenged by Craig & Schulkes (1985), who argued that Kuin and Martens had underestimated the coupling between chromosphere and corona, and that an improved treatment leads to stable loops.

Again in 1982, several nonlinear hydrodynamic flux tube codes (for a review of such codes see Ulmschneider & Muchmore 1986) became available, which were used to study loop stability (Craig et al. 1982; Oran et al. 1982; Peres et al. 1982; An et al. 1983; Mariska et al. 1986). The results of these calculations typically indicate that the TR is very robust and nonlinearly stable. If a loop is forced into instability, then the nonlinear evolution is determined by the chromospheric evaporation/coronal condensation process (An et al. 1983).

5 CONCLUSIONS AND PROSPECTS

Waves are responsible for much of the dynamics and for some of the inhomogeneity of the outer Solar atmosphere. We have now a basic understanding of the mean structure of the chromosphere and TR, as illustrated by the Vernazza et al. (1981) models (Sect. 2.1). In particular, we know why in such models the temperature gradient is flat in the chromosphere and steep in the TR, and we think we know what determines the location of the TR and thus the vertical extent of the chromosphere. We also know, however, that the real Solar atmosphere fluctuates about the mean in space (inhomogeneity), velocity space (steady flows), and time (dynamics) to such an extent that the value of an average atmosphere model becomes questionable (Sect. 2.2).

The inhomogeneity of the outer atmosphere leads to the possibility of a multitude of wave phenomena (Sect. 3), of which we are presently only beginning to understand the most idealized cases. We do have a sound knowledge of the behavior of adiabatic waves in flux tubes, magnetic interfaces, and slabs. Nonadiabatic waves have been studied only in the thin flux tube approximation and for highly simplified treatments of the nonadiabatic processes, radiation and conduction. Since the thin flux tube approximation breaks down in the chromosphere, we ultimately have to study thick tubes. We also have to aim for a full treatment of nLTE radiative transfer in the few most important chromospheric lines, and we will have to include the effects of nonequilibrium ionization and nonlocal thermal conduction.

Observationally, we should strengthen ongoing efforts to detect chromospheric wave phenomena by means of time series. Moreover, it is of crucial importance that we find ways to improve the measurements of the chromospheric canopy, where the field lines of neighbor flux tubes merge (Sect. 2.1). Theoretically, we need to understand better what the mergence of the field lines implies for the transmission and reflection at the canopy of waves in the ambient medium, and for the propagation and dissipation of waves in the flux tubes. We need to understand, and take into account in model calculations, the mechanical interaction between the ambient medium and a (nonlinear) wave in a (thick) flux tube (Sect. 3).

Molecules are important for both the energetics and the stability of the lower chromosphere (Sect. 4.1). We should try to clarify observationally whether a hot chromosphere exists only inside flux tubes that are embedded in a cool atmosphere.

The thermal stability of the TR and corona (Sect. 4.2) is still debated in the literature. A strong tendency towards neutral stability becomes apparent, however. The stability of a model depends most sensitively on the assumed physics of the lower TR and of those regions that strongly influence this layer. Of particular importance for stability are radiation transfer in Lα, the mass exchange between chromosphere and TR, and the heating mechanism. Proposers of new heating mechanisms should check whether their mechanism produces a stable or unstable corona.

Let me return to waves. I have reviewed only very briefly (Sect. 3) some properties of the basic wave modes of a thin flux tube: longitudinal, transverse, and torsional waves. Propagating waves can be excited efficiently in all three modes. Unlike the longitudinal mode, the transverse and torsional modes are not much affected by radiative damping in the photosphere. Various dissipation mechanisms are conceivable. We should, however, keep in mind that we need not necessarily heat the whole outer atmosphere of the Sun by waves of any kind. Other heating mechanisms (reconnection) are possible as well. But waves are certainly present; they have been observed. They carry fluxes of mass, momentum, and energy with them. These fluxes produce flows; they control the extent of the atmosphere and the observed line profiles; and they lead to energy redistribution and/or heating. In all these ways, waves affect the structure of the atmosphere - but how exactly they do so can only be understood by means of ever improving calculations.

REFERENCES

An, C.-H., Canfield, R.C., Fisher, G.H. & McClymont, A.N. (1983). Ap.J. 267, 421.
Antiochos, S.K. (1979). Ap.J. 232, L 125.
Antiochos, S.K. (1984). Ap.J. 280, 416.
Antiochos, S.K. & Noci, G. (1986). Ap.J. 301, 440.
Antiochos, S.K., Shoub, E.C., An, C.-H. & Emslie, A.G. (1985). Ap.J. 298, 876.
Anzer, U. & Galloway, D.J. (1983). M.N.R.A.S. 203, 637.
Anzer, U. & Galloway, D.J. (1985). In: Chromospheric Diagnostics and Modelling, ed. B.W. Lites, p. 199.
Athay, R.G. (1981). Ap.J. 250, 709.
Athay, R.G. (1985a). Solar Phys. 100, 257.
Athay, R.G. (1985b). In: Theoretical Problems in High Resolution Solar Physics, ed. H.U. Schmidt, MPA 212, p. 205.
Ayres, T.R. (1981). Ap.J. 244, 1064.
Ayres, T.R. (1985). In: Chromospheric Diagnostics and Modelling, ed. B.W. Lites, p. 259.
Ayres, T.R. & Testerman, L. (1981). Ap.J. 245, 1124.
Ayres, T.R., Testerman, L. & Brault, J.W. (1986). Ap.J. 304, 542.
Bohn, H.U. & Stein, R.F. (1985). In: Chromospheric Diagnostics and Modelling, ed. B.W. Lites, p. 228.
Bonnet, R.M., Bruner, E.C., Jr., Acton, L.W., Brown, W.A. & Decaudin,M. (1980). Ap.J. 237, L 47.
Brückner, G. (1986). These proceedings.
Chiuderi, C., Einaudi, G. & Torricelli-Ciamponi, G. (1981). Astron. Ap. 97, 27.
Couturier, P., Mangeney, A. & Souffrin, P. (1979). In: IAU Symp. 91, eds. M. Dryer & E. Tandberg-Hanssen, Reidel, p. 127.
Craig, I.J.D. & McClymont, A.H. (1981). Nature 294, 333.
Craig, I.J.D. & Schulkes, R.H.S.M. (1985). Ap.J. 296, 710.
Craig, I.J.D., Robb, T.D. & Rollo, M.D. (1982). Solar Phys. 76, 331.
Deubner, F.-L. (1984). In: Small Scale Dynamical Processes in Quiet Stellar Atmospheres, ed. S.L. Keil, p.2.
Deubner, F.-L. (1985a). In: Chromospheric Diagnostics and Modelling, ed. B.W. Lites, p. 279.
Deubner, F.-L. (1985b). In: Theoretical Problems in High Resolution Solar Physics, ed. H.U. Schmidt, p. 160.
Dufton, P.L., Kingston, A.E. & Keenan, F.P. (1984). Ap.J. 220, L 35.

Edwin, P.M. & Roberts, B. (1986a). Wave Motion 8, 151.
Edwin, P.M. & Roberts, B. (1986b). These proceedings.
Einaudi, G. (1984). In: The Hydromagnetics of the Sun, ESA SP-220, p. 147.
Feldman, U. (1983). Ap.J. 275, 367.
Ferriz-Mas, A. & Moreno-Insertis, F. (1986). Astron. Ap., in press.
Field, G.B. (1965). Ap.J. 142, 531.
Foing, B., Bonnet, R.-M. & Bruner, M. (1986). Astron. Ap. 162, 292.
Foukal, P. & Smart, M. (1981). Solar Phys. 69, 15.
Giovanelli, R.G. (1975). Solar Phys. 44, 299.
Giovanelli, R.G. (1978). Solar Phys. 59, 293.
Giovanelli, R.G. (1980). Solar Phys. 68, 49.
Giovanelli, R.G., Livingston, W.C. & Harvey, J.W. (1978). Solar Phys. 59, 40.
Giovanelli, R.G. & Jones, H.P. (1982). Solar Phys. 79, 267.
Hammer, R. (1982a,b). Ap.J. 259, 767 and 779.
Hammer, R. (1983). Adv. Space Res. Vol. 2 No. 9, p. 261.
Hammer, R., Linsky, J.L. & Endler, F. (1982). In: Four Years of IUE Research, eds. Y. Kondo, J.M. Mead & R.D. Chapman, NASA CP-2238, p. 268.
Hammer, R. & Linsky, J.L. (1984). In: Fourth European IUE Conference, ESA SP-218, p. 25.
Hasan, S.S. & Kneer, F. (1986). Astron. Ap. 158, 288.
Hasan, S.S. & Schüßler, M. (1985). Astron. Ap. 151, 69.
Hearn, A.G. & Vardavas, I. (1981a,b). Astron. Ap. 98, 230 and 241.
Hearn, A.G., Kuin, N.P.M. & Martens, P.C.H. (1982). Astron.Ap. 125, 69.
Herbold, G., Ulmschneider, P., Spruit, H.C. & Rosner, R. (1985). Astron. Ap. 145, 157.
Heyvaerts, J. (1984). In: The Magnetohydrodynamics of the Sun, ESA SP-220, p. 123.
Hollweg, J.V. (1982). Ap.J. 257, 345.
Hollweg, J.V. (1985). In: Chromospheric Diagnostics and Modelling, ed. B.W. Lites, p. 235.
Hollweg, J.V., Jackson, S. & Galloway, D. (1982). Solar Phys. 75, 35.
Hood, A.W. (1985). In: Solar System Magnetic Fields, ed. E.R. Priest, Reidel, p. 80.
Hood, A.W. & Priest, E.R. (1978). Astron. Ap. 87, 126.
Jones, H.P. (1985). In: Chromospheric Diagnostics and Modelling, ed. B.W. Lites, p. 175.
Jones, H.P. & Giovanelli, R.G. (1983). Solar Phys. 87, 37.
Jordan, C. (1980). Astron. Ap. 86, 355.
Joselyn, J.A., Munro, R.H. & Holzer, T.E. (1979). Ap.J.Suppl. 40, 793.
Kneer, F. (1980). Astron. Ap. 87, 229.
Kneer, F. (1983). Astron. Ap. 128, 311.
Kneer, F. (1984). In: Small Scale Dynamical Processes in Quiet Stellar Atmospheres, ed. S.L. Keil, p. 110.
Kneer, F. (1985). In: Chromospheric Diagnostics and Modelling, ed. B.W. Lites, p. 252.
Kneer, F. (1986). In: Small Scale Magnetic Flux Concentrations in the Solar Photosphere, eds. W. Deinzer, M. Knölker & H.H. Voigt, Vandenhoeck & Ruprecht, Göttingen, p. 147.
Kneer, F. & v. Uexküll, M. (1985). Astron. Ap. 144, 443.

Kuin, N.P.M. & Martens, P.C.H. (1982). Astron. Ap. 108, L 1.
Krall, K.R. & Antiochos, S.K. (1980). Ap.J. 242, 374.
Mariska, J.T. (1984). Ap.J. 281, 435.
Mariska, J.T. (1986). Ann. Rev. Astron. Ap. 24, 23.
Mariska, J.T. & Boris, J.P. (1983). Ap.J. 267, 409.
Mariska, J.T. & Hollweg, J.V. (1985). Ap.J. 296, 746.
Mariska, J.T., Klimchuk, J.A. & Antiochos, S.K. (1986). Bull. Am. Astron. Soc. 18, 708.
Martens, P.C.H. & Kuin, N.P.M. (1983). Astron. Ap. 123, 216.
McClymont, A.N. & Canfield, R.C. (1983). Ap.J. 265, 497.
McClymont, A.N. & Craig, I.J.D. (1985a,b). Ap.J. 289, 820 and 834.
Mein, N. (1977). Solar Phys. 52, 283.
Muchmore, D. (1986a). Astron. Ap. 155, 172.
Muchmore, D. (1986b). Private communication.
Muchmore, D. & Ulmschneider, P. (1985). Astron. Ap. 142, 393.
Nordlund, A. (1985). In: Theoretical Problems in High Resolution Solar Physics, ed. H.U. Schmidt, MPA 212, p.1.
Noyes, R.W. & Hall, D.N.B. (1972). Ap.J. 176, L 89.
Oran, E.S., Mariska, J.T. & Boris, J.P. (1982). Ap.J. 254, 349.
Owocki, S.P. & Canfield, R.C. (1986). Ap.J. 300, 420.
Parker, E.N. (1953). Ap.J. 117, 431.
Parker, E.N. (1982). Geophys. Astrophys. Fluid Dyn. 22, 195.
Peres, G., Rosner, R., Serio, S. & Vaiana, G.S. (1982). Ap.J. 252, 791.
Pneuman, G.W., Solanki, S.K. & Stenflo, J.O. (1986). Astron. Ap. 154, 231.
Priest, E.R. (1981). In: Solar Active Regions, ed. F.Q. Orrall, Boulder Ass. Univ. Press, Ch. 9.4.
Priest, E.R. (1982). Solar Magnetohydrodynamics. Reidel, Dordrecht.
Rabin, D. & Moore, R. (1984). Ap.J. 285, 359.
Rae, I.C. & Roberts, B. (1982). Ap.J. 256, 761.
Roberts, B. (1983). Solar Phys. 87, 77.
Roberts, B. (1985). Phys. Fluids 28 (11), 3280.
Roberts, B. (1986). In: Small Scale Magnetic Flux Concentrations in the Solar Photosphere, eds. W. Deinzer, M.Knölker & H.H. Voigt. Vandenhoeck & Ruprecht, Göttingen, p. 169.
Roberts, B. & Mangeney, A. (1982). MNRAS 198, 7p.
Roberts, B. & Webb, A.R. (1978). Solar Phys. 56, 5.
Rosner, R. (1981). Nature 294, 611.
Rosner, R., Tucker, W.H. & Vaiana, G.S. (1978). Ap.J. 220, 643.
Roussel-Dupré, R. (1980a,b). Solar Phys. 68, 243 and 265.
Schmieder, B. (1978). Solar Phys. 57, 245.
Schmitz, F., Ulmschneider, P. & Kalkofen, W. (1985). Astron. Ap. 148, 217.
Schüßler, M. (1984). Astron. Ap. 140, 453.
Shoub, E.C. (1983). Ap.J. 266, 339.
Skumanich, A., Smythe, C. & Frazier, E.N. (1975). Ap.J. 200, 747.
Spiegel, E.A. (1957). Ap.J. 126, 202.
Spruit, H.C. (1981). In: The Sun as a Star, ed. S.D. Jordan, NASA SP-450, p. 385.
Spruit, H.C. (1982). Solar Phys. 75, 3.
Spruit, H.C. (1984). In: Small Scale Dynamical Processes in Quiet Stellar Atmospheres, ed. S.L. Keil, p. 249.

Spruit, H.C. & Roberts, B. (1983). Nature 304, 401.
Staiger, J. (1985). Ph.D. Thesis, Univ. Freiburg.
Stein, R.F. (1985). In: Chromospheric Diagnostics and Modelling, ed.
    B.W. Lites, p. 213.
Stein, R.F. (1986). Private communication.
Tavakol, R.K. & Tworkowski, A.S. (1981). Solar Phys. 71, 203.
Thomas, J.H. (1985). In: Theoretical Problems in High Resolution Solar
    Physics, ed. H.U. Schmidt, MPA 212, p. 126.
Trujillo Bueno, J. & Kneer, F. (1986). These proceedings; and Astron.
    Ap., in press.
Ulmschneider, P. (1970). Solar Phys. 12, 403.
Ulmschneider, P. (1985). In: Theoretical Problems in High Resolution
    Solar Physics, ed. H.U. Schmidt, MPA 212, p. 150.
Ulmschneider, P. (1986). Adv. Space Res., in press.
Ulmschneider, P. & Muchmore, D. (1986). In: Small Scale Magnetic Flux
    Concentrations in the Solar Photosphere, eds. W. Deinzer,
    M. Knölker & H.H. Voigt, Vandenhoeck & Ruprecht, Göttingen,
    p. 191.
Vernazza, J.E., Avrett, E.H. & Loeser, R. (1981). Ap.J.Suppl. 45, 635.
Webb, A.R. & Roberts, B. (1980). Solar Phys. 68, 87.
Zähringer, K. & Ulmschneider (1986). These proceedings.
Zweibel, E. (1980). Solar Phys. 66, 305.

Nr. 274   Mitteilungen aus dem Kiepenheuer-Institut

NON-LINEAR ALFVEN-WAVES AND THE FINE-STRUCTURE OF MAGNETIC FIELDS IN THE PHOTOSPHERE

Eberhart Jensen
Institute of Theoretical Astrophysics
University of Oslo

Abstract: It is assumed that photospheric magnetic fields carry Alfven-waves excited in the convection zone. Postulating an Alfven-wave flux-density required to heat the outer solar atmosphere to be of the order of F_A = 5E5 cgs, it is shown that magnetic fields below 25-40 gauss will be destroyed by non- linear effects in the photospheric layers where magnetic fields are measured.

It has previously been suggested that non-linear Alfven-waves are responsible for the generation of the turbulent velocity-field characterizing the main body of quiescent prominences (Jensen, 1983, 1986). The observed magnetic fields in these prominences are found to have values within a limited interval. According to our hypothesis this may be understood as an effect of non-linear waves. If the carrier field is too strong the waves do not become sufficiently non-linear to generate turbulence. If the field is too weak it will be destroyed by dissipation.
In the following we shall apply a similar line of reasoning to photospheric magnetic fields.
Many observers have stressed the absence of weak magnetic fields in the photosphere (Frazier and Stenflo, 1972, Livingston and Harvey, 1975, Stenflo and Lindegren, 1977, Stenflo, 1982). In a recent paper Stenflo arrives at a value for the lower limit of the photospheric field in the interval 10G<B<100G.
Consider a vertical magnetic field, B, carrying the wave-flux, F_A, at a location where the mass-density is ρ. The amplitudes in velocity, ΔV, and in the magnetic field, ΔB, in the wave are then given by:

$$\Delta V = 2\pi^{1/2} \rho^{-1/4} F_A^{1/2} B^{-1/2} \qquad (1)$$

$$\Delta B = 4\pi^{3/4} \rho^{1/4} F_A^{1/2} B^{-1/2} \qquad (2)$$

We further introduce the relative magnetic amplitude in the waves, $R = \Delta B/B$. Then the second of the equations above takes the form,

$$R = 4\pi^{3/4} \rho^{1/4} F_A^{1/2} B^{-3/2}$$

Solved with respect to the carrier field of the waves, we get

$$B(R) = 2^{4/3} \pi^{1/2} \rho^{1/6} F_A^{1/3} R^{-2/3} \qquad (3)$$

We shall consider what happens at a depth in the photosphere where spectral lines used for observations of magnetic fields are formed i.e. at a continuum optical depth of $\tau_5 = 3E-2$. At this depth we choose a value of the density $\rho = 7E-8 \, gcm^{-3}$. Equation (3) now determines the magnetic field corresponding to a given value of the Alfven-wave flux F_A, and of the relative magnetic amplitude R. For a value of R of the order of unity the wave becomes strongly non-linear and it is assumed that the wave will dissipate and be unable to carry the flux imposed from below. The limiting value of the magnetic field strength obtained in this way is designated by $B_L(R)$. Choosing R=1 we get the lowest value for the limiting field strength below which the field is destroyed by non-linear effects. It is further assumed that excitation of Alfven-waves by turbulent eddies in the convection zone takes place all over the sun, creating a wave-flux, F_A.

The result of the discussion is shown in Fig. 1. The limiting field strength, $B_L(R)$, calculated from Equ. 3 is here plotted as a function of the wave flux for two values of the amplitude ratio, R=1 and 1/2. If we choose F_A = 5E5 cgs. as a "canonical" value the lower limit becomes 23 G with R=1 and 40 G for R=1/2. It should be pointed out that these limits are rather insensitive to the value chosen for the density, the only parameter where model data for the photosphere

Fig. 1. This is a double logarithmic plot to illustrate Equation (3). The limiting magnetic field strength for the carrier field, B_L, is shown as a function of the Alfven-wave flux density F_A for two values of the relative amplitude $R = \Delta B/B$, 1/2 and 1. In the photosphere at a depth where lines of moderate strengths are formed, the mass density, ρ, is taken to be $7E-8 \, gcm^{-3}$. The lowest line refers to conditions in quiescent prominences with $\rho = 1.3E-13 \, gcm^{-3}$. The velocity amplitudes in the waves, in km/s, are indicated along the lines.

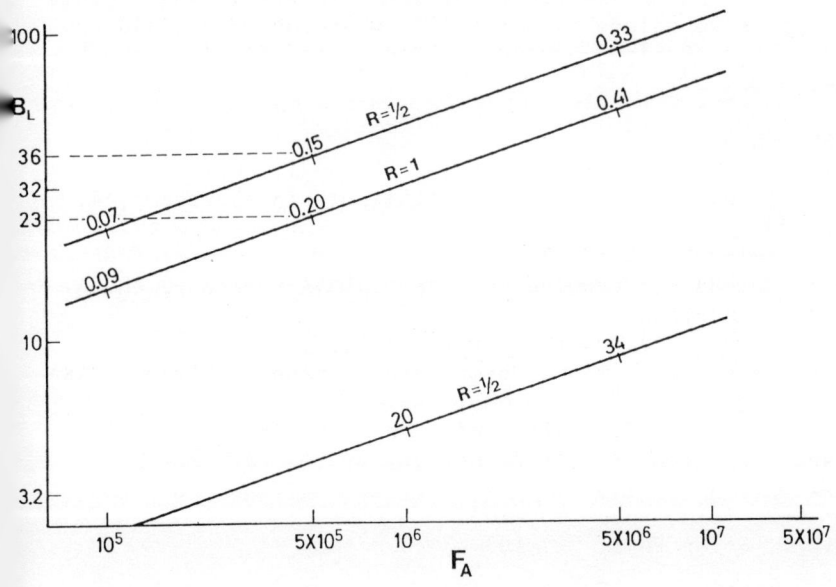

enters. A change in density by a factor of two corresponds to a 12% change in $B_L(R)$.

We note that the velocity amplitudes indicated along the straight lines are small, well below the velocity of sound, which at the depth we consider is C_s = 6.5 km/s. This is in contrast to the case of quiescent prominences, where the velocities set up are partly in the supersonic range.

It is of interest to see how the damping length for Alfven waves, L_A, as given by Wentzel (1977) fits into this picture. His equation is:

$$L_A = \frac{\tau}{64\pi^3 \alpha} \frac{B^4}{\rho F_A}$$

where τ denotes the wave period, and α is a constant of the order of unity. Referring to Fig. 1 a parameter combination R = 1, F_A = 5E5cgs, ρ = 7E-8gcm^{-3}, τ = 100s giving B = 23G, corresponds to a damping length L_A = 4km. For R = ½, with ρ and F_A the same but with τ = 300s and B = 36G we get L_A = 73km.

The simple arguments presented here thus lead to lower limits that tend to be consistent with Stenflo's observational results.

Local increase in F_A may be found in active regions. An increase in F_A by two orders of magnitude to 5E7 would, for instance, give B_L = 110G.

One may ask whether other pieces of evidence may be found that corroborate the existence of such effects on the sun. It would seem that in addition to the case of quiescent prominences already mention we may possibly count one more. That is the recent beautiful observations at Big Bear Solar Observatory (Hermans and Martin, 1986) of the disappearence of tiny filaments. They observed small dark arches getting fuzzy in the upper part and then gradually fading. This could be what happens to a loop with a fairly weak magnetic field when F_A increases locally with time, provided the magnetic field in the footpoints exceeds B_L.

REFERENCES

Frazier, E.N. and Stenflo, J.O., 1972, Solar Phys. <u>22</u>, 402.
Harvey, J.W., 1985, in E.A. Muller (ed.) Highlights in Astronomy, <u>4</u>, II, 223.
Jensen, E., 1983, Solar Phys. <u>89</u>, 275.
Jensen, E., 1986, Corona and Prominence Plasma, NASA workshop, ed. A. Poland, in press.
Livingston, W. and Harvey, J.W., 1975, Bull. AAS, <u>7</u>, 346.
Hermans, L.M. and Martin, S., 1986, Corona and Prominence Plasma, NASA workshop, ed. A. Poland, in press.
Stenflo, J.O. and Harvey, J.W., 1975, Solar Phys. <u>95</u>, 99.
Stenflo, J.O. and Lindegren. L., 1977, Astron. and Astrophys. <u>59</u>, 367.
Stenflo, J.O., 1982, Solar Phys. <u>80</u>, 209.
Wentzel, D.G. 1977, Solar Phys. <u>52</u>, 163.

POLOIDAL MODE COUPLING OF ALFVEN CONTINUUM MODES IN 2D
CORONAL LOOPS

S. POEDTS
Research Assistant of the Belgian National Science Foundation
KU Leuven, Celestijnenlaan 200B, 3030 Heverlee, Belgium

M. GOOSSENS
KU Leuven, Celestijnenlaan 200B, 3030 Heverlee, Belgium

1 INTRODUCTION

Resonant absorption and phase mixing of Alfvén waves are two candidates for the magnetic heating of the solar corona which have attracted attention in recent years (Ionson 1978; Heyvaerts and Priest 1983). They are based on the existence of a continuous spectrum in linear ideal magnetohydrodynamics (MHD) which implies that in ideal MHD individual magnetic surfaces can oscillate independently at their own frequency (see, e.g., Appert et al. 1974; Goedbloed 1975). The range of driving frequencies for resonant absorption and phase mixing, the position of the resonant magnetic surface for a given driving frequency, the spatial confinement of resonant absorption and the efficiency of phase mixing are determined by the continuous spectrum of linear ideal MHD. It is therefore necessary to study the continuous spectrum of linear ideal MHD in order to understand resonant absorption and phase mixing. In the context of heating of the solar corona, the continuous spectrum, resonant absorption and phase mixing have been studied only for one dimensional (1D) plasmas. Study of the continuous spectrum of 2D solar loops is obviously required to see how 2D effects change the continuous spectrum and influence resonant absorption and phase mixing.

2 THE EQUILIBRIUM AND COORDINATES

The continuous spectrum is determined for static 2D models for coronal loops with a purely poloidal magnetic field. A Cartesian system of coordinates (x, y, z) is introduced so that the equilibrium quantities depend on x and z but not on the ignorable coordinate y. The poloidal magnetic field is described with a flux function $\psi(x, z)$ which is chosen as

$$\psi(x, z) = \psi_0 \left(\frac{x^2}{a^2} + \frac{z^2}{b^2}\right) . \tag{1}$$

The cross-sections of the magnetic surfaces with a plane y = constant are circles ($a = b$) or ellipses ($a \neq b$). A system of magnetic coordinates (θ, ψ, y) is introduced, defined as

$$x = a \left(\frac{\psi}{\psi_0}\right)^{1/2} \cos\theta \quad , \quad z = b \left(\frac{\psi}{\psi_0}\right)^{1/2} \sin\theta \quad , \quad y = y. \tag{2}$$

ψ is the continuous label of the magnetic surfaces and θ is a poloidal angle-like variable on a magnetic surface. The system of coordinates (θ, ψ, y) is orthogonal only if $a = b$. The geometrical effects are contained in the elements of the metric tensor and do not appear explicitly in the equations for the continuous spectrum.

The pressure satisfies the equation of hydrostatic equilibrium and is a linear function of ψ alone,

$$p(\psi) = P_0 \left[1 + \frac{1}{\beta_0} (1 + \frac{1}{c^2})(1 - \frac{\psi}{\psi_0})\right] \tag{3}$$

where $P_0 = p(\psi_0)$, $\beta_0 = 2\mu P_0/B_0^2$ and $B_0 = 2\psi_0/a$. Since gravity has been neglected, the density does not occur in the equation of hydrostatic equilibrium. Density is chosen as

$$\rho = \rho_0(\psi) , \quad \text{or} \quad \rho = \rho_0(\psi) \exp(-Kz) \tag{4}$$

with K a positive constant.

3 ALFVEN CONTINUOUS SPECTRUM

The continuous spectrum of a 2D static plasma with a purely poloidal magnetic field consists of two uncoupled parts which can be referred to as an Alfvén continuum and a slow continuum since the solutions are polarized in the magnetic surfaces either perpendicular to the magnetic field lines (Alfvén continuum) or along the magnetic field lines (slow continuum) (Poedts et al. 1985; Goossens et al. 1985). The perturbed quantities are Fourier analyzed with respect to the ignorable coordinate y and are put proportional to exp(iny). The improper normal dependence $\delta(\psi - \psi_c)$ is separated in the solutions, and $\xi^y(\theta)$ denotes the proper dependence of the contravariant y-component of the displacement vector in the magnetic surface $\psi = \psi_c$. The present paper is limited to a brief presentation of results for the Alfvén continuum. The equation that governs the Alfvén continuous spectrum is

$$\lambda \rho(\psi_c, \theta) \xi^y(\theta) + \frac{d^2 \xi^y(\theta)}{d\theta^2} = 0 , \tag{5}$$

where
$$\lambda = \frac{\mu}{(B^\theta)^2} \sigma^2 . \tag{6}$$

B^θ is the contravariant component of the magnetic field and σ is the eigenfrequency (time dependence exp(iσt)).

Equation (5) is a non-singular eigenvalue problem on a given magnetic surface, when supplemented with proper boundary conditions. The boundary conditions adopted here are $\xi^y(0) = \xi^y(\pi) = 0$. The eigensolutions obtained under these boundary conditions are mutually orthogonal and can be used to represent any function in $[0, \pi]$ after a simple change of variable. In particular, they can be used to represent the solution of Eq. (5) with non-homogeneous boundary conditons (see, e.g., Section IV of Heyvaerts and Priest 1983 for a similar situation). The Alfvén continuous spectrum is obtained as follows. We solve Eq. (5) for a given toroidal wave number n on a given magnetic surface ψ_c and obtain a discrete set of eigenvalues $\{\lambda_A(\psi_c)\}_m$, where the index m refers to the different oscillatory solutions. When the flux surface $\psi = \psi_c$ is varied the discrete eigenvalue yields the Alfvén continuous spectrum for given m, $\{\min \lambda_A(\psi), \max \lambda_A(\psi)\}$.

Equation (5) does not contain any geometrical effects and non-circularity does not affect Alfvén waves. We therefore present results for circular cross-sections with a density profile that decreases exponentially with height z as given by (4 b). For small values of K a perturbation scheme has been used. It is found that a given zeroth-order solution with poloidal wave number M

$$(\lambda^0)_M = \frac{M^2}{\rho_0}, \qquad (\xi^{y0})_M = \eta \sin M\theta , \qquad (7)$$

acquires a first order correction to the eigenfunction and the eigenfrequency. The correction to the eigenfunction involves all poloidal wave numbers m with the same parity as M. The dependence of the eigenfuction $\xi^y(\theta)$ on θ cannot be specified by one wave number but several wave numbers are required. For Alfvén continuum modes we even need an infinite number of wave numbers. This phenomenon is called poloidal wave number coupling and is caused by the variation of the equilibrium quantities with respect to θ. Density stratification in poloidal magnetic surfaces causes, at least in principle, a complete poloidal wave number coupling for poloidal wave numbers with the same parity as M. In practice, a finite number of wave numbers suffices to represent the eigensolution. The analytical results for small K already suggest that density variations along magnetic field lines have a strong effect on the Alfvén continuous spectrum. The perturbation scheme can only be applied for small values of K. Equation (5) has been solved by numerical integration for increasing values of K.

Numerical results are displayed on Figures 1 and 2. A linear function has been taken for $\rho_0(\psi)$. The Alfvén continuum for a spatial eigenfunction with five oscillations (M = 5) is depicted on Fig. 1 for K = 1.6. This value of K corresponds to a variation of density from the bottom to the top of a magnetic field line by a factor of 5. The range of ψ-values corresponds to a density variation across the loop by a factor of 5. In Fig. 1 we have also shown the Alfvén continuum for the circular loop with constant density $\rho_0(\psi)$ along field lines. Fig. 1 shows that density stratification along magnetic field lines has a profound effect on the Alfvén continuum. The continuum eigenvalues are shifted to larger values, but more importantly the range of continuum eigenvalues and hence $d\sigma_A(\psi)/d\psi$ is increased by about a factor 3. Consequences for resonant

Fig. 1 : Alfvén continuum for M= 5, K = 0 (Z) and K = 1.6 (E)

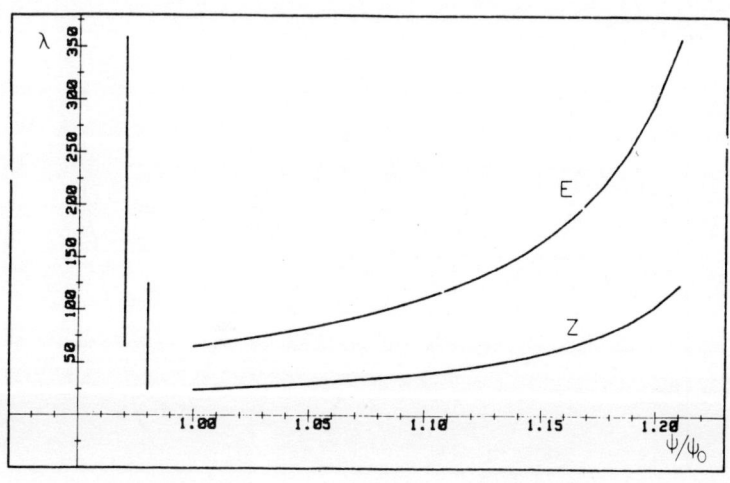

absorption and phase mixing are a substantial increase in the range of
driving frequencies, a change of the position of the resonant surface for
a given driving frequency, a pronounced spatial confinement of resonant
absorption and an increase of the efficiency of phase mixing which is
proportional to $d\sigma_A(\psi)/d\psi$ as indicated by Sakurai (1985). In Fig. 2
the eigenfunction is depicted. This eigenfunction has to be compared to
$\sin 5\theta$ which is the eigenfunction of the corresponding loop with constant
density along the magnetic field lines. The amplitude of the eigenfunction increases from bottom to top so that the spatial derivative of
velocity with respect to ψ increases from bottom to top. The efficiency
of phase mixing is increased compared to a loop with constant density
along the magnetic field lines. The difference between the actual
eigenfunction and $\sin 5\theta$ clearly illustrates that the eigenfunction
cannot be represented by one poloidal wave number, namely $M = 5$, but that
several poloidal wave numbers are required (in principle we need to have
an infinite number of poloidal wave numbers).

The results reported here show that the Alfvén continuum of
static 2D solar loops can be changed substantially compared with the 1D
plasma case. This can have important consequences for resonant absorption
and phase mixing. In particular the present results indicate that
density variation along magnetic field lines increases the efficiency of
phase mixing of Alfvén continuum waves.

REFERENCES
Appert, K., Gruber, R., Vaclavik, J. (1974). Phys. Fluids **17**, 1471
Goedbloed, J.P. (1975). Phys. Fluids **18**, 1258
Goossens, M., Poedts, S., Hermans, D. (1985). Solar Physics **102**, 51
Heyvaerts, J., Priest, E.R. (1983). Astron. Astrophys. **117**, 220
Ionson, J. (1978). Astrophys. J. **226**, 650
Poedts, S., Hermans, D., Goossens, M. (1985). Astron. Astrophys. **151**, 16
Sakurai, T. (1985). Proceedings of the MPA/LPARL Workshop, Theoretical
        Problems in High Resolution Solar Physics, 263

Fig. 2 : Eigenfunction of Alfvén continuum ($M = 5$, $K = 1.6$)
on $\psi/\psi_0 = 1$.

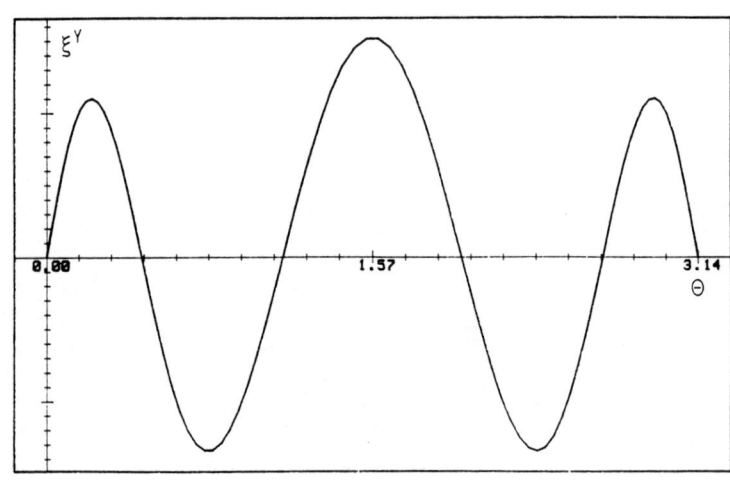

RADIATIVE RELAXATION IN SMALL SCALE STRUCTURES

JAVIER TRUJILLO-BUENO and FRANZ KNEER
Universitäts-Sternwarte
Geismarlandstr. 11
3400 Göttingen, F.R.G.

INTRODUCTION

When one aims at understanding the role of fine-scale magnetic fields on the structure and dynamics of the solar atmosphere, Multidimensional Radiative Transfer (MRT) has to be taken into account. Energy exchange by radiation must be included in gas dynamic studies (e.g. Hammer, Schüßler; these proceedings). Also, to interpret observations of a hightly inhomogeneous plasma like the solar atmosphere, we must learn about the effects which photons produce when moving around and interacting with different structures before escaping.

This contribution has both, comparative and demonstrative purposes with respect to some of the effects of MRT on the energy exchange by radiation in small-scale structures.

We start with Spiegel's work (1957) on this subject. He formulated the radiative relaxation problem of thermal perturbations and obtained an analytic expression for the relaxation time of small sinusoidal temperature fluctuations in an infinite, homogeneous, grey medium in LTE. The formula has been widely used (see e.g. ref.1 and 9) to simulate radiative relaxation in the solar atmosphere, i.e. in a physical system which possesses a surface through which photons escape and which is stratified. Thus several questions arise: What are the effects of a surface and stratification on the relaxation process? Can spectral lines compete with the continuum radiation? Are non-LTE lines as important as LTE lines? What can we say about the vertical versus horizontal migration of photons during radiative relaxation processes?

There are of course more questions which we will mention at the end. It is however our strategy to find out the important effects by means of a step by step investigation. In this contribution the effects of perturbations in the Planck function will be considered. More details and information can be found in Kneer (1986) and in a series of papers (e.g. Trujillo-Bueno and Kneer, 1987). Here, the main objective is physical insight.

FORMULATION OF THE PROBLEM

We follow Unno and Spiegel (1966) by simplifying the energy equation and concentrating on the radiative transfer effects. Accordingly, the time behaviour of a thermal perturbation applied to a system in radiative equilibrium (RE) is written as

$$\rho C_p \frac{\partial T}{\partial t} = 4\pi \int_0^\infty \chi_\nu (J_\nu - S_\nu) \, d\nu \quad , \tag{1}$$

with ρ the density and C_p the specific heat per gram at constant pressure. The transfer of radiation is a complicated process: The mean intensity J_ν is a non-local function of the emitting and absorbing conditions throughout the whole atmosphere. These conditions enter the radiative heat equation (1) via the opacity χ_ν and source function S_ν. Because these functions are in part determined by the radiation field itself, the radiative relaxation problem is in addition non-linear. Both, non-localness and non-linearity act together in a complicated way damping (or amplifying !) the fluctuations.

By applying small temperature disturbances of the form $T = \bar{T} + \Delta T(Z) \cos kx$ with wavenumber $k = 2\pi/\Lambda$, fluctuations in all physical quantities appear. We assume that the amplitude of the Planck function B_ν remains independent of height Z. Thus for a k=0 perturbation the fluctuating part of the problem reduces to that of a plane-parallel isothermal atmosphere. This facilitates gaining of insight and some analytical work.

After linearizing (1) one is left with the equation which governs the time evolution of $\Delta T(Z)$. Then a <u>local radiative relaxation time</u> $t_r = 1/n$ can be defined, with n the growth rate of the perturbation. t_r is a time scale for the speed of radiative processes in comparison with dynamic time scales. In linear analysis contributions due to Planck function and opacity fluctuations can be investigated separately. As mentioned, only the former effects will be considered. Therefore $t_r = 1/n_{\Delta B}$.

SPECIFIC PROBLEMS
a) Grey continuum in LTE

For this case the growth rate of the perturbation is

$$n(K,Z) = \frac{16\sigma \bar{T}^3}{C_p} \frac{\bar{\chi}}{\bar{\rho}} (1 - \frac{\Delta J}{\Delta B}) \qquad (2)$$

It depends on non-perturbed quantities describing the background atmosphere like $\bar{\chi}/\bar{\rho}$ and on the amplitude of the frequency-integrated mean intensity fluctuations. For an infinite and homogeneous medium Spiegel (1957) found $\Delta J/\Delta B = \bar{\chi}/k \cot^{-1}(\bar{\chi}/k)$.

In the presence of a surface and stratification $\Delta J/\Delta B$ has to be calculated by numerically solving the transfer equation in two dimensions. To mimic the gravitational stratification in the Sun, we choose

$$\bar{\chi}_c = \bar{\chi}_{5000}(\tau_c = 1) e^{-Z/\mathcal{H}} \text{ and } \bar{\rho} = \bar{\rho}(\tau_c = 1) e^{-Z/2\mathcal{H}} \quad ,$$

where $\mathcal{H}$ is the opacity scale height (~60km for continuum radiation). For $\bar{T}(Z)$ we use a grey RE model atmosphere with $T_{eff} = T_{eff}^\odot$. For C_p we adopt the value $1.6 \cdot 10^8$ erg $K^{-1}g^{-1}$ throughout the atmosphere.

b) LTE and non-LTE spectral lines.

The contribution to $n_{\Delta B}$ due to one spectral line of strength $q = \chi_\ell/\chi_c$ and non LTE parameter ε (i.e. the collisional destruction probability is found to be

$$n_{\Delta B} = \frac{1}{C_p} \frac{\bar{X}_{5000}(\tau_c=1)}{\bar{\rho}} \frac{\bar{B}_\nu}{\bar{T}} \frac{h\nu}{k\bar{T}} \Delta\nu_D \Delta L(Z,k). \tag{3}$$

The function $\Delta L(Z,k)$ is the amplitude of the fluctuations of the energy losses by radiation per unit volume and time in units of $\bar{X}_{5000}$ ($\tau_c=1$) ΔB_ν $\Delta\nu_D$, and is given by

$$\Delta L(Z,k) = 4\pi e^{-Z/\lambda} \left[\epsilon_q \left(1 - \frac{\Delta \bar{J}_\nu}{\Delta B_\nu}\right) + \int_{line} \left(1 - \frac{\Delta J_\nu}{\Delta B_\nu}\right) dV \right]. \tag{4}$$

Frequencies are measured in units of the Doppler width ($V = \frac{\Delta\nu}{\Delta\nu_D}$). The relaxation time depends again on the background atmosphere. We use the same model as for the continuum calculations. Now line-defining parameters enter the equation, like ϵ, σ_ℓ and $\Delta\nu_D$. The two terms within brackets in equation (4) contain information on the amplitudes of the mean intensity. The first one comes from the spectral line itself for which the assumptions of two level atoms and complete redistribution have been considered. The second one has to do with the background continuum on which the line is superimposed. The chosen stratification for this problem is similar to the grey continuum case. Geometric distances are to be measured in units of the scale height.

The two-dimensional transfer problem has been solved by using the procedures outlined in Kneer and Heasley (1979) and Kneer (1981).

RESULTS AND DISCUSSION

Figure 1 has comparative purposes. We first discuss radiative relaxation by continuum radiation in a grey, LTE stratified background atmosphere. For a k=0 (i.e. $\Lambda=\infty$) perturbation an infinite relaxation time is obtained with Spiegel's formula. However, a surface does allow relaxation even for a k=0 disturbance. The depth variation of $1/t_r$ for this case is given by the dotted line. Only at optical depths $\tau_c = \exp(-Z) > 1$, the relaxation time grows rapidly approaching the conditions of an infinite medium.

As the wavenumber increases, the agreement between Spiegel's formula and our calculation improves. Note, however, that for k=0.2 (i.e. $\Lambda \simeq 1800$ km ~ granulation scale) we differ by an order of magnitude around $\tau_c=1$. For $k \geq 1$ the coincidence between the two calculations is good. At $\tau_c=1$ and $k \simeq 1$ relaxation takes place within a few seconds.

Near the surface the optically thin limit (i.e. $\Delta J \to 0$) is rapidly reached. There the opacity is low, the non-local effects extreme and the smoothing of the mean-intensity fluctuations very efficient. However, at large optical depths the use of the optically thin limit (k=∞ curve) would considerably overestimate the growth rate of the perturbation.

With respect to the coincidence between the two calculations for $k \geq 1$, two comments should be made. First, it is well known that the evolution in time of an arbitrary perturbation can be found by superposing Fourier components. As time progresses, and the disturbance smooths out, the dominant wavenumber tends to zero. Consequently, a superposition performed by using Spiegel's components would lead to

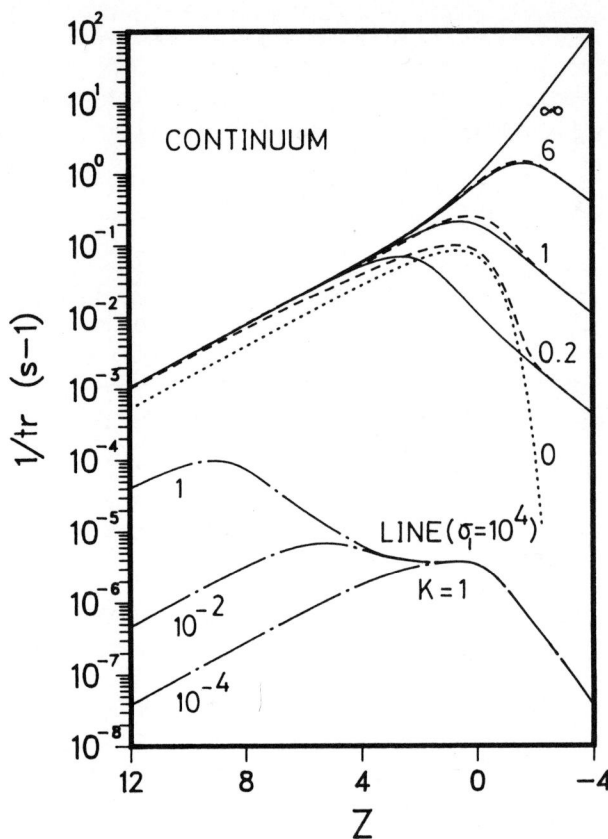

Fig. 1: Growth rate $n_{\Delta B}$ as a function of height Z. Parameter is the wavenumber k. Full lines: $1/t_r$ calculated with Spiegel's formula. Dashed: $n_{\Delta B}$ for grey continuum transfer in a stratified LTE atmosphere. Dashed-dotted: contribution of one spectral line of strength 10^4 for three values of ε. Optical depth unity in the continuum is at $Z=0$.

a very slow decay in time in comparison with the situation in which a surface is present. Secondly, Spiegel (1957) demonstrated that, to first order, opacity perturbations have no effect on the relaxation. This, however, is only true for situations in which an average radiative flux does not exist (Rybicki 1965). In a stratified atmosphere the growth rate due to opacity fluctuations is proportional to the average radiative flux and to the temperature sensitivity of the absorption coeffitient. It can be shown (Kneer, 1986) that the physical conditions in the solar atmosphere are such that the growth rate for opacity perturbations may be comparable to that of the Planck function fluctuations. We will expand on this in a forthcoming study.

Relaxation due to one spectral line of strength $\sigma_l = 10^4$ is also shown in Figure 1 for three non-LTE parameters ε. k=1 has been chosen for comparison with the continuum radiation effects. As it can be seen from Figure 1 and eq. (4), lines are an essential ingredient for the energy balance in small scale structures: When they are very strong (e.g. Ca K line) or very abundant they compete with the continuum in their dissipative effect.

Comparing strong lines with different values of ε we find that, near the surface, LTE lines (ε =1) have much more influence on the energy budget than non-LTE lines. This has to be expected form line-blanketing stellar atmospheres calculations. The stronger the non-LTE effects, the smaller the chance of energy deposition per scattering event and, correspondingly, the greater the relaxation time. However, at optical depths larger than the thermalization length, LTE and non-LTE lines have the same amount of radiative loss.

The influence of MRT on the energy exchange by spectral lines is shown in Figure 2. There the amplitude of the radiative loss function (i.e. $\Delta L(Z,k)$) is given for various wavelengths of the horizontal perturbation.

Fig. 2: Height dependence of the amplitude of the energy loss function due to one spectral line of strength 10^4 for three values of ε in units of $X_{5000}(\tau_C=1)\Delta B_\nu \Delta \nu_D$. Bottom to top: k= 0,0.2, 0.6, 2,4,6,8.

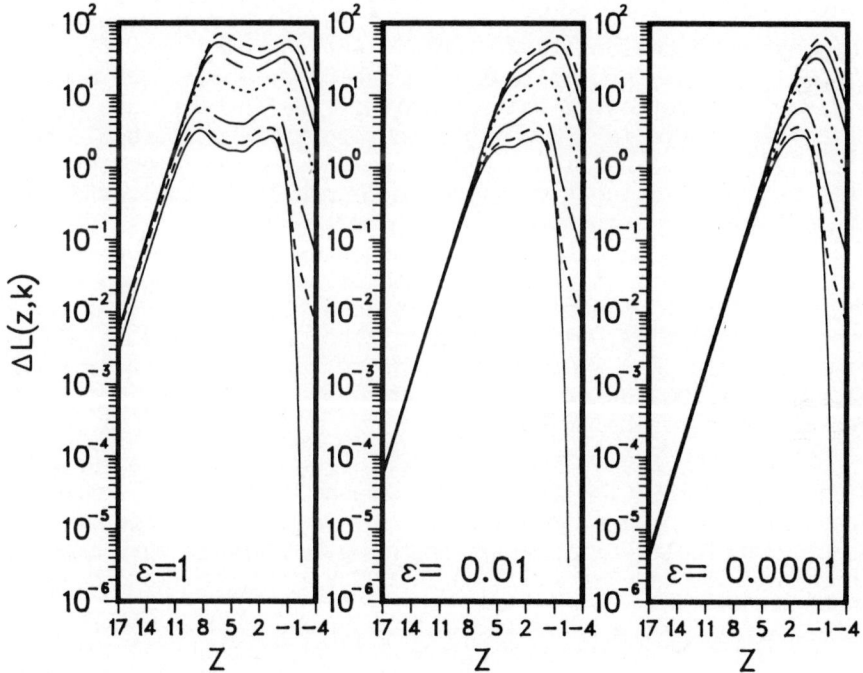

Generally, $\Delta L(Z,k)$ increases with increasing k. Near the surface the amplitudes of $J\nu$ are already small for k=0 and the optically thin limit (i.e. $\Delta J\nu$=o for k→∞) is rapidly reached. Consequently the k-dependence is smallest there.

The height Z where the radiation field thermalizes decreases with increasing wavenumber due to the possibility of horizontal photon exchange. This can be seen in the shift of the first maxima of $\Delta L(Z,k)$ towards lower Z values.

At larger depths $\Delta S_\nu \simeq \Delta B_\nu$, and the dependence of $\Delta L(Z,k)$ on k is largest, also due to horizontal radiative exchange. There the net radiative losses occur in the wings of the line, where the opacity is lowest and photons are still untrapped. These photons smooth out the fluctuations of J_ν, increasing $\Delta L(Z,k) = 4\pi \int X_\nu (\Delta S_\nu - \Delta J_\nu) \, d\nu$ with increasing k.

At still larger optical depths even wing photons are trapped due to absorption and destruction by the background continuum. Then the radiative loss function drops rapidly with decreasing height Z.
An important result to remark is the following: Already at structural lengths of three to four scale heights the radiative energy exchange deviates from the plane parallel case by an order of magnitude.

VERTICAL VERSUS HORIZONTAL MIGRATION OF PHOTONS DURING RADIATIVE RELAXATION PROCESSES

From the foregoing discussion it has become clear that horizontal transfer of photons is very efficient and damps the fluctuations of the mean intensity of the radiation field. We would like however to have a visual indication of this efficiency. We go back to a grey, RE, stratified atmosphere in LTE which is horizontally perturbed. Then the system evolves to reestablish equilibrium and photons migrate within a characteristic relaxation time given in Figure 1. Is horizontal migration of photons as important as the vertical? The answer is given in Figure 3. There, the amplitudes of the components of the radiative flux perturbations are shown as function of height. Although they are the result of "exact" numerical calculations, their behaviour can be understood by analytical manipulation of the multidimensional Eddington approximation (i.e. $\vec{F} = -(4\pi/3\mathbf{X}) \vec{\nabla} J$). We do not solve here this mathematical exercise. On the contrary we just turn to discussing the main features of the numerical results.

Clearly for k=0 only vertical flux exists. The curve k=0 indeed shows the radiative flux which happens to be in an isothermal, plane-parallel atmosphere. However, for k>0 a net transport of photons in the horizontal direction arises becoming relatively more important as k increases. Already at k=0.6 ($\Lambda \simeq 10 \hat{=} 600$ km) ΔF_x has become larger than ΔF_z for $Z \leq 4$.

There is other behaviour which merits physical interpretation. It refers to the existence of the maxima and their position. They indicate a transition region between an optically thick regime where the opacity is high and an optically thin limit where it is low. As the characteristic length of the structures is diminished, one has to go deeper in order to find higher opacities capable of forcing the photons to be destroyed within the same structures in which they were created.

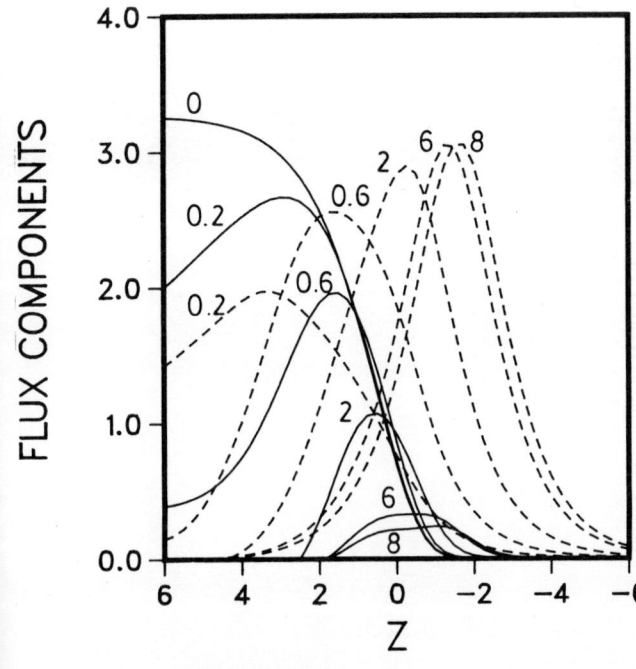

Fig. 3: Amplitudes of the flux components in dimension of ΔB as a function of height Z. Parameter is K. Full lines: vertical flux. Dashed: horizontal flux. Optical depth unity is at Z=0.

CONLUSIONS

Radiative relaxation is an important dissipative process in stellar atmospheres. Like on the Sun, the plasma of such physical systems may be very inhomogeneous. Abrupt changes of the thermodynamic properties within a few scale heights in the horizontal direction may exist which force radiative energy exchange to occur not only towards the stellar surface but into all directions. Thus, the effects of MRT have to be investigated.

To gain physical insight into its nature and to sort out the important effects is the motivation of this work. In this contribution the radiative relaxation problem of small thermal perturbations has been addressed with special emphasis on the effects of fluctuations in the Planck function.

In a future paper the role of opacity fluctuations and the possibility of radiative instabilities driven by molecules will be investigated. In realistic situations it should be considered that small-scale structures are height-dependent and that they are not small perturbations. In any case we are convinced that to gain first physical insight by means of studying basic problems is worthwhile.

ACKNOWLEDGEMENTS: J.T.B. would like to thank all friends at the Instituto de Astrofísica de Canarias for offering hospitality and facilities during several stays at this Institute. Support by the Deutsche Forschungsgemeinschaft through grant Kn 152/3-1 is gratefully acknowledged.

REFERENCES

Hasan, S.S.: 1985, Astron. Astrophys. 143, 39
Kneer, F.: 1981, Astron. Astrophys. 93, 387
Kneer, F.: 1986, in "Small-Scale Magnetic Flux Concentrations in the Solar Photosphere", Deinzer, Knölker, Voigt (eds.), Göttingen
Kneer, F., Heasley, J.: 1979, Astron. Astrophys. 79, 14
Rybicki, G.B.: 1965, Ph.D. Thesis, Harvard University
Spiegel, E.A.: 1957, Astrophys. J. 126, 202
Trujillo-Bueno, J., Kneer, F.: 1987, Astron. Astrophys. (in press)
Unno, W., Spiegel, E.A.: 1966, Publ. Astron. Soc. Japan 18, 85
Webb, A.R., Roberts, B.: 1980, Solar Phys. 68, 87

SOME STATISTICAL PROPERTIES OF THE MAGNETIC FIELD IN PROMINENCES

Kim I.
IZMIRAN, Moscow Region, SU-142092 Troitsk, USSR

Koutchmy S.
Institut d'Astrophysique CNRS, 98bis Bd Arago, F-75014 PARIS

Stellmacher G.
Institut d'Astrophysique CNRS, 98bis Bd Arago, F-75014 PARIS

Stepanov A.I.
IZMIRAN, Moscow Region, SU-142092 Troitsk, USSR

Magnetic field observations based on the Zeeman effect, measuring the longitudinal component $B_\parallel$ of the field with an entrance hole of 6.4 arcsec diameter, were obtained for a great number of prominences during 1979 to 1985 using the spectrally scanning magnetograph installed at the coronograph ($\emptyset$=53 cm) of the high altitude Observatory at Kislovodsk (USSR).

The observed fieldstrengths $B_\parallel$ show a two-peaked distribution with maxima near 8 and 20 gauss and a high field tail near 45 gauss (Fig.1)

When separating the observations into two subsets corresponding to the latitudes occupied by the prominences, one sees that the first peak near 8 gauss relates to quiet region prominences at higher latitudes $\phi > 35°$, while the second peak near 20 gauss is characteristic for active region prominences at latitudes $\phi \leq 35°$ (Fig.2). (Such narrow optimum ranges of fieldstrengths seem to be required for the formation of prominences, see An et al.,1986).

In its latest version (Nikolsky et al. 1985) the magnetograph allows to measure together with $\{B_\parallel, W_\lambda, \Delta\lambda_{1/2}\}$ also the line of sight velocity $v_\parallel$. Figure 3 shows the distribution of the observed fieldvalues $B_\parallel$ and the simultaneously measured line of sight velocities $v_\parallel$ over β, the angle between the filament axis and the line of sight. Fieldvalues and velocities statistically show a similar distribution with maximum values around $\beta \simeq 25°$. It can be supposed that $B_\parallel^{max} \simeq B$, the mean pitch-angle α of the field will then be $\alpha \simeq \beta = 25°$ (see Tandberg-Hanssen and Anzer 1970). In prominences magnetic forces will dominate mechanical forces, the flows will follow the fieldlines as indicated in Figure 3.

Figure 1. Histogram showing the distribution of the measured field values $B_\parallel$.

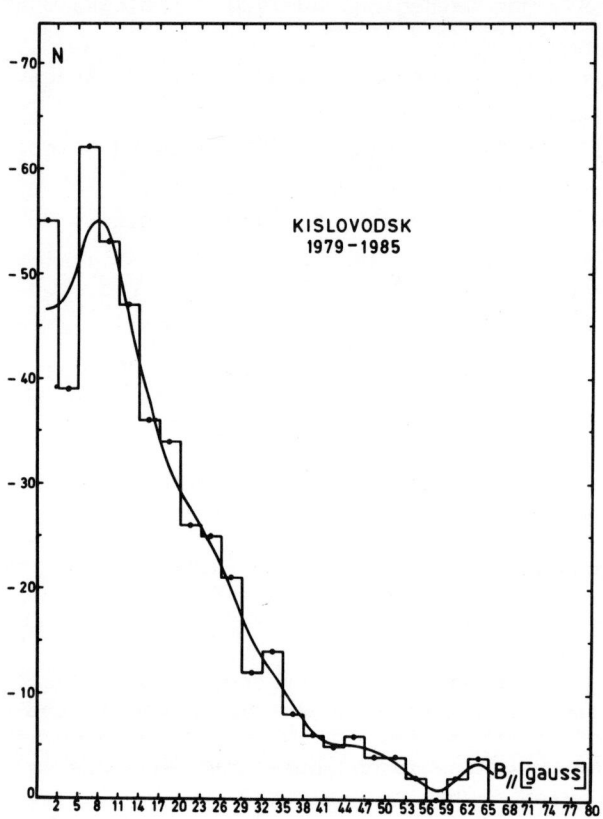

REFERENCES

Nikolsky, G.M., Kim, I.S., Koutchmy, S., Stepanov, A.I., Stellmacher, G: 1985, Astron. Journal, **62**, 1147.

Tandberg-Hanssen, E., Anzer, U : 1970, Solar Phys., **15**, 158.

An, C.H., Bao, J.J., Wu, S.T. : 1986, Proceedings of Coronal and Prominence Plasma Physics Meeting, Berkeley Springs, West Virginia, April 8-10, 1986.

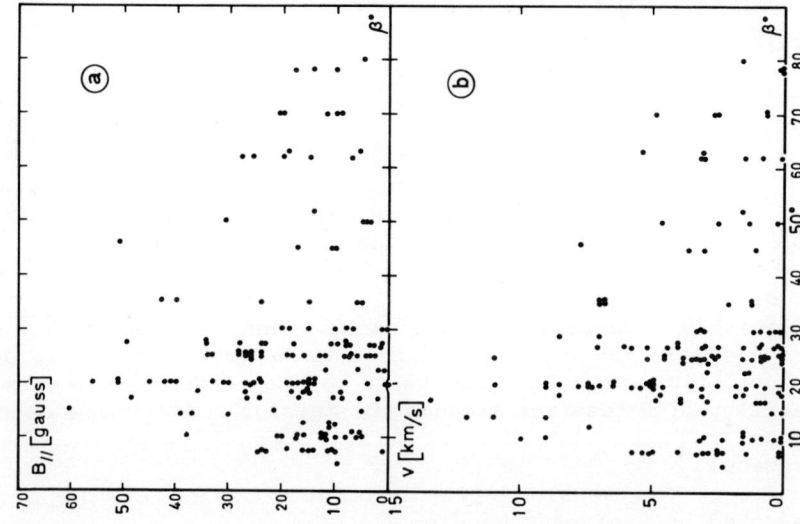

Figure 2. As Figure 1: a) for active region prominences, latitudes φ ≤ ±35°, b) for high latitude prominences, latitudes φ > ±35°.

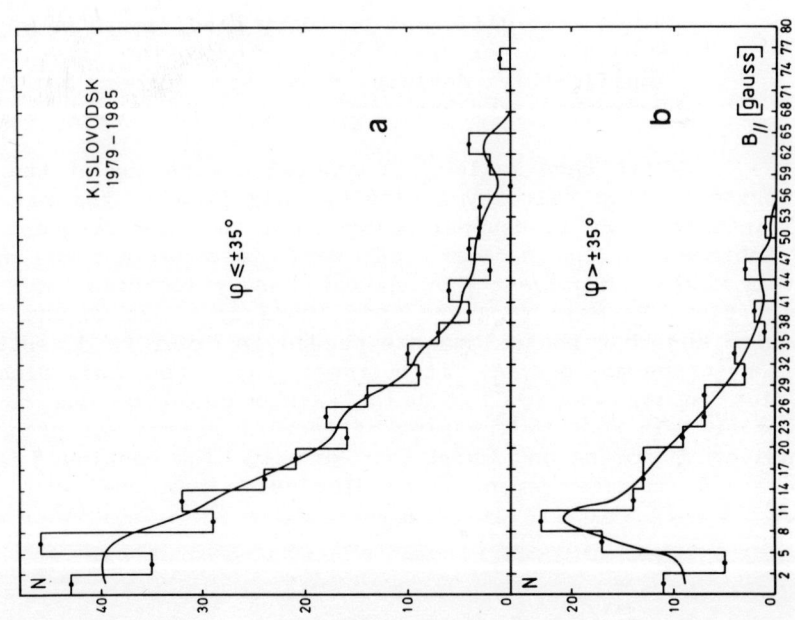

Figure 3. a) observed field values B∥ versus β the angle between the line-of-sight and the filament axis, b) observed line shifts v∥ versus β.

JOINT DISCUSSION ON TOPICS OF SESSIONS 9 AND 10, SUMMARIZED
BY THE CHAIRMEN

B. Roberts
University of St. Andrews

P. Mein
Observatoire de Paris

Sessions 9 and 10 were concerned with the dynamical nature of flux concentrations in the photosphere, the waves that such flux tubes support, and their influence on the chromosphere and corona. In addition to the extensive surveys presented by the reviewers, M. Schüssler and R. Hammer, contributing authors discussed the behaviour of nonlinear, radiatively damped, longitudinal and transversal waves (P. Ulmschneider) and nonlinear Alfven waves (E. Jensen) that propagate in isolated flux tubes. Two aspects of coronal effects were considered, the stability of force-free arcades (U. Anzer), and the mode coupling of the Alfven and slow continuum modes in coronal loops (S. Poedts and M. Goossens).

The session opened with discussion of the first review, presented by Manfred Schüssler :

F. Kneer :
What is the lower boundary condition in your models ?

M. Schüssler :
We used two different boundary conditions, namely (1) fixed temperature or (2) fixed flux. The results did not significantly deviate from each other as far as the observable intensity structure is concerned.

There then followed a general discussion of the problem of the phase-shifted velocity-intensity relations in the network and in cell interiors. F. L. Deubner pointed out that low frequencies (1000s) are enhanced in the network (cf. Damé 's observations) and that if gravity waves are present then upward phase velocities implied downward energy flow. P. Mein remarked that 1000s is close to the lifetime of spicules and that phase-lags are needed in order to disentangle waves and convective motions. A. Title, remarking on the small size (<100 km) of flux tubes, wondered if brightenings could be due to waves. M. Schüssler felt that brightening of facular points was probably due to transient phenomena and added that we need high continuum intensity to understand FTS measurements. The discussion continued with :

E. Wiehr :
You showed Hasan's fluxtube oscillations at the level 50 km below $\tau = 1$. What do these oscillations look like at observable layers ?

M. Schüssler :
You have to take into account the Wilson depression within the flux tube of about 100 - 150 km. Thus the level shown refers to $\tau \leqslant 0.1$ within the tube. However, the τ-scale should be calculated for these simulations.

H. Spruit :
In my '76 and '77 models, I found that flux tubes can produce a net excess heat flux (taking into account the dark ring around them) and got numbers up to 60% of the mean solar flux (per unit area of cross section of the tube). But these numbers were rather sensitive to details of the model, and especially to the flux tube size. What numbers for the excess flux did you obtain in your numerical models ?

M. Schüssler :
The excess mean intensity for $\mu = 1$ is between + 5 % for small tubes (d $\leqslant$ 150 km) and - 10% for larger tubes (d $\geqslant$ 500 km). However, our domain of calculation (850 km x 850 - 2000 km) is much smaller than your's. Consequently, we could not very accurately follow the propagation of the flux disturbance into the convection zone. On the other hand, the observations seem to show that there is no much propagation and we just have a local re-distribution.

C. Zwaan :
You present the hot-wall model and the hot-cloud model as alternatives. We know, of course, that hot-cloud model is also true : there is strong chromosphere heating, that reaches down to the temperature minimum. Should we not try to glue the hot-wall model and the hot-cloud model together ?

M. Schüssler :
I agree. My point was that we may need overlapping of hot-clouds for the continuum contrast near the limb, too.

The discussion then turned to the review presented by R. Hammer.

E. Priest (comment to R. Hammer) :
It is most unlikely that thermal conduction is classical. For example, small-scale plasma instabilities may change the $T^{5/2}$ temperature dependence and may change its magnitude by a factor of a hundred.

R. Hammer :

Such effects can drastically affect the structure of the atmosphere and the type of wave and stability calculations that I have been reviewing. So far they have not yet been included in such calculations, to the best of my knowledge.

R. Rosner :

You mentioned that non-local thermal conduction, e.g. penetration of transition region matter by coronal electrons, could strongly affect energy balance at the transition region base. As I understand recent results of F. Shab, and of S. Owocki and R. Canfield, such effects are in fact not important, at least for active regions ; however, plasma diagnostics may be strongly affected by the fact that coronal electrons could produce enhancements in the lower transition region of the abundance of species with high excitation potentials. These calculations are classical in the sense that they do not account for collective effects, viz., the excitation of plasma waves, and the resulting enhanced electron scattering.

R. Hammer :

I have restricted myself to the discussion of quiet regions. For these, non-local conduction may be very important for the energetics, according to Owocki and Canfield.

P. Ulmschneider (in comment to R. Hammer) :

You mentioned that mechanical heating is not able to balance the strong Lyman α radiation losses. We have to keep an open mind for this. Longitudinal (acoustic) waves emerging out of the $\tau_{Ly\alpha} = 1$ surface will suffer large radiation damping. Thus more energy can contribute significantly to the $L_{y\alpha}$ cooling.

E. Priest (to Hammer) :

Which of the various wave modes in a flux tube do you think are the most important in the solar atmosphere ?

H. Spruit :

I think transversal waves are probably the most important. Transversal displacements are going on all the time by granulation. One can measure these and their time scales. The transversal tube wave also has the advantage that it propagates energy easily at low frequency, and that there is no radiative damping until it becomes nonlinear.

The question of the importance of the various tube waves was discussed further by the floor. M. Schüssler commented upon the importance of torsional waves and their generation by downflows, which readily induce rotational flows around flux tubes and thus generate Alfven waves. Spruit agreed with this comment. It was further remarked that Alfven waves have no propagation cut-off. B. Roberts pointed out that, if impulsively generated, longitudinal waves will be readily excited and then their cut-off frequency is simply manifest as the frequency of oscillation of their wake which trails behind the wavefront. Longitunal waves may also be induced by the nonlinear development of the transversal wave. The discussion of the generation of waves was concluded on a point of clarification :

F. L. Deubner (to P. Ulmschneider) :
> What is the period of the shaking action that was used in your calculations ?

P. Ulmschneider : It was 45 seconds.

The discussed then turned to coronal effects.

E. Priest (to U. Anzer) :
> Some two-ribbon flares have no prominence eruption, but the magnetic configuration may still be very similar with a magnetic island. My feeling is that the ballooning modes may be important for creating small-scale structure and filamentation in the corona since they are localized modes and may be driven by resistivity or small pressure gradients. In contrast, the eruption of an arcade to give a two-ribbon flare must be a global mode.

U. Anzer :
> According to the stability diagram of Hood one needs substantial pressure gradients to drive ballooning instabilities. Since the coronal magnetic fields are very close to a force-free situation ballooning modes are probably not very important in the corona.

E. Priest (to M. Goossens) :
> To the eyes of a theorist like myself, this is just as beautiful as Alan Title's movies ! Is the Alfven continuum more important or less important than the slow continuum for coronal loops ?

M. Goossens :
> The relative importance of the Alfvén continuum and the slow continuum for coronal heating depends on the range of frequencies involved and on the way of excitation. Consider a 2D equilibrium with a poloïdal magnetic field so that Alfvén continuum modes and slow continuum modes are clearly distinct and neglect gravity. The slow continuum then coincides with the Alfven continuum when the plasma is incompressible ; i.e., in the limit $c^2 / V_A^2 \to \infty$ and collaspses on to zero in the limit $V_A^2 / c^2 \to \infty$ (c^2 is the square of the velocity of sound and V_A^2 is the square of the Alfvén velocity). The ranges of the slow continuum and the Alfven continuum frequency are comparable for plasmas where $c^2 / (c^2 + V_A^2)$ is not much less then 1. Alfven continuum waves are excited by motions in the magnetic surfaces perpendicular to the magnetic field lines while slow continuum waves are excited

by motions in the magnetic surfaces but parallel to the magnetic field lines. For 2D plasmas with mixed poloïdal and toroïdal magnetic fields things become more complicated since we cannot divide the continuum waves in Alfven and slow waves.

B. Roberts (to M. Goossens) :
Do you think that the enhancement in resonant absorption you have explained to exist when density stratification is allowed for would be particularly important in the transition region ?

M. Goossens :
The mechanism discussed here to enhance phase mixing requires variation of density (more precisely of Alfven velocity) both <u>normal</u> to the magnetic surfaces and <u>in</u> the magnetic surfaces along the magnetic field lines. If these two conditions are fulfilled in the transition region and if there is a driving agent to shake the field lines, then this mechanism is very likely to be important in the transition region.

APPEARANCE AND DISAPPEARANCE OF MAGNETIC FLUX AT THE SOLAR SURFACE

E R Priest
Applied Mathematics Department
St Andrews University, KY16 9SS, Scotland

Abstract. Theory and observation of the emergence and submergence of magnetic flux and the role of magnetic reconnection are briefly reviewed, together with a new unified theory for fast steady-state reconnection that has recently been proposed. New flux appears on a wide variety of scales due to magnetic buoyancy or convective motions. However, only a small fraction of the flux which emerges is likely to escape. Active regions seem to disappear by the fragmentation and cancellation of flux. Cancelling magnetic features (discovered by Sara Martin and coworkers) probably represent sites where magnetic flux is disappearing on small scales by reconnection submergence rather than by simple submergence. On large scales flux is also escaping in coronal transients and is probably submerging below quiescent prominences.

1 INTRODUCTION

The emergence and submergence of magnetic flux may appear at first sight to be a very simple process — you just carry flux up through the solar surface and bring it back down. However, the Sun appears to work in more complex ways involving the interaction between convective motions, magnetic buoyancy and magnetic reconnection. In the 1970's there was a great interest in the process of appearance of flux and its relation to solar flares (eg Priest, 1984), but more recently several authors have started studying its disappearance.

The appearance of flux may be caused by convective motions bringing it to the photosphere, as reviewed by Weiss in these proceedings, or by magnetic buoyancy (see § 2 here). The basic theory of reconnection is summarised in § 3, but the main part of this review is a discussion in § 4 of the disappearance of flux on small scales in cancelling magnetic features, sunspots and active regions, as well as on large scales between active regions, in prominences and in coronal transients.

2 EMERGENCE OF FLUX BY MAGNETIC BUOYANCY
Parker (1958) pointed out that an isolated isothermal

magnetic flux tube would be lighter than its surroundings and so would rise. Denoting quantities inside and outside the flux tube by i and e, respectively, lateral pressure balance gives

$$p_e = p_i + B_i^2/(2\mu), \qquad (1)$$

where the perfect gas law implies that $p_e = R\rho_e T_e$ and $p_i = R\rho_i T_i$. The buoyancy force is therefore

$$(\rho_e - \rho_i)g = \frac{B_i^2 g}{2\mu R T_e} - \rho_i g(1 - \frac{T_i}{T_e}) . \qquad (2)$$

Thus, for an isothermal situation ($T_i = T_e$) the buoyancy force dominates the magnetic tension force ($B_i^2/\mu L$) when the radius of curvature (L) exceeds twice the scale height ($\Lambda = RT/g$). This implies that very small loops will sink because tension dominates, whereas loops of size a few arcsec (such as the intra-network field) may go up or down due to slight imbalances. Furthermore, loops which are cool enough ($T_i < T_e$) with respect to their surroundings will not be buoyant at all when the second term on the right of (2) dominates and so they will sink.

A second role of buoyancy is to make a stratified equilibrium magnetic field unstable to the formation of rising tubes (the magnetic buoyancy instability, Parker 1966). Straight field lines are unstable if the magnetic field decreases with height faster than the density (ie d/dz(B/ρ) < 0). However, if the field lines are allowed to curve so that plasma can drain from the crests to the troughs, the instability condition is less restrictive (ie dB/dz < 0). Modifications due to rotation, curvature and diffusion have been made, but the rise-time of flux is rather uncertain – the general view is that flux tubes rise very much more quickly than the solar cycle period and so need to be stored below the base of the convection zone (Galloway and Weiss, 1981). They may still emerge by magnetic buoyancy instability and a nonlinear calculation of the unstable modes has recently been completed (Cattaneo and Hughes, 1986).

The equilibrium shape of a buoyant flux tube anchored at two footpoints a distance 2W apart has been estimated by Parker (1979) and Moreno-Insertis (1986). With the notation of Figure 1a, a balance between tension and buoyancy gives

$$- B_i^2 \, d\theta/dz = (\rho_e - \rho_i)\mu g \cot\theta \qquad (3)$$

or, for a uniform temperature T, and ($\rho_e - \rho_i$) given by (2),

$$d\theta/dz = \tfrac{1}{2}\Lambda^{-1}\cot\theta . \qquad (4)$$

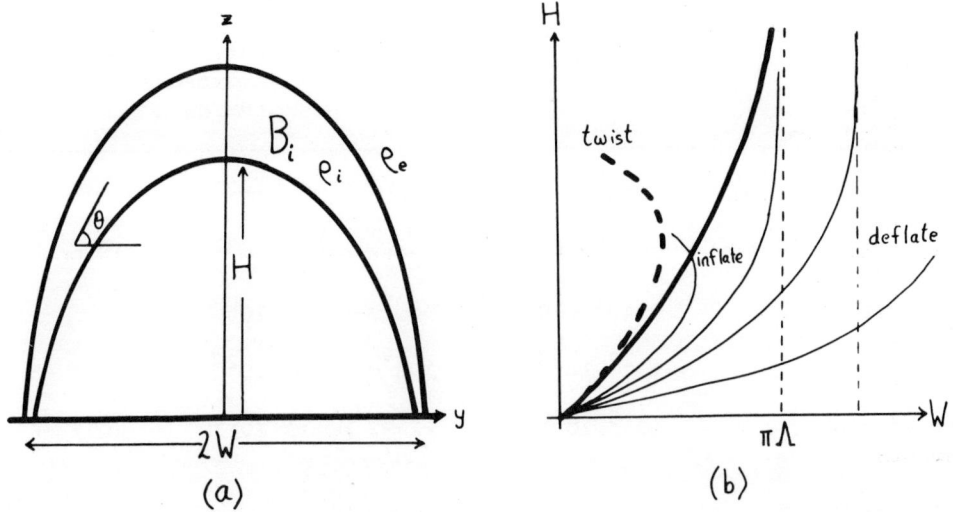

Figure 1(a) Notation for a buoyant tube. (b) The height (H) as a function of the footpoint separation (2W).

Since $dz/dy = \tan\theta$, this may be integrated to give

$$\tan^2(\tfrac{1}{2}W/\Lambda) = \exp(H/\Lambda) - 1, \tag{5}$$

the solution being sketched as the thick solid curve in Figure 1b. The effects of twisting the tube or of having an external magnetic field (and therefore allowing the tube to be inflated ($p_i > p_e$) or deflated ($p_i < p_e$)) have been calculated by Browning and Priest (1984, 1986) and are sketched as the dashed and thin solid curves, respectively, in Figure 1b. It can be seen that usually the rise of a flux tube can be caused by increasing the footpoint separation whereas its sinking can be caused by decreasing the separation.

The emergence of flux is observed to take place on a variety of scales. Active regions stand out on a full-disc magnetogram, but ephemeral regions bring up ten times the flux of active regions, while intra-network elements bring up a hundred times the flux of ephemeral regions. The speed of rise is something of a puzzle, however. Chou and Wang (1987) have recently measured the separation velocity of new poles in emerging flux regions and ephemeral regions to be between 0.2 and 1 kms^{-1}, while the magnetic fluxes, mean field strength and sizes range between 3×10^{18} Mx and 7×10^{20} Mx, 66 G and 1000 G, and 3 Mm and 22 Mm, respectively. They find that the separation velocities predicted from a simple theory are an order of magnitude larger than the observed values.

Emerging flux regions may produce an arch-filament system in Hα when

TABLE Properties of Emerging Flux

	Flux(Mx)	Lifetime	Rate Emergence (Mx/day)	Size(Mm)
Large active region	$5\times10^{21} - 4\times10^{22}$	months	10^{22}	
Small active region	$2\times10^{20} - 5\times10^{21}$	weeks–days	10^{22}	3–100
Ephemeral region	$2\times10^{18} - 10^{20}$	hours–day	10^{23}	
Network	$10^{18} - 10^{20}$	1–3 days	10^{23}	<0.7 –5
Intra-network elements	$10^{16} - 10^{18}$	<1 –2 hours	10^{26}	<0.7 –5

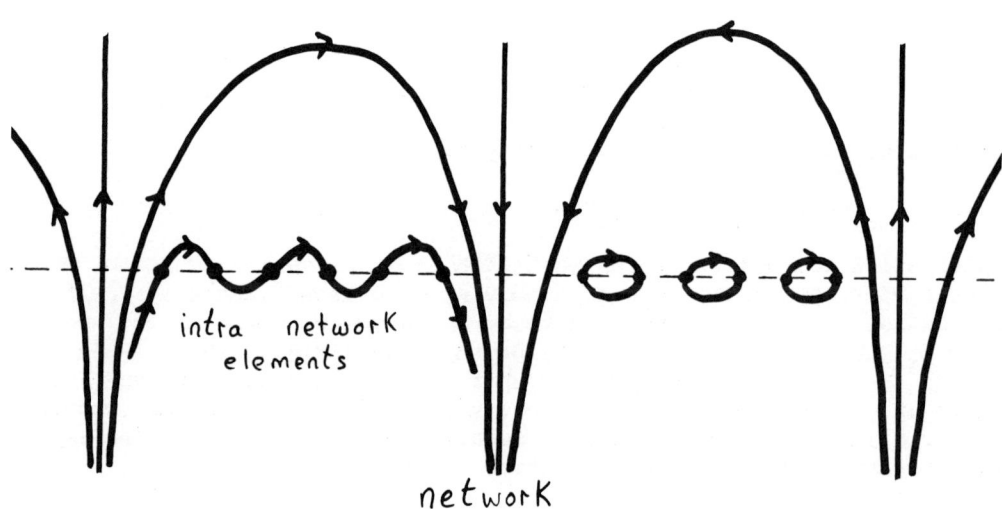

Figure 2. Possible magnetic connections of intra-network elements.

the flux exceeds $0.5 \times 10^{20} - 1.0 \times 10^{20}$ Mx and white-light pores when the flux exceeds 10^{20} Mx (Chou and Wang, 1987). Sometimes flares are produced. In the soft X-ray corona, active regions show up very bright and some ephemeral regions produce X-ray bright points. The properties of flux emergence on different scales have been summarised by Martin (1986), Zirin (1986) and Zwaan (1986) and are given in the Table above.

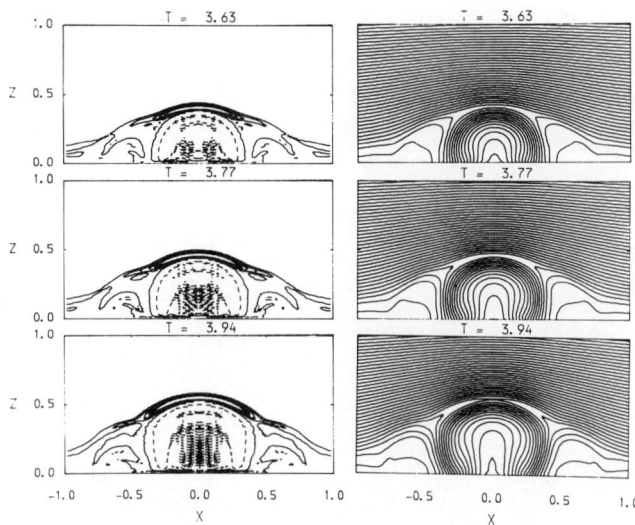

Figure 3. Current density contours and magnetic field lines in a numerical experiment (Forbes and Priest, 1984).

Intra-network elements are of mixed polarity and move quickly. They appear through local dipole emergence and fragmentation of network elements, and then they continually merge with network elements. However, little is known about them. How do they form? How deep do they extend? How do they interact with granulation? Are they connected above and below the photosphere in the way that is sketched in Figure 2a, Figure 2b or in some more complex manner?

During the emergence of new magnetic field ($\underline{B}_e$) it pushes up against the overlying field ($\underline{B}_0$) and the part $\frac{1}{2}(\underline{B}_0 - \underline{B}_e)$ can reconnect. A numerical experiment on this process has been conducted by Forbes and Priest (1984) who solve the resistive MHD equations. The way the flux reconnects can be seen in Figure 3.

3 BASIC RECONNECTION THEORY

Magnetic reconnection theory has been split into two main parts, as reviewed by Priest (1985). The first is the linear tearing mode instability of a current sheet or sheared magnetic field, whereby the equilibrium goes unstable to the breaking and reconnection of magnetic field lines. The second is the fast nonlinear steady state of Petschek-Sonnerup reconection, in which the current sheet bifurcates into pairs of slow-mode shock waves which stand in the flow and convert incoming magnetic energy to the heat and kinetic energy of hot fast plasma streams. The age of detailed numerical experimentation on reconnection, which has recently begun, has shown that in its nonlinear development the tearing mode can evolve to a

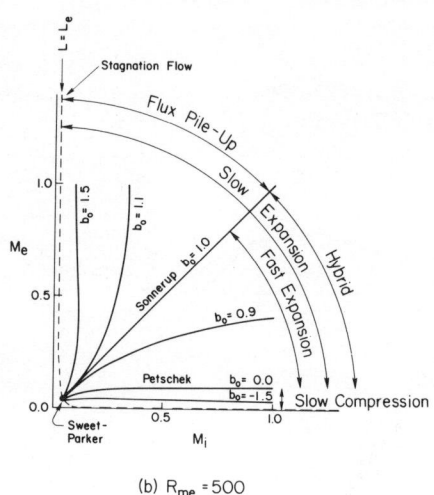

(b) $R_{me} = 500$

Figure 4. Reconnection rate (M_e) as a function of M_i for a unified reconnection theory (Priest and Forbes, 1986).

fast quasi-steady regime, but often fast reconnection is rather different from the classical modes. Analytically, a new theory has been discovered which includes the classical modes as special cases. The classical theory of fast steady reconnection consists of several distinct models. The Sweet-Parker (1958) model consists of a simple current sheet into which oppositely directed magnetic field lines are carried, reconnected and then ejected at the Alfven speed. Petschek's (1964) mechanism has a central Sweet-Parker current sheet from which the standing slow-mode shocks propagate, whereas Sonnerup's (1970) model has extra sets of discontinuities in the inflow. The inflow region for Petschek's model is a <u>fast-mode expansion</u>, in the sense that as the plasma approaches the central sheet the magnetic field strength and pressure decrease while the streamlines converge. In contrast, the inflow region for Sonnerup's model has discrete <u>slow-mode expansions</u>, which make the plasma pressure decrease while the field strength increases and the flow diverges. Sonnerup and Priest (1975) also discovered a stagnation-point flow solution in which the field lines are carried in by a diverging flow and annihilated.

The new unified theory (Priest and Forbes, 1986) has many more families of solutions labelled by a parameter b, which is determined by the nature of the inflow, in particular how converging or diverging it is (Figures 4 and 5). When b < 0 there is a family of <u>slow compressions</u> with strongly converging flows, whereas when b > 1 there

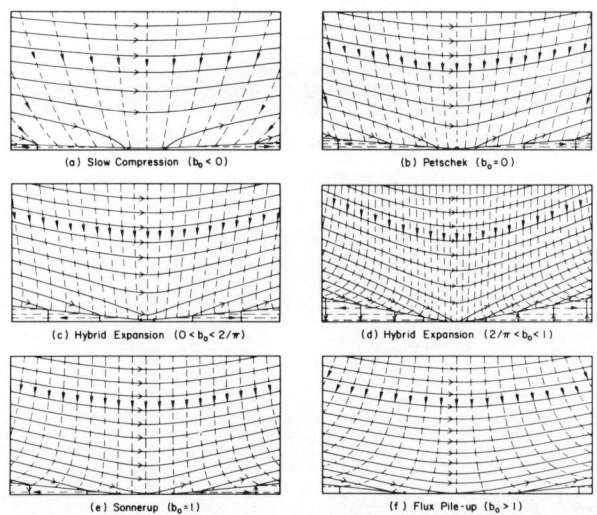

Figure 5. Magnetic field lines (solid) and streamlines (dashed) for several reconnection solutions (Priest and Forbes, 1986).

is a <u>flux pile-up regime</u> of solutions with strongly diverging flows, in which the magnetic field strength increases and the central diffusion region is long. When $0 < b < 1$ a <u>hybrid regime</u> exists with a fast expansion on the axis and slow expansions at the sides of the inflow region. These three families of solutions are separated by the Petschek solution at b=0 with a weak fast-mode expansion and a Sonnerup-like solution with a weak slow-mode expansion.

Figure 4 shows the external Alfven Mach number (M_e), or reconnection rate, calculated at the inflow to the region as a function of the inflow Alfven Mach number (M_i) at the inflow to the central current sheet for several values of b and an external Lundquist number (R_{me}) of 500. It can be seen that for Petschek's solution (b=0) the maximum reconnection rate is about 0.1, but that for other solutions with $b > 0$ the reconnection rate is larger. In particular, when $b \geq 1$, the magnetic field can reconnect at any rate up to the Alfven speed within the limitations of the theory.

4 DISAPPEARANCE OF FLUX

After flux has appeared, how does it disappear again? Just as it emerges on a variety of length-and-time scales, so it may submerge on several scales too. Ephemeral regions disappear in hours, whereas the large-scale polar-cap field takes years to reverse. In this section we start with small scales and then move to large ones.

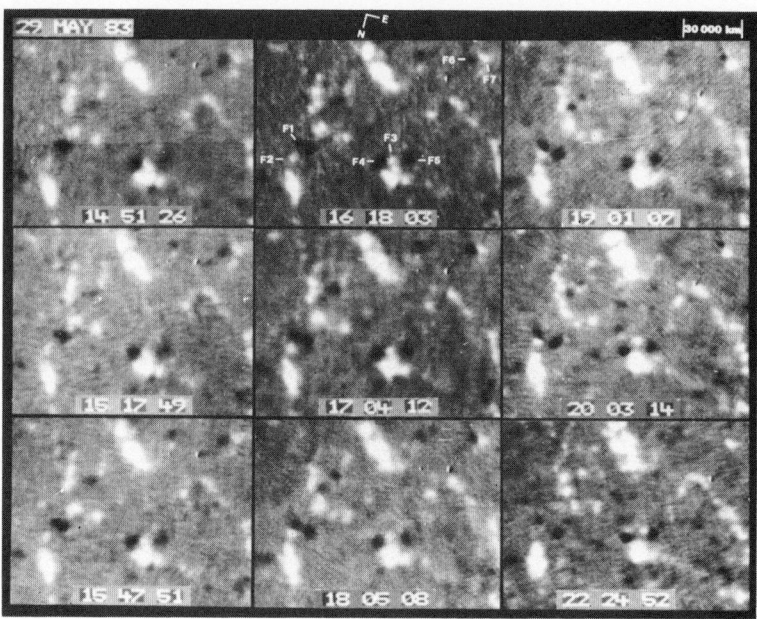

Figure 6. Cancelling magnetic features (Martin, 1985).

General comments on flux disappearance have previously been made by Zwaan (1978, 1986) and Spruit et al. (1986).

4.1 Cancelling Magnetic Features

Pictures of the line-of-sight magnetic field in the photosphere, such as Figure 6, show that the solar surface is covered with magnetic fragments. Some of them lie along the network and many form bipolar pairs which often produce X-ray bright points and He 10830 dark points. It used to be thought that X-ray bright points are caused by emerging flux, but at Big Bear Observatory Sara Martin has used the videomagnetograph to follow the time-development of positive and negative polarity fragments. She finds that two-thirds of the positive-negative pairs are so-called cancelling magnetic features (Martin, 1984; Martin ey al., 1985; Livi et al., 1985; Hermans and Martin, 1986). For example it can be seen how the light and dark fragments F1 and F2 in Figure 6 representing opposite polarities approach one another, come into contact and disappear. Similarly, fragments F6 and F2 approach and disappear. Many examples of this process have been presented with the flux in both polarities decreasing; sometimes both fragments disappear, whereas sometimes, if the initial flux in one fragment is larger than the other, part of the first fragment remains after the cancellation.

Most of the emergence of flux is in the centre of a supergranule cell and most of the cancellation takes place at the edge, at a rate of

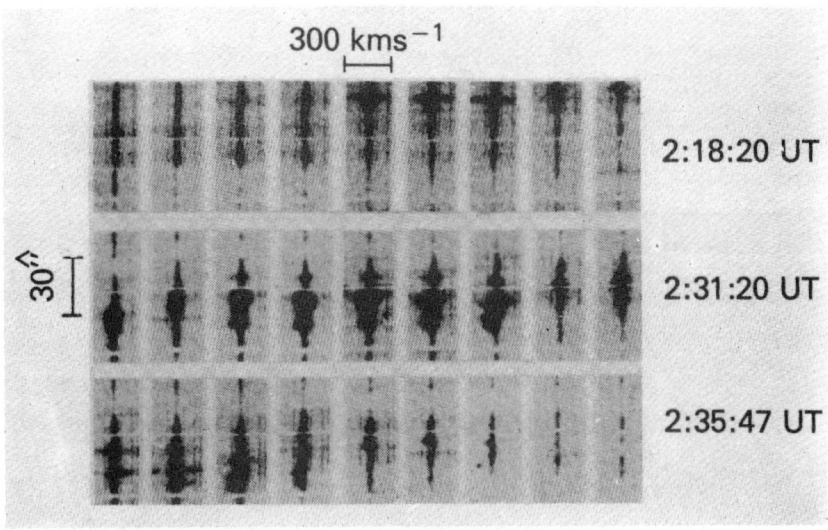

Figure 7. Turbulent CIV spectra (Brueckner *et al.*, 1986).

10^{18} Mx per cell per hour. Presumably flux is being carried up by the supergranule flow pattern and then transported to the edge of the cell, where it tends to cancel with whatever flux is at that part of the network it reaches. According to Harvey (1985), two-thirds of the X-ray bright points are associated with cancelling magnetic features rather than emerging flux regions. As well as giving rise to X-ray bright points, cancelling magnetic features also produce subflares and macrospicules (Marsch, 1978). In addition, they often are associated with very tiny erupting filaments, which occur at a rate of 500-600 per day, three-quarters of them producing subflares (Hermans and Martin, 1986). In the transition region one sees CIV brightenings in UVSP data at X-ray bright points (Porter et al, 1986) and often there are turbulent events and jets up to 300 kms^{-1} (Brueckner et al, 1986). For example, Figure 7 shows turbulent CIV spectra from the HRTS instrument on Spacelab 2 above an emerging flux region at several locations and three different times.

How can one explain flux disappearance in a cancelling magnetic feature if it is not a purely instrumental effect? Several possibilities may occur to the reader. If the opposite polarity fragments are initially connected by a flux tube, then simple submergence of the tube would make the two poles come together and disappear (Figure 8a). However, this is most unlikely because the process of appearance of flux looks quite different from its disappearance. When bipolar flux appears, the two poles of an emerging flux region separate and never come back to one another and cancel, as one would expect if a curved tube was lifted up and then just brought back down. Furthermore, the magnetic gradient between

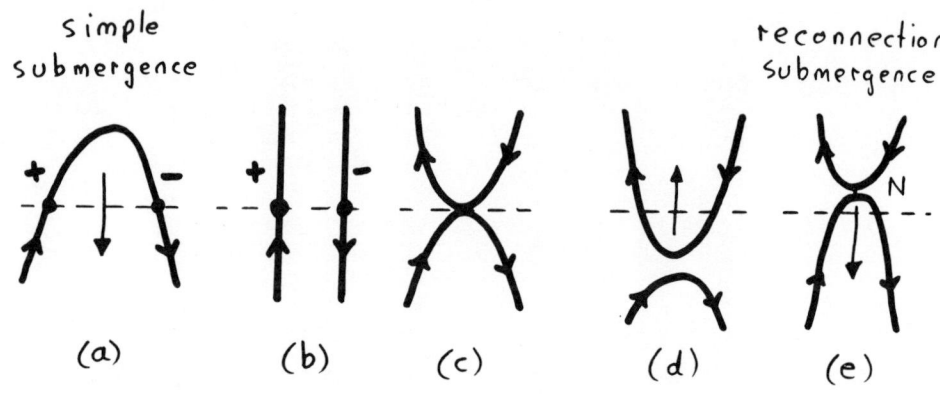

Figure 8. Explanations for cancelling magnetic features.

the two poles is relatively weak and in Hα they are often seen to be joined by fibrils. In contrast, during the disappearance of flux the two poles of a cancelling magnetic feature have quite different origins. Also, the magnetic gradient is much stronger with the peak field closer to the interface, and in Hα they are often separated by a filament perpendicular to the line joining the poles. Furthermore, as mentioned above, cancelling features are associated with much more energetic phenomena, such as X-ray bright points, macrospicules and subflares. All this suggests that the two poles may not be initially part of the same magnetic flux tube (Figure 8b) and that the cancellation involves reconnection of initially separate magnetic fields, with a resulting release of magnetic energy.

Reconnection exactly at the photosphere (Figure 8c) is most unlikely. This level is certainly special in the sense that the magnetic diffusivity (η) has its highest value (2×10^4 m^2/s), so that the magnetic field diffuses most easily through the plasma here, since the plasma is only slightly ionised. However, the typical magnetic Reynolds number is still very large (2×10^4) and so reconnection proceeds in a very similar way to elsewhere, but with a slightly larger diffusion region. The situation shown in Figure 8d with reconnection below the photosphere is also unlikely since a large mass of plasma would be trapped on the upper field line and so would severely impede its rise.

The most likely situation, in our view, is what may be called <u>reconnection submergence</u> (Figure 8e) with reconnection taking place above the photosphere so that the lower field line moves down as the separate poles approach. Of course, Doppler measurements would show whether or not the material between the poles is indeed falling. The

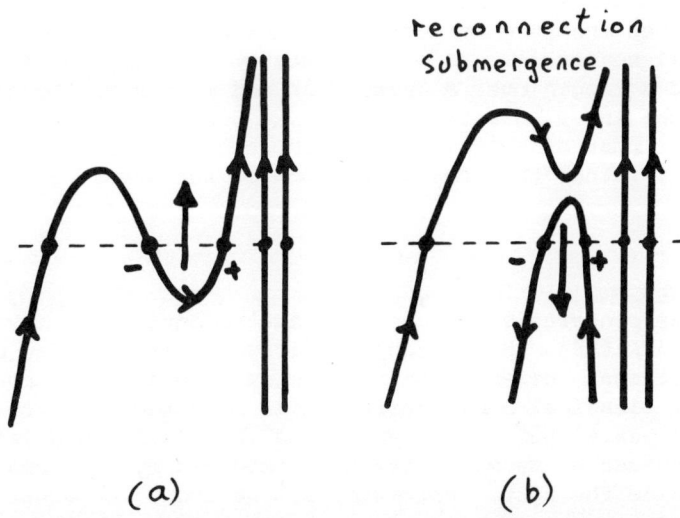

Figure 9. Collision of an ephemeral region with network.

difference between this and simple submergence (Figure 8a) is that here the submergence is driven by the reconnection process and the field line curvature produced by reconnection, whereas presumably for simple submergence it is subphotospheric behaviour which is pulling the field lines down. What makes reconnection submergence likely is that reconnection (at the Alfven speed (V_A) or some fraction thereof) proceeds much faster in the corona where $V_A \approx 10^3$ kms^{-1} than in the photosphere where $V_A \approx 1$ kms^{-1}. The neutral point N probably forms in the low corona and moves to the low chromosphere as the two polarities approach, since its height is roughly equal to the separation of the poles. He 10830 dark points and therefore X-ray bright points are observed to begin when the magnetic fragments are 5 arcsec apart and they continue throughout the cancellation (Harvey, 1986).

Often an ephemeral region is observed to bump into the network, where one pole cancels with a network element. Such a cancelling magnetic feature may represent the simple emergence of U-shaped field lines if the ephemeral region and the network are already connected below the photosphere before the collision (Figure 9a). It is more likely, however, that the two flux systems are not initially connected and then the cancellation would again be caused by cancellation submergence (Figure 9b) as the two systems reconnect.

Creation of the inner network field has been discussed by Spruit et al. (1986). They consider a large-scale U-loop with a field strength of 10^3G at a depth 10 Mm which rises and expands to produce a field of only 10G when it reaches the photosphere. It spreads laterally as it rises and either granulation or magnetic buoyancy forms horizontal undulations which break through the surface and reconnect to form a series of closed loops in the interior of a supergranule cell. Diffusion is then enhanced by magnetoconvection or turbulence producing small scales of $1-10^3$ m. The result is a region of low magnetic field strength over a large area and with mixed polarity, as one sees for the inner network field. Like Schussler (1980), these authors suggest that ephemeral regions represent the weak field of rising U-loops, rather than small versions of the deeply anchored active regions.

4.2 Sunspots

The properties of sunspots have recently been reviewed by Moore and Rabin (1985). Most sunspots disappear after a few days, for some unknown reason, whereas some of them decay over a few months. The decay from classical ohmic diffusion is much too slow, taking typically 1000 years, but it was suggested that flux could leak slowly from a spot because of an eddy diffusion produced by weak small-scale convection inside the spot. For example, the induction equation in cylindrical polars, namely

$$\frac{\partial B_z}{\partial t} = \frac{\tilde{\eta}}{R} \frac{\partial}{\partial R} \left(R \frac{\partial B_z}{\partial R} \right),$$

admits simple solutions for the decay of flux in a cylindrical flux tube of the form

$$B_z = \frac{\phi_o}{4\pi\tilde{\eta} t} \exp\left(-\frac{R^2}{4\tilde{\eta} t} \right)$$

The resulting rate of decrease of flux agreed with the observed decrease if an eddy diffusivity ($\tilde{\eta}$) of $2 \times 10^7 \, m^2 s^{-1}$ is adopted.

However, the problem with this suggestion is that the magnetic flux is not observed to spread out and increase the flux of the surrounding active region (Wallenhorst and Howard, 1982; Liggett and Zirin, 1983; Rabin et al., 1984). It looks as if the flux tubes sink below the surface, although the way in which this is accomplished may be quite complex. A decaying spot is observed to be surrounded by a moat cell, across which so-called moving magnetic features are seen to transport flux at a rate of $6 \times 10^9 - 8 \times 10^{11}$ Wb h^{-1} (Vrabec, 1974; Harvey and Harvey, 1973). Meyer et al. (1974, 1979) have suggested that flux strands are carried across the moat by the annular circulation and pulled downwards, but the cause of the fragmentation of the boundary of the spot into strands is unclear. Perhaps it is an interchange instability or magnetoconvection or stimulated by granulation.

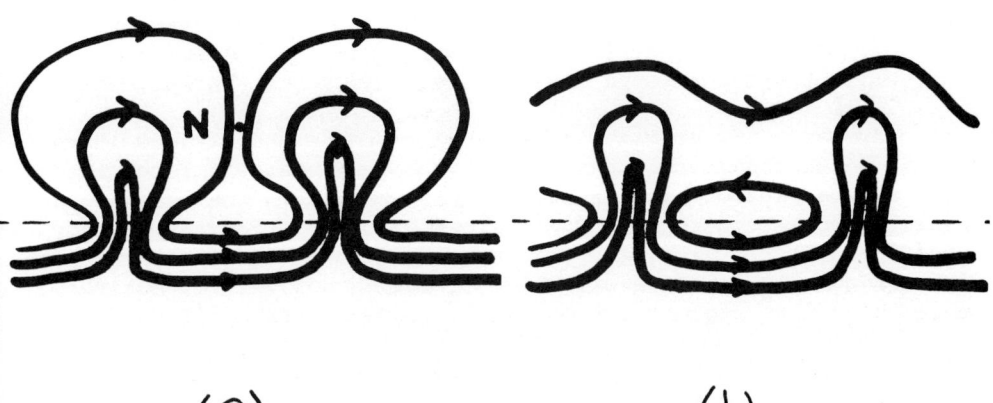

Figure 10. Escape of flux of active regions.

4.3 Active regions

In a most important piece of work Martin *et al.* (1985) observed an active region decaying by flux fragmentation (perhaps due to magnetoconvection) and cancellation. Twenty-one subflares were produced and all were located at sites of cancelling magnetic features (c.f. Marsch, 1978; Moore *et al.*, 1977). One large two-ribbon event occurred, and it too started at and was brightest at a cancelling magnetic feature.

An interesting phenomenon is an active region complex, where many active regions appear in turn at the same location (Gaizauskas *et al.*, 1985). Such active regions do not spread out but disappear in place, so that either the flux is pulled back down before reemerging as a new active region (Parker, 1986) or U-loops are reconnecting below the photosphere to create a closed flux loop which then escapes through the surface by cancellation (Spruit *et al.*, 1986) A clear example of flux disappearance within an active region has been observed by Rabin *et al.* (1984), in which 4×10^{20} Mx of flux in an area of 30 arcsec across disappeared overnight rather than spread out.

4.4 Escape of large-scale flux

Magnetic buoyancy is thought to cause a large-scale azimuthal flux tube stretched out by differential rotation to rise to the surface as an active region. For a straight tube of cross-sectional area A_0, field strength B_0 and containing plasma of dnesity ρ_0 at the base of the convection zone, conservation of flux (BA) and mass (ρA) imply that the tube becomes weak and broad at the surface, where the area and field strength are

$$A_1 = A_0 \rho_0 / \rho_1 \; , \; B_1 = B_0 \rho_1 / \rho_0 \; .$$

However, the magnetic diffusivity is so small that there is no direct way for the whole flux tube to free itself of the plasma within it and escape. Only that part of the flux tube which breaks and reconnects can escape. There is a parting of the ways, with the plasma remaining behind, while the reconnected flux escapes into space (Parker, 1984).

Magnetic buoyancy instability is thought to create undulations in the flux tube so that different parts of it rise and emerge through the surface as a series of bipolar regions (Figure 10a). The breaking and reconnection of the outermost field lines allows the flux at the top to escape and at the same time creates closed loops between the active regions (Figure 10b). However, this process is not very effective because the active regions are so widely spaced. Parker (1984) estimates that, of the 10^{21} Mx per day that emerges, only 3×10^{19} Mx per day escapes. The fraction of flux that escapes can be estimated as roughly $2b/(\pi a)$ by modelling the active regions as pairs of monopoles separated by a distance b, where a is the distance between two active regions.

Thus the surface of the Sun is not open but is very much of an inhibiting barrier for the escape of flux. The consequences for dynamo theory have not been worked through: for example, does it imply that the solar cycle is a recycling of flux? Another important question is: what happens to the closed flux between active regions? Perhaps it fragments and cancels, or is it pulled back down by convective motions? What is the role here of topological pumping or the migration of dynamo waves?

Howard and Labonte (1981) have stressed that most of the magnetic flux in the active region belts disappears in situ, with only 1% migrating towards the poles. Also, they point out that most of the flux disappears very fast, with 10% being recycled per day, in other words, the average rate of flux emergence is so high that it would double the total flux on the Sun in only 10 days. By comparison the increase in flux from sunspot minimum to sunspot maximum is only a factor of 3-4 (according to Mount Wilson data) or 10 (according to Kitt Peak data, Livingston, private correspondence), so that at any one time flux emergence and removal are nearly balancing one another throughout the solar cycle.

4.5 Prominences

The motions in prominences are rather puzzling, but the presence of upflows in prominences when viewed on the disc (Malherbe et al., 1983) may have implications for magnetic flux transport. For example, for Kippenhahn – Schluter prominences inside active regions the upflows suggest that the prominence is a sight of large-scale flux emergence. On the other hand most quiescent prominences are thought to have a magnetic configuration with Kuperus – Raadu topology, as shown

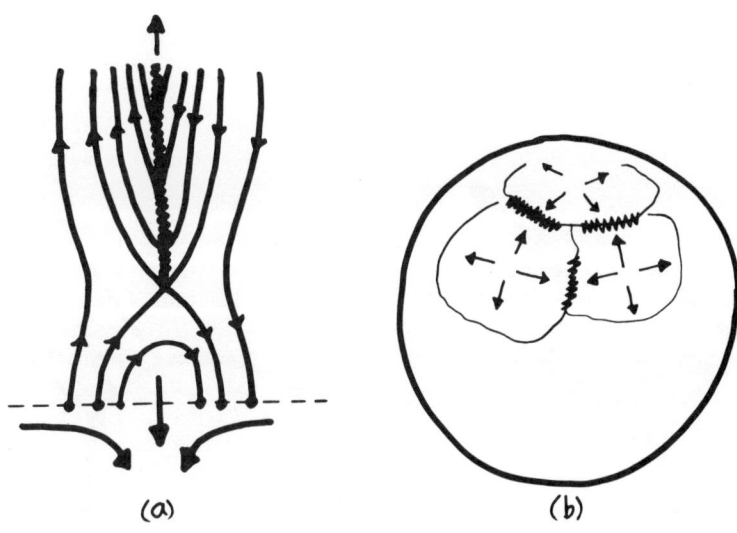

Figure 11. Flux loss at prominences.

in Figure 11a in a section normal to the long axis of the the prominence (Leroy et al., 1983). In view of the reversal in horizontal field direction at the neutral point in this model, the upflows in a prominence would imply (for a steady state in which the plasma and magnetic field are frozen together) that the region below the prominence represents a site of large-scale submergence. This is consistent with the suggestion of Malherbe and Priest (1983) that quiescent prominences lie at the boundaries of giant cells, as indicated in Figure 11b, and therefore they represent large-scale fault lines in the solar surface. Giant cell motions of the footpoints of the prominence fields are typically 0.03 - 0.1 kms^{-1} and, because of the weakening of the field with height, would produce the observed upflows of 0.5 kms^{-1} in the prominence itself. For a field strength of 10G over a length of 100 Mm, this would suggest that 5×10^{20} Mx per day of flux is escaping out of the top of each prominence and the same amount (but opposite in sign) is submerging below the prominence.

4.6 The Corona

Magnetic flux escapes on a large-scale from the corona during prominence eruptions. Usually, these produce a coronal transient or coronal mass ejection; 70% of such ejections are associated with erupting prominences. Occasionally, one finds evidence of reconnection below the escaping structure (Figure 12), although it probably always takes place when the coronal transient magnetic field is stretched out enough. Such a disconnection is required to prevent an observed build-up in magnetic flux in the

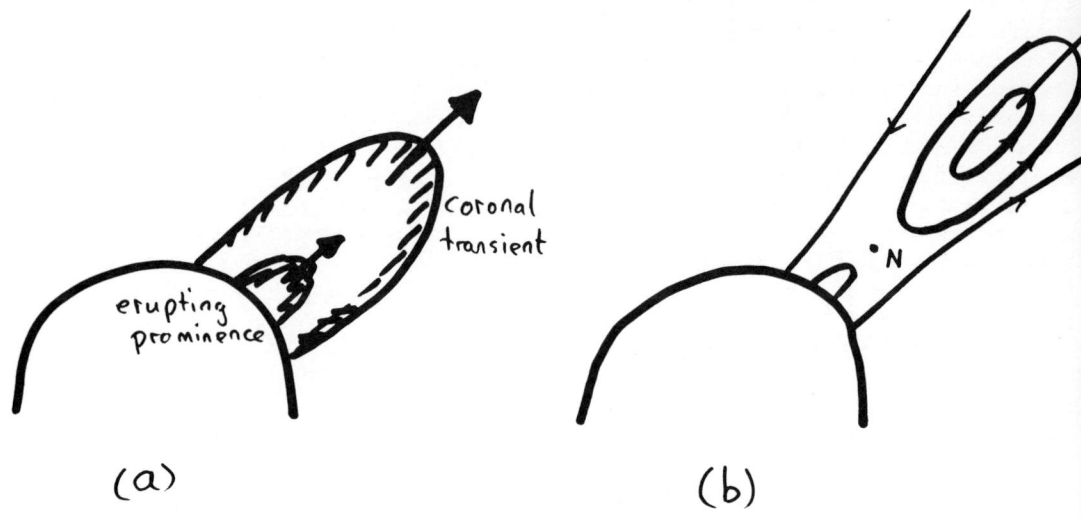

Figure 12. Flux escape in coronal transients.

solar wind, and it probably gives rise to a large plasmoid or magnetic cloud that is sometimes seen in the solar wind. There is typically one coronal mass ejection every two days and each contains about 10^{21} Mx of flux, so this again represents a flux loss of 5×10^{20} Mx per day. Perhaps coronal mass ejections are therefore caused when the build-up of magnetic flux above a prominence by escape out of its top becomes so great that the overlying configuration loses equilibrium; the removal of the overlying stabilising field lines from the prominence would then cause it to erupt. This would be an alternative to the standard picture of coronal mass ejections being a secondary response to a prominence eruption when the prominence field goes unstable.

On a small scale we have seen that cancelling magnetic features produce X-ray bright points and possible small plasmoids or HRTS jets. Ahmad and Webb (1978) noticed plumes in soft X-ray emission overlying bright points in coronal holes. They estimated the outflow speed to be 100 kms^{-1}, and more recently Holt and Mullan (1987) have discovered a corresponding outflow of 1-2 kms^{-1} down at the denser CaIIK level. This would be enough to supply the mass flux of the high-speed solar wind and it is possible that magnetic flux is escaping too in plumes in the form of tiny plasmoids. As far as coronal heating is concerned, a magnetic flux escape from the photosphere and destruction in the corona at a rate of 10^{21} Mx per day for an active region (or 10^{20} Mx per day for a quiet region) would provide the necessary heat.

5 CONCLUSIONS

The emergence and submergence of magnetic flux is a complex process and occurs on a variety of scales. Submergence in particular is sometimes simple submergence, but is often probably reconnection submergence, although the relative role of these two possibilities is at present uncertain. Inner network magnetic field appears to be under a delicate balance between magnetic tension and buoyancy and so it may easily float upwards or fall down. Only a small fraction of active region or ephemeral region flux could escape if it passively sat in one place. But active regions appear to disappear by a process of fragmentation followed by cancellation. Ephemeral regions, on the other hand, appear to be very active and to hunt out other magnetic fragments and cancel with them; this process is helped by the supergranular motion carrying them to the edge of a cell where they may collide with fragments of the network.

Cancelling magnetic features most of the time probably represent flux disappearing by reconnection submergence with reconnection occurring rapidly in the corona in response to photospheric footpoint motions. The transfer of flux in coronal transients and prominences is important for the global circulation of large-scale flux. Furthermore, the appearance and disappearance of flux at the solar surface is not only of interest in its own right but its study has implications both for the interior of the Sun and the dynamo problem and also for the evolution of the coronal magnetic field and solar wind. The new telescopes on Tenerife should play a key role in helping us understand this process much better in future. However, it is crucial to concentrate not just on the photosphere but to understand the subtle connections with the overlying chromosphere and corona.

ACKNOWLEDGEMENT

I am most grateful to Henck Spruit, Aart van Ballegooijen, Ces Zwaan and Sara Martin for invaluable discussions on the physics of flux disappearance.

REFERENCES

Ahmad, I. and Webb, D. (1978). X-ray analysis of a polar plume. Solar Phys. 58, 323-336.

Browning, P.K. and Priest, E.R. (1984). The magnetic non-equilibrium of buoyant flux tubes. Solar Phys. 92, 173-188.

Browning, P.K. and Priest, E.R. (1986). The shape of buoyant coronal loops and the eruption of coronal transients and prominences. Solar Phys., 106, 335-351.

Brueckner, G., Bartoe, J., Cook, J., Dere, K. and Socker, D (1986). HRTS results from Space Lab 2. Adv. Space Res., in press.

Chou, D.-Y. and Wang, H. (1987). The separation velocity of emerging flux. Solar Phys., submitted.

Forbes, T.G. and Priest, E.R. (9184). Numerical simulation of reconnection in an emerging flux region. Solar Phys. 94, 315-340.

Gaizauskas, V., Harvey, K., Harvey, J., and Zwaan, C. (1985). Large-scale patterns formed by solar active regions during the ascending phase of cycle 21. Ap. J. 265, 1056-1065.

Galloway, D.J. and Weiss, N.O. (1981). Convection and magnetic fields in stars. Astrophys. J. 243, 945.

Harvey, K. (1985) Aust. J.Phys.

Harvey, K. and Harvey, J.W. (1973). Observations of moving magnetic features near sunspots. Solar Phys. 28, 61-76.

Harvey, K. (1986). Proc. 2nd Workshop on High-Resolution Solar Physics.

Hermans, L.M. and Martin, S.F. (1986). Small-scale Eruptive Filaments on the Quiet Sun. Proc. Coronal and Prominence Plasmas Workshop, in press.

Holt, R. and Mullan, D (1987). Shifts of CaII K line in HeI 10830 dark points. Solar Phys., in press.

Howard, R. and Labonte, B. (1981). Surface magnetic fields during the solar activity cycle. Solar Phys. 74, 131-145.

Leroy, J. Bommier, V. and Sahal-Brechot, S. (1983). Magnetic field in prominences of the polar crown. Solar Phys. 83, 135-142.

Liggett, M. and Zirin, H. (1983). Naked sunspots. Solar Phys. 84, 3-12.

Livi, S.H.B., Wang, J. and Martin, S.F. (1985). The Cancellation of Magnetic Flux I On the Quiet Sun. Aust. J. Phys. 38, 855-873.

Malherbe, J., Schmieder, B., Ribes, E., and Mein, P. (1983). Dynamics of Solar Filaments. Astron. Astrophys. 119, 197.

Malherbe, J and Priest, E (1983). Current sheet models for solar prominences I. Astron. Astroph. 123, 80-88.

Marsch, K. (1978). Ephemeral region rates and diffusion of the network. Solar Phys. 59, 105-115.

Martin, S.F. (1984). Dynamic signatures of quiet Sun magnetic fields. Proc. Symp. on Small-Scale Dynamical Processes in Quiet Stellar Atmospheres. (ed. S. Keil), Nat. Solar Obs. Sac. Peak, p 30.

Martin, S.F. (1984). Emergence, growth and decay of magnetic flux. Proc. 2nd Workshop on Problems in High-Resolution Solar Physics, Boulder, Sept. 1986.

Martin, S.F., Livi, S.H.B., Wang, J. and Shi, Z. (1985). Proc. Workshop on Vector Magnetic Fields, Alabama, NASA Washington, p 403.

Martin, S.F., Livi, S.H.B. and Wang, J. (1985). The Cancellation of Magnetic Flux II In a Decaying Active Region. Aust. J. Phys. 38, 929-959.

Moreno-Insertis, F. (1986). Nonlinear time-evolution of flux tubes in the convection zone. Astron. Astrophys., submitted.

Moore, R., Tang, F., Bohlin, D. and Golub, L. (1977). Ap. J. 218, 286.

Moore, R. and Rabin, D. (1985). Sunspots. Astron. Astrophys. 23, 239-266.

Meyer, F., Schmidt, H., Simon, G. and Weiss, N. (1979). Astron. Astrophys. 76, 35.

Meyer, F., Schimidt, H., Weiss, N. and Wilson, P. (1974). The growth and decay of sunspots. Mon. Not. Roy. Astron. Soc. 169, 35.

Parker, E.N. (1958). Astrophys. J. 128, 664.

Parker, E.N. (1966). Astrophys. J. 145, 811.

Parker, E.N. (1979). Cosmical Magnetic Fields, Oxford University Press, England.

Parker, E.N. (1984) Magnetic buoyancy and the escape of magnetic fields from stars. Astrophys. J. 281, 839-845.

Parker, E.N. (1986). Astrophys. J., in press.

Petschek, H.E. (1964). Magnetic field annihilation. AAS-NASA Symp. on Physics of Solar Flares, 425-439.

Porter, J.G., Reichmann, E., Moore, R. and Harvey, K. (1986). Magnetic location of CIV events in the quiet network. Proc. of Coronal and Prominence Plasmas Workshop.

Priest, E.R. (1984). The role of newly emerging flux in the flare process. Adv. Space Res. 4, 37-48.

Priest, E.R. (1985). The MHD of current sheets. Rep. Prog. Phys. 48, 955-1090.

Priest, E.R. and Forbes, T.G. (1986). New models for fast steady-state magnetic reconnection. J. Geophys. Res. 91, 5579-5588.

Rabin, D., Moore, R., and Hagyard, M. (1984). A case for submergence of magnetic flux in a solar active region. Astrophys. J. 287, 404-411.

Schussler, M. (1980). Nature 288, 150.

Sonnerup, B.U.O. (1970). Magnetic reconnection in a highly conducting incompressible fluid. J. Plasma Phys. 4, 161-174.

Sonnerup, B.U.O. and Priest, E.R. (1975). Resistive MHD stagnation-point flows at a current sheet. J. Plasma Phys. 14, 283-294.

Spruit, H., Title, A. and Van Ballegooijen, A. (1986). Is there a weak mixed polarity background field? Theoretical arguments. Solar Phys., submitted.

Sweet, P.A. (1958). The neutral point theory of solar flares. IAU Symp. 6, 123-134.

Vrabec, D., (1974) Streaming magnetic features near sunspots. Chromospheric Fine Structure (ed. R.G.Athay) D. Reidel, 201-231.

Wallenhorst, S. and Howard, R. (1982). On the dissolution of sunspot groups. Solar Phys. 76, 203-210.

Zirin, H. (1986). Caltech science objectives on HRSO. Proc. 2nd Workshop on Problems in High-Resolution Solar Physics, Boulder, Sept. 1986.

Zwaan, C. (1978). On the appearance of magnetic flux in the solar photosphere. Solar Phys. 60, 213-240.

Zwaan, C. (1986). Elements and patterns in solar magnetic structure. Proc. 2nd Workshop on Problems in High-Resolution Solar Physics, Boulder, Sept. 1986.

TRANSITION ZONE FLOWS IN SUNSPOT

O. Kjeldseth-Moe, N. Brynildsen, O. Engvold, P. Maltby
Institute of Theoretical Astrophysics, Univ. of Oslo

J.-D. F. Bartoe and G.E. Brueckner
US Naval Research Laboratory, Washington, D.C.

Introduction

Downflows in the transition region over sunspots have been observed with the High Resolution Telescope and Spectrograph on rocket flights in 1975 and 1978. Brueckner (1981) and Nicolas et al. (1982) have described this phenomenon where supersonic flow velocities are seen in the 100 000 K region. However, rocket flights could only give a snapshot of the conditions in the sunspot transition region. Any information on the duration of the downflow phenomenon and reliable indications as to how common it was had to wait until one could make the observations from a satellite.

Observations

The High Resolution Telescope and Spectrograph - HRTS - flew as part of Spacelab 2 from 29 July to 6 August 1985. The instrument, mission and scientific results so far have been described by Bartoe et al. (1986), Brueckner et al. (1986) and Cook (1985).

An extensive series of observations was made of the sunspot in active region NOAA AR 4682 on 7 occasions over a period of 4 days. Some of the observing programs repeated the observations of the entire sunspot every 5 minutes. Two-dimensional images were built up by stepping the HRTS slit across the spot and active region in steps as small as 2 arc seconds. Pointing accuracy was determined by the Instrument Pointing System - IPS, which for the observations used in this investigation, was stable to 2 arc seconds. This value may thus be regarded as the spatial resolution.

The HRTS spectral range, 1190-1680 Å contains emission lines from ions at temperatures from 5000 K to 250 000 K covering the chromosphere and lower transition region. A few weak coronal lines are also present. The high spectral, spatial and time resolutions make HRTS an excellent instrument for the study of dynamical phenomena.

The observations described here use spectra from two different observing programs. In one program, executed 3 August, the slit stepped across the spot 4 times at intervals of 2 arc seconds. The small frame option was used (see Brueckner et al., 1986) giving spectra in a 14 Å range including the C IV resonance lines at 1550 Å. This observing sequence thus serves as a high resolution mapping of the sunspot at the temperature of formation of the C IV lines, 100 000 K. The interval between repeated rasters is 4.5 minutes, and the entire sequence allows the study of short time variations over a 20

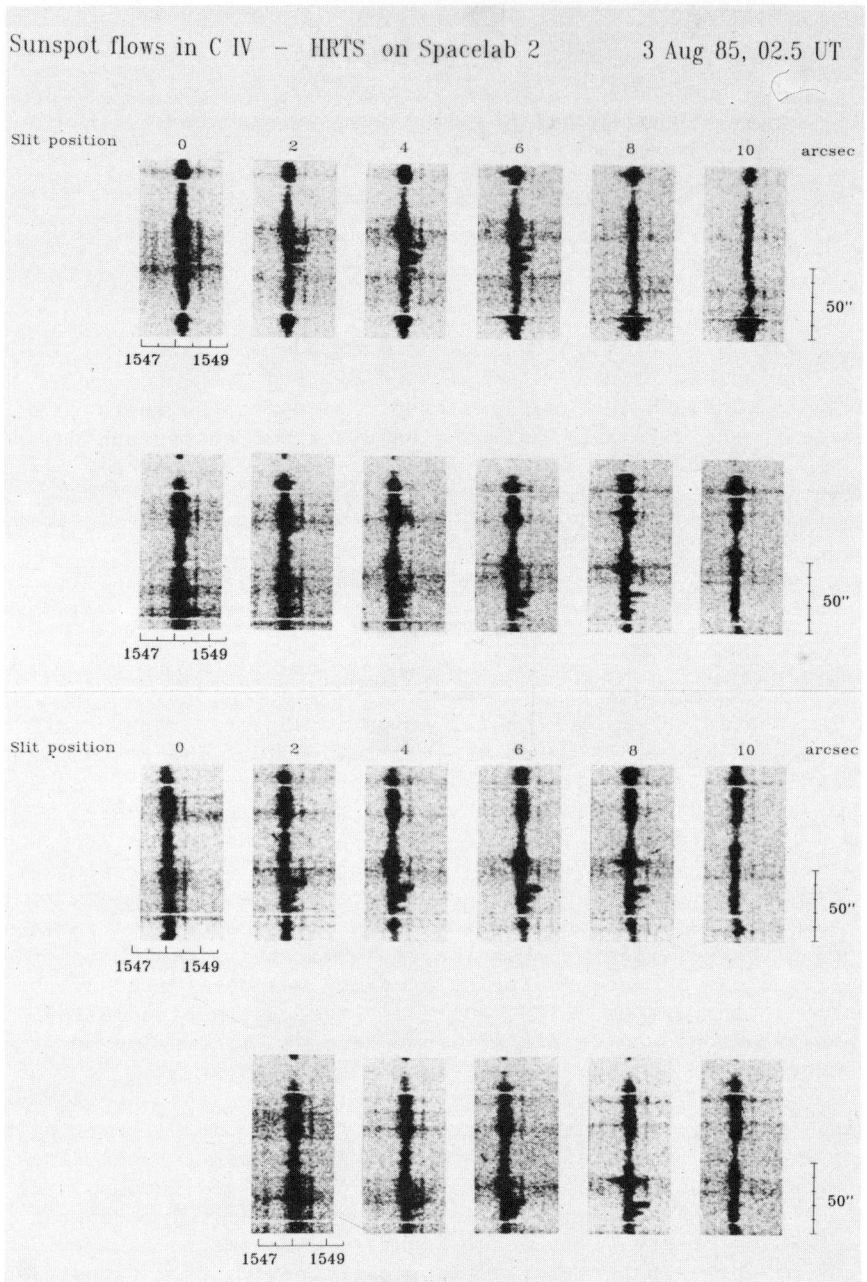

Figure 1. Downflow in C IV 1548 Å stepping across a sunspot four times in increments of 2 arc seconds. The four rows of panels each represent data from one of the four rasters of the slit across the spot.

minute time period. The second program, run on 1 August, consisted of a raster of full frame spectra covering the entire HRTS spectral range. This program is designed for studying the observed phenomena at a range of temperatures.

Results

Figure 1 shows the four rasters across the sunspot in C IV on 3 August. Only the 1548 Å line is included in the figure. The downflow in the sunspot is clearly visible as 2-4 regions of limited extent with a strong redshift in the C IV line. The size of the downflow regions is 4-6 arc seconds. As one steps across the spot the number of redshifted areas varies between 2 and 4 and the position along the slit also appears to change slightly. This may indicate that the individual downflow regions are components of a larger region with a complex structure. While there are minor differences between the flow patterns on the four successive rasters, the main structure did not change over the 20 minute time span the observations lasted.

The flow velocity has been measured by making least squares fits of gaussian profiles to the observed lines. The veloci-

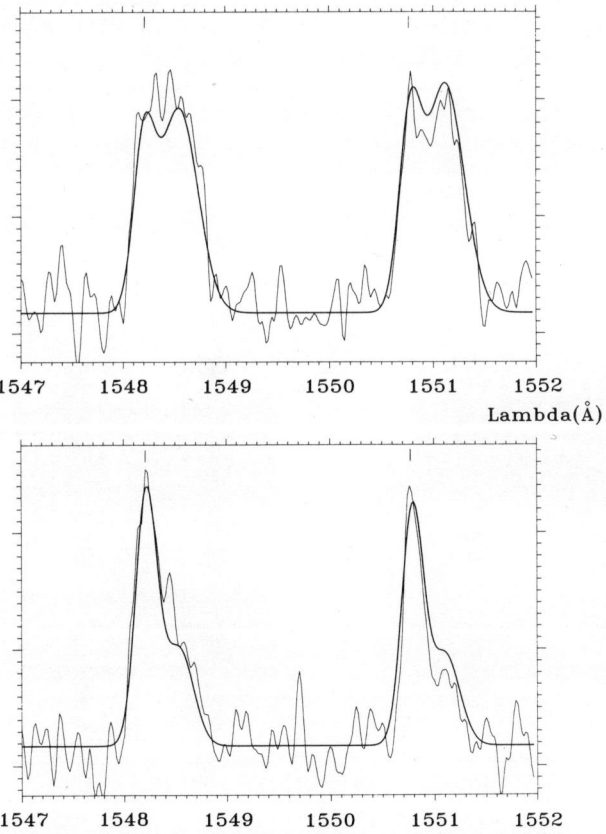

Figure 2. Line profiles of C IV 1550 Å in regions with high and low downward masse flux. The abscissa is proportional to film density. The markers above the lines are indicating the laboratory wavelengths for the C IV lines.

ties are of the order of 50 km/s to 70 km/s in the C IV lines and show no strong variation within the downflow regions or between different regions. The method of measuring the flows is hampered by the presence of lines from Si I at a wavelength distance from the unshifted C IV lines corresponding to the redshift caused by the downflow. The Si I lines have, however, been allowed for in the fitting procedures. The correspondence between observed and fitted profiles is apparent from Figure 2.

Figure 2 also shows two other interesting features. One is the presence of a line component with zero flow velocity. Clearly an appreciable fraction of the gas in the line of sight is at rest and not participating in the downflow. This is always the case at all downflow locations.

The two panels in Figure 2 furthermore show the line profiles at two locations. The upper panel depicts the C IV lines in one of the strong downflow regions while the lower panel shows the line profile in an adjacent region located between two areas with prominent downflow. We note that downflows are present also in this second region. The velocities are slightly smaller, but still amount to approximately 50 km/s. The main difference is the intensity of the redshifted component of the C IV lines, i.e. the downward mass flux, which is considerably smaller than in the region with more prominent downflow.

With the observed strong downflows one may ask the question of where the corresponding upward flow of gas occurs. A coronal loop above the sunspot would be depleted of gas in a few minutes

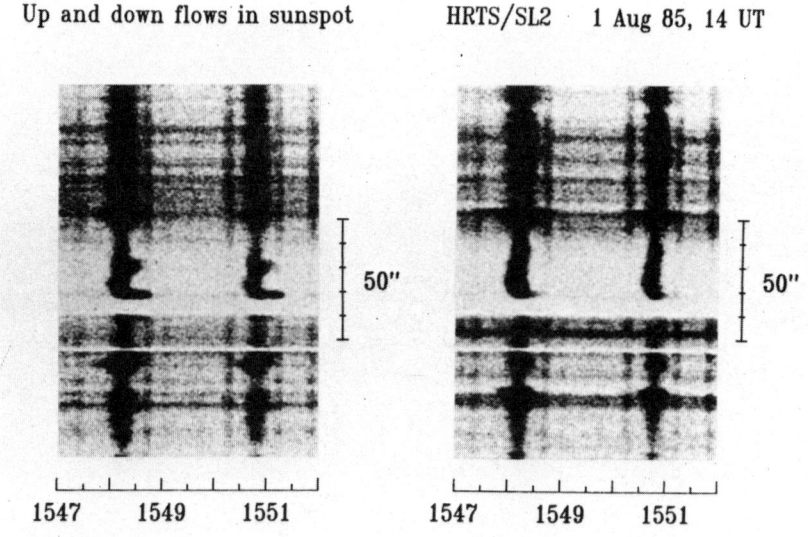

Figure 3. Downflow (left) and upflow (right) in C IV within the same sunspot.

unless supplied by an upward flow. The HRTS spectra indicate that at least part of the required mass may be supplied from areas inside the spot itself. Evidence for this is obtained from some of the panels in Figure 1 where a weak and extended blueshift of the C IV line is seen outside, but near the position of the strong downflow.

The effect is more prominent in Figure 3, which shows an observation of the C IV lines made on 1 August. The left panel shows two typical redshifted regions in the sunspot on this day. (The gap in the data below the lower region is due to light being blocked off by a speck of dust on the slit of the spectrograph.) The panel to the right shows the same lines from an adjacent region 3 arc seconds away. The redshift has disappeared and has been replaced by a blueshift of the C IV line extending over the greater part of the spot. In this case the whole line seems to be shifted, i.e. there are no co-existing zero velocities. Flow velocities are of the order of 10-20 km/s. These observed upflows represent a new discovery not seen in earlier spectra of transition zone lines from sunspots. It still remains to estimate the total upward and downward mass fluxes to determine if the observed upflow within the spot exactly compensates the downflows.

Conclusions

The downflow in the transition region over sunspots first detected on HRTS rocket flights, seems to be a general phenomenon. Although details in the flow pattern appear to change over a period of minutes, the phenomenon itself persists for days. While the mechanism producing the observed redshifts is not understood it seems most likely that they are produced by actual downflow of gas in thin filamentary structures. This may be inferred from the co-existence within the same spatial resolution element of tubes with strong downflows and tubes where the gas is at rest. Other explanations such as fast waves are possible, but appear less likely. Thus the line profiles result from an unresolved fine structure in a similar fashion that Evershed effect in the photosphere produces "flag"-like line profiles in visual lines.

Acknowledgement

We wish to acknowledge support from the Norwegian Research Council for Science and the Humanities under contract D.10.08.044.

References

Bartoe, J-D F, Brueckner G E, Cook J W, Dere K P, Morrison M D, Prinz D K, Socker D G, and VanHoosier M E; (1986) AIAA 24th Aerospace Science Meeting, Reno, Nevada.

Brueckner G E; (1981) in Solar Active Regions, Colorado Associated University Press, ed F Orrall, pp 113-127.

Brueckner G E, Bartoe J-D F, Cook J W, Dere K P, and Socker D G; (1986) Advances in Space Research (in press), 1986.

Cook J W; (1985) in Theoretical Problems in High Resolution Solar Physics, ed H U Schmidt, Max Planck Institut für Physik und Astrophysik, pp 308-310.

Nicolas, K R, Kjeldseth-Moe O, Bartoe J-D F, and Brueckner G E; (1982) Solar Physics 81, 253.

OVERSHOOT OF THE SOLAR GRANULATION

Anastasios Nesis
Kiepenheuer-Institut für Sonnenphysik
7800 Freiburg, FRG

Introduction

The dynamics of the solar granulation is closely coupled with the dynamics of the layers which are directly above the continuum ($\tau_5=1$). Although these layers are convectively stable (above $\tau_5 \approx 0.8$), they show a velocity field induced by the decaying granular motions. Therefore, they are called "overshoot" layers.

The dynamics of the overshoot layers was inferred from investigations of the vertical and horizontal velocities and their variation with height in the photosphere (Nesis, 1985a). The extension of the convective motions into the deep photosphere was determined by means of coherence analysis (Durrant and Nesis, 1981; Nesis, 1985a). Using the measured rms velocity, a model variation of the vertical and horizontal velocity (V_v and V_h) with height in the photosphere was obtained (Nesis, 1985b). The results indicated that there is a layer of about 200 km extension above the continuum ($\tau_5=1$, which corresponds to a geometrical height z=0 km) in which the horizontal and vertical velocities decrease with height. Furthermore, we found that the horizontal and vertical velocity fields are not coherent in the overshoot layers. This indicates that vertical and horizontal velocity fields coexist in these layers, but that they cannot be regarded as components of the overshoot velocity (Nesis, 1985a; Nesis et al., in preparation).

To understand the dynamics of the overshoot layers it is necessary to determine the contribution of the horizontal and the vertical velocity fields to the line-of-sight velocity which we measure by means of the Doppler-shift of the absorption lines. This can preferrently be done by observations at $\cos\theta = 0.6$ where the horizontal and vertical velocities contribute approximately to an equal extent. The measurement at $\cos\theta=0.6$ has an additional advantage: the measured horizontal velocity includes the velocity field just above the continuum (up to ca. 100 km above $\tau_5=1$), which is not the case by measuring at the limb of the solar disk. Here we cannot measure deeper than ca. 100 km above $\tau_5=1$ because of the center-to-limb variation of the sun.

A quantitative description of the contribution of V_v and V_h to the line-of-sight velocity requires a model formulation for the velocity fields as given by Keil (1978). But, here we have to take into consideration that in the overshoot layers exist secondary motions such as gravity waves and turbulence which make the situation more complicated. Therefore, we approached the problem qualitatively by asking only for the predominance of the vertical and horizontal velocities in the line-of-sight velocity, instead of asking for their contribution to this velocity. This was done by comparing the results of the coherence analysis for the spectra taken at $\cos\theta=0.6$ with those of the spectra taken at the center and the limb, respectively, of the solar disk.

Material and Methods

Spectrograms taken with the balloon-borne Spektrostratoskop in the wavelength range 5120-5214 Å were used. The spectrograms contained absorption lines of different strength, and, thus, it was possible to measure the velocity fields (by means of the Doppler-shift of the absorption lines) at different heights (residuals intensity levels I-IV) in the overshoot layers. Spectrograms taken at the center ($\cos\theta=1$) and near the limb ($\cos\theta=0.2$) of the solar disk gave information about the pure vertical velocity V_v, and about the pure horizontal velocity V_h, respectively. Spectrograms were also taken at $\cos\theta=0.6$. The analysis of the spectrograms included photometry, determination of the velocity field, and power- and coherence analysis. The procedure is described in detail in Durrant and Nesis (1981, 1982). The comparison of the cross-spectra was performed by convolution:

$$h(t) = \int_{-\infty}^{\infty} f(t')g(t-t')dt'$$

f(t): Input
g(t): Filter-Function
h(t): Output

Results

Fig. 1 shows the cross spectra between the velocity variations at the higher layers (level I-IV) of the photosphere and those of the deeper layers (level 0) at three positions $\mu=1.0$, $\mu=0.6$ and $\mu=0.2$ ($\cos\theta \equiv \mu$) of the Solar disc. In the following, we compare the velocity cross-spectra at positions $\mu=0.6$ with those at the positions $\mu=1.0$ and $\mu=0.2$ respectively. In doing so, we have to take into consideration that the residual intensity levels I-IV at the different positions μ do not represent the same geometrical height in the photosphere. For example, level I at $\mu=1.0$ lies below level I at $\mu=0.6$, and this again below level I at $\mu=0.2$. The same holds for the levels II, III and IV. Let us assume in a first approximation, that level I at $\mu=0.6$ is about at the same height as level II at $\mu=1.0$; level I at $\mu=0.2$ is at the same height as level II at $\mu=0.6$; and correspondingly for the other levels. Therefore, in Fig. 1 for $\mu=0.6$ (dashed lines) the cross-spectra for the levels I, II and III are drawn together with the cross-spectra for the levels II, III and IV, respectively, at $\mu=1.0$. Accordingly, the cross-spectra for the levels II, III and IV at $\mu=0.6$ (dashed lines) are drawn together with the cross-spectra for the levels I, II and III, respectively, at $\mu=0.2$.

The comparison of the cross-spectra at $\mu=0.6$ and those at $\mu=1.0$ shows the following (Fig. 1, Ia left column): Within the interval $0<X<9(2\rlap{.}{''}9)$ all values of the cross-spectrum at $\mu=0.6$ lie above the values of the cross-spectrum at $\mu=1.0$. This is the case for all layers. From

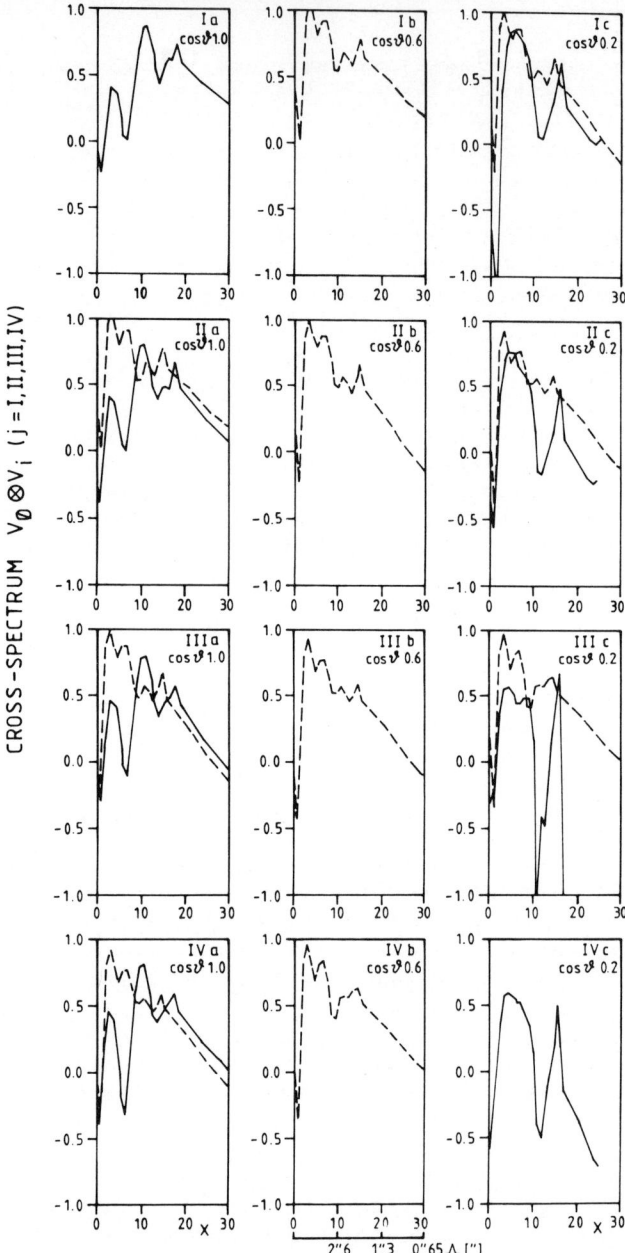

Fig. 1. Cross-spectra of the velocity fields of the deepest layers (V_o) and the velocity fields of higher layers (V_{I-IV}). Left column: results obtained in a measurement at $\cos\theta=1.0$ (center of the solar disk), corresponding to the pure vertical velocity field; right column: measured at $\cos\theta=0.2$ (limb of the solar disk) corresponding to the pure horizontal velocity field; mid column: measured at $\cos\theta = 0.6$. The four rows stand for the levels I-IV (I being the lowest level).
Dashed lines: see text.

$X=9(2\rlap{.}''90)$ up to $X=20(1\rlap{.}''30)$, both cross-spectra show on average the same values.

The comparison of the cross-spectra at $\mu=0.6$ with those at $\mu=0.2$ shows for the levels I-III approximately the same behaviour within the interval $0<X<9(2\rlap{.}''90)$ (Fig. 1. Ic right column). Between $X=10$ and $X=15$ the values of the cross-spectra at $\mu=0.2$ lie on average below those of the cross-spectra at $\mu=0.6$. Both cross-spectra, however, show a similar decline with the exception of the value at $X=16(1\rlap{.}''90)$ where the cross-spectra at $\mu=0.2$ show a peak.

The fact that the cross-spectra values at $\cos\theta=1.0$ are suppressed with respect to those at $\cos\theta=0.6$ at lower wave numbers $X<9(>2\rlap{.}''90)$, and the finding that the cross-spectra values at $\cos\theta=0.2$ are lower than those at $\cos\theta=0.6$ at higher wave numbers $10<X<15(2\rlap{.}''60>\Lambda>1\rlap{.}''80)$ implies the existence of a filter process. Under the premise that a filter process can be described by convolution, we convoluted the cross-spectrum at $\cos\theta=0.6$ with the cross-spectra at $\cos\theta=1.0$ and 0.2. Here, we assumed that the cross-spectrum at $\cos\theta=0.6$ is always the input.

The dashed line in Fig. 2a shows the result of the convolution between the cross-spectra at $\cos\theta=0.6$ in level I and at $\cos\theta=1.0$ in level II. It gives information about the predominance of the vertical velocity field at $\cos\theta=0.6$. Here, the values of the convolution are small within the range $0<X<9(>2\rlap{.}''90)$, increase rapidly within the range of $8<X<12$ $(3\rlap{.}''8>\Lambda>3\rlap{.}''3)$ and remain high up to $X=20(1\rlap{.}''30)$. The result of the convolution between the cross-spectra at $\cos\theta=0.6$ in level II and at $\cos\theta=0.2$ in level I, which gives information about the predominance of the horizontal velocity field at $\cos\theta=0.6$, shows a different behaviour (solid line in Fig. 2a). Here, the values of convolution increase rapidly to higher wave numbers within the interval $3<X<8(9\rlap{.}''00>\Lambda>3\rlap{.}''30)$ and decline moderately in the range $9<X<15$ $(3\rlap{.}''30>\Lambda>1\rlap{.}''8)$. Figs. 2b and 2c show the result of the convolution for the higher levels in the photosphere. Here, the behaviour of the curves was similar as in Fig. 2a. However, the difference between the solid and the dashed curves is less pronounced.

Discussion

The velocity field at $\cos\theta=0.6$ is determined by vertical as well as horizontal motions. High values of convolution indicate that the filter function, i.e. the cross-spectra at $\cos\theta=1.0$ and 0.2, has the same value or a larger value than the input function, i.e. the cross-spectrum at $\cos\theta=0.6$. In Fig. 2 (dashed lines), the high values of convolution at the higher spatial wave numbers indicate that at $\cos\theta=0.6$ the vertical V_v motions are predominantly connected with structures smaller than $2\rlap{.}''5$. The high values of convolution of the cross-spectra at $\cos\theta=0.6$ with those at $\cos\theta=0.2$ (solid lines in Fig.2) at lower wave numbers indicate that the horizontal motions at $\cos\theta=0.6$ are predominantly connected with spatial structures larger than $2\rlap{.}''6$. At the higher levels (Figs. 2b and 2c), the relative difference between the solid and the dashed curves was less pronounced at lower spatial

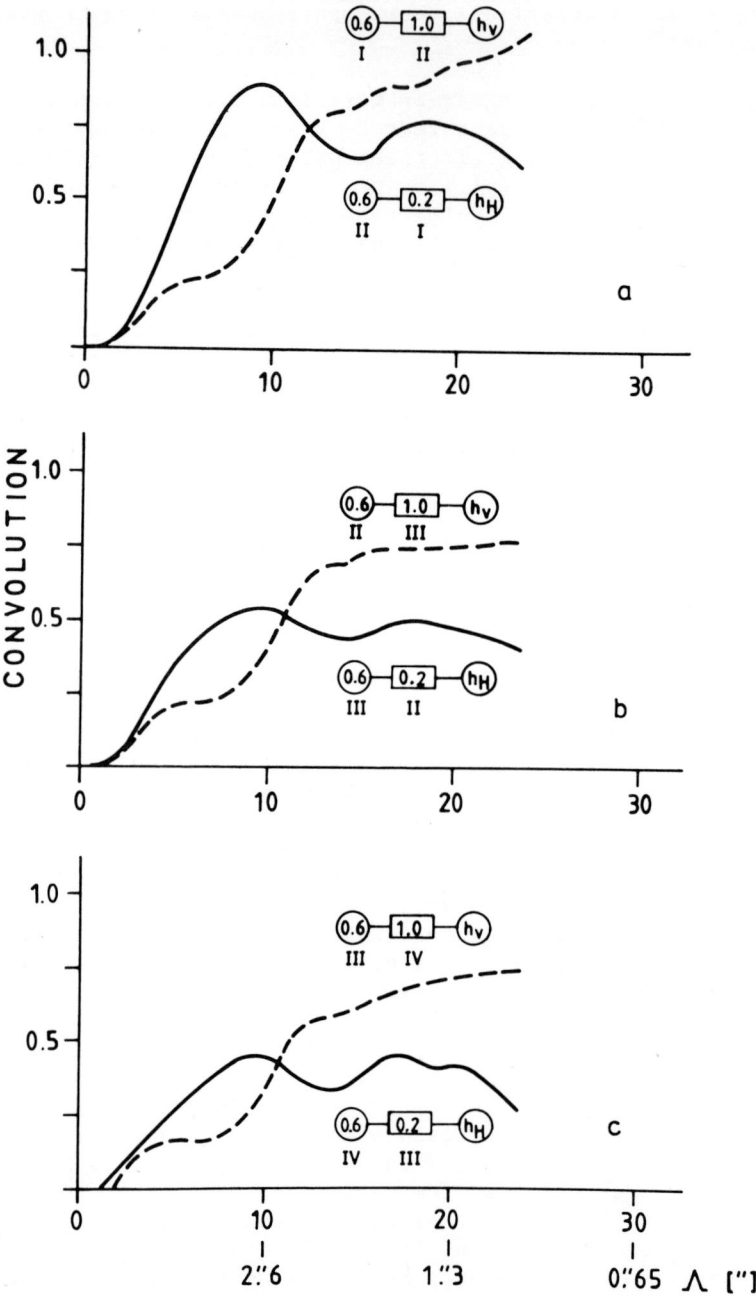

Fig. 2. Convolution between the cross-spectra at $\cos\theta=0.6$ (input function) and the cross-spectra at $\cos\theta=1.0$ (dashed line), and the cross-spectra at $\cos\theta=0.2$ (solid line), respectively, a,b,c: results for the lower levels (I, II), medium levels (II, III) and highest levels (III, IV), respectively.

wave numbers. This indicates that at cosθ=0.6 the horizontal velocity structures disappear gradually with height.

The high values of convolution for X>10(<2".60) (solid and dashed curves in Fig. 2) indicate that at cosθ=0.6 horizontal as well as vertical velocities are connected with spatial structures smaller than 2".5.

Our results can be interpreted as follows: At the deepest layers which are just above the continuum, the velocity field is dominated by horizontal motions which are connected with structures larger than 2".6. This is in accordance with models for the convective overshoot by Hurlburt et al. (1984). By correlating the continuum intensity variations with the vertical velocity variations we have found in previous investigations (Mattig, Mehltretter, Nesis, 1969) that the structures smaller than 3" belong to the convective phenomenon (granulation), whereas structures larger than 3" can be attributed to oscillations. In K-Ω diagrams a low frequency power (in structures >3") could indicate the existence of gravity waves. Our finding that the horizontal velocity V_h is connected with structures 3" fits well in this situation. On the other hand we know that the 5 min oscillations do not show any horizontal velocity (Stix and Wöhl, 1974), and the short period oscillations are vertically directed. Thus the horizontal velocity V_h could be due to gravity waves. Furthermore, our finding that the horizontal velocity decreases with the height in the overshoot layers could demonstrate the existence of large vortices with a high turn-over time. Mehltretter (1978) found larger granula having a life time larger than 8 min. To decide whether gravity waves or vortices are involved is, however, difficult.

References
Durrant, C.J.& Nesis, A.(1981), Astron. Astrophys. **95**, 221.
Durrant, C.J.& Nesis, A.(1982), Astron. Astrophys. **111**, 272.
Hurlburt, L.N., Toomroe, J. & Massaguer, M.J.(1984), Ap.J. **282**, 557.
Keil, S.L.(1978), Astron. Astrophys. **70**, 169.
Mattig, W., Mehltretter, J.P. & Nesis, A.(1969), Solar Phys. **10**, 254.
Mehltretter, J.P.(1978), Astron. Astrophys. **62**, 311.
Nesis, A.(1985a), Thesis, TU Berlin.
Nesis, A.(1985b), in High Resolution in Solar Physics, ed. R. Muller, (Lectures Notes in Phyics 233), Springer, Heidelberg, p. 249.
Nesis, A., Durrant, D.J. & Mattig, W.(1986), in preparation.
Stix, M. & Wöhl, H.(1974), Solar Phys. **37**, 63.

Mitteilungen aus dem Kiepenheuer-Institut Nr. 275

JOINT DISCUSSION ON TOPICS OF SESSION 11, SUMMARIZED BY THE CHAIRMAN

W. Deinzer
Universitätssternwarte Göttingen, F.R.G.

The discussion concentrated primarily on the invited review by Eric Priest:

1) M. Schüßler and J.O. Stenflo pointed out that the observed decreasing circular polarisation does not necessarily imply disappearing magnetic flux. If for instance a large fluxrope is considered which breaks up into smaller fragments, they are hotter and the spectral lines are weakened by a large factor (maybe 2.5) as compared with the unfragmented case. This results in a decrease of the integrated circular polarisation by this factor and looks like an apparent disappearance of magnetic flux. There have been reports of flux disappearance in unipolar regions not accompanied by cancellations which might be interpreted in this way.

2) Further remarks were concerned with the disappearance of magnetic flux via reconnection. Is there a special attraction mechanism between heteropolar flux tubes (Garcia)? Are not funny large scale velocity fields needed to set up a reconnection configuration (Zwaan)? A. Title is disturbed about Zirin's observations where motions of about sound speed are needed to bring the far edges of different flux tubes together.

M. Schüßler referred to Sheeley, who showed that differential rotation and the action of granular and supergranular flow -described by a "turbulent diffusivity" - are sufficient to reproduce the evolution of the magnetic field on the largest spatial scales. A. Title wanted meridional flows to be included.
H. Spruit remarked "with a seaserpent you can make single poles disappear" (Spruit, Title, van Ballegooijen preprint).

3) Concerning submergence of magnetic flux E. Priest (in answer to U. Anzer) pointed out that in the case of disconnected poles reconnection is needed to form a single flux tube before submergence is possible; furthermore reconnection helps submergence by causing part of the reconnected field to be ejected down. C. Zwaan mentioned that Sara Martin has never seen an ephemeral region to converge again and then to disappear; hence simple submergence cannot be the mechanism in all cases of flux disappearance.

4) Finally there were some clarifying remarks about the con-

cept "turbulent diffusivity". Turbulence by itself does not dissipate magnetic flux; it only brings flux tubes together closely enough and thus generates sufficiently strong field gradients to make Ohmic losses effective (M. Schüßler, R. Rosner).

There was some discussion on Maltby's talk: F. Kneer pointed out that the asymmetric doubling of the CIV-line could be due to self-reversal in optically thick layers. P. Maltby replied, that the lines had been treated as optically thin; but even if the lines were optically thick velocities of the given order were required.

Further remarks concerned the mass flux and the correlation between upflows and umbral dots (Garcia, Koutchmy).

Referring to A. Nesis's paper N. Weiss raised the question, whether the scale of the horizontal velocity indicates that it is driven by exploding granules. Nesis postponed a definite answer to future observations with the Teneriffa telescopes.

CROSS CORRELATION OF IMAGES OF SOLAR
FINE STRUCTURE AND POSSIBLE APPLICATIONS

- A PROGRESS REPORT

Tron André Darvann
Institute of Theoretical Astrophysics, University of
Oslo, P.O.Box 1029, Blindern, N-0315 Oslo 3, Norway

INTRODUCTION
Cross correlation is used for high precision detection of time changes in solar images. The maximum cross correlation coefficient of two images is a measure of similarity and best superposition of the images. The technique has been successfully carried out on granulation images for high precision tracking (von der Lühe 1983). The accuracy of the method for various frame sizes and resolution of the data is studied further by Andreassen et al (1984), and Darvann et al (1986). The method is currently being used to study time evolution in images of solar prominences. The objective of this presentation is to discuss various applications of the method.

Figure 1. Plots of displacement vectors. Length of smallest vectors is 0.12 arcsec. (The arrows have a 3x magnification relative to the scale in the diagrams.)
1a) Visual seeing quality 0. Mean diameter of isoplanatic areas 8.1 arcsec. RMS fluctuation of amplitudes 0.072 arcsec.

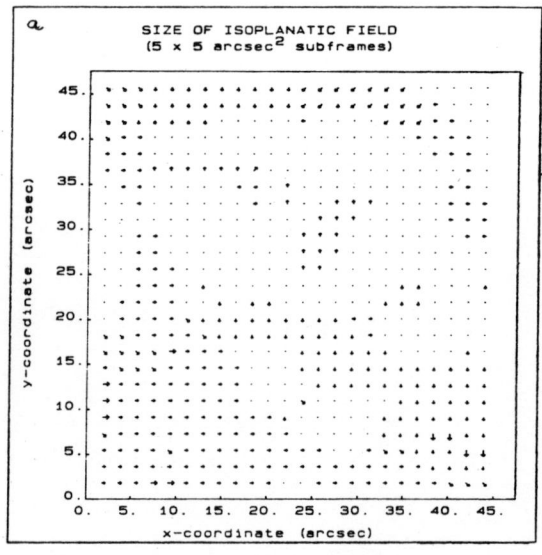

APPLICATIONS OF THE CROSS CORRELATION METHOD
Measurements of solar seeing.

The quality of seeing is measured from photographically recorded granulation images obtained with the 50 cm vacuum telescope of the Royal Swedish Academy of Sciences on La Palma, Canary Islands (Wyller and Scharmer 1985). We first calculate the cross correlation of the full images, thus neglecting the local image distortion caused by seeing. The relative displacement of local areas (image distortion) is then calculated by cross correlating smaller areas successively within the full frames. The time separation between the two correlated images must be short enough so that no real changes in the granulation pattern has occured.

In order to measure the true effect of the seeing the subframes should be comparable to or (rather) smaller than the angular coherence area of the seeing (isoplanatic area) and at the same time be large enough to enclose sufficiantly many granules.

Figures 1a, b, c show the image distortion detected in pictures of different seeing quality, – the best ones having resolution of about 0.3 arcseconds. Areas consisting of displacement vectors with equal amplitude and direction represent isoplanatic areas.

The method is presently being applied for seeing measurements during the LEST site testing campaign at La Palma. The mean size of coherence area, and the average image distortion has been found to correlate well with other measurements of seeing (Kusoffsky et al., 1986).

Fig. 1b,c. Visual seeing quality b) 1.5, c)4. Mean diameter of isoplanatic areas b) 3.5 arcsec, c) 2.2 arcsec. RMS fluctuation of amplitudes b) 0.23 arcsec, c) 1.1 arcsec.

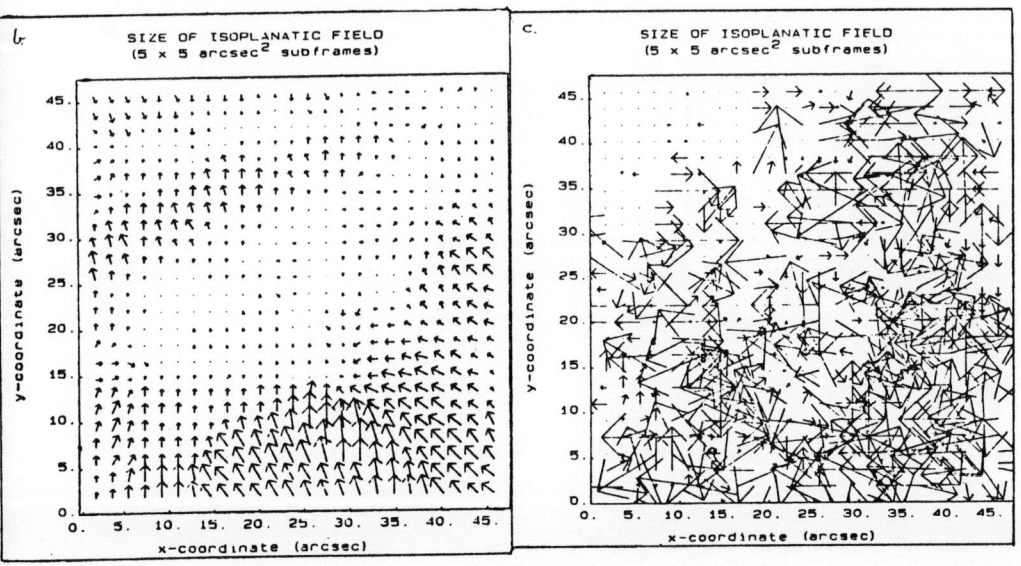

Figure 2. The reference image must be updated before the crosscorrelation coefficient drops below 0.5.

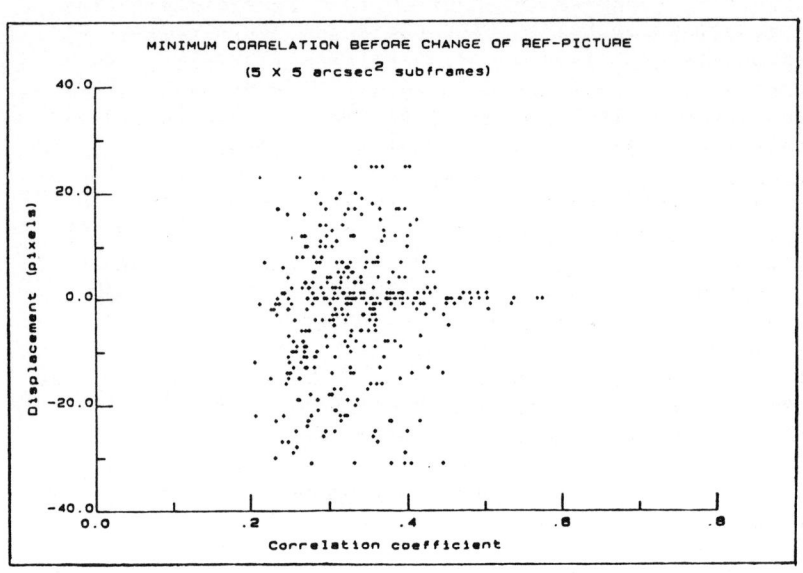

Figure 3. Granulation decay curve.

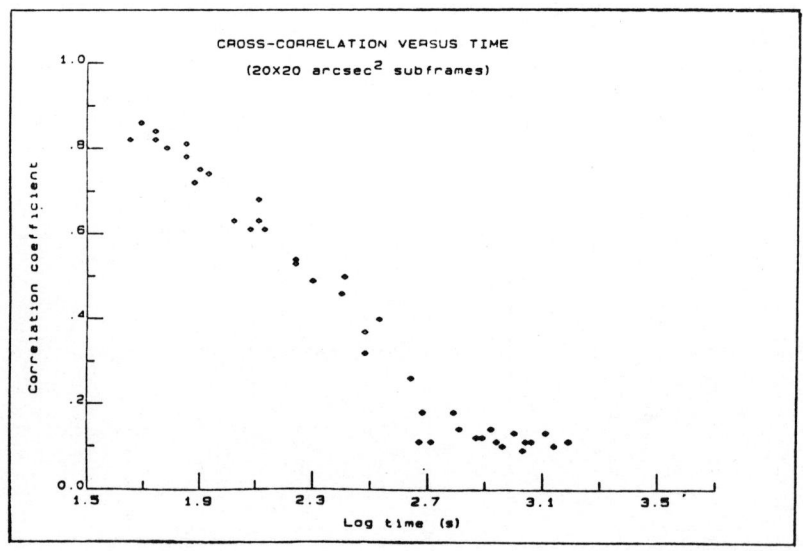

High precision correlation tracking

A correlation tracker using solar granulation is recently developed by von der Lühe and collaborators at National Solar Observatory at Sacramento Peak (von der Lühe 1983). The use of solar granulation for high precision tracking has been investigated further by the use of high quality time series of granulation (Darvann et al. 1986). The objective is to see how the accuracy of the method depends on such as (i) the area of the correlated images, e.g. the number of granules, (ii) the time difference between the two images, and (iii) the number of bits per data point. The observations for the study were made with the 50 cm refractor of Pic du Midi Observatory.

The tracking error signal is derived by cross correlating an image in the time series with successively later ones. The precision in the superposition of two images decreases (i) in proportion to decreasing overlap of the correlated images, and (ii) with increasing time difference of the images.

The study shows that the background noise level of the CC decreases with increasing frame size. Images of 5 x 5 $arcsec^2$ which typically contain 7 - 15 granules seem adequate for tracking. The maximum area of a tracker must be comparable to, or less than, the isoplanatic area of the seeing. The reference image has to be updated before the correlation coefficient falls below a critical value set by the lifetimes of the individual granules (see Fig. 3), and by the background noise level of the CC. Figure 2 shows that in the case of a 5x5 $arcsec^2$ subframe size the reference image must be updated before the peak CC drops below 0.5. This requires that reference images are refreshed within 4 minutes or less.

Time evolution of solar fine structure

The decay curve of the maximum cross correlation coefficient in a timeseries of granulation images is a measure of the lifetime of the granules. Decay curves have been plotted using the observations from Pic du Midi. Figure 3 shows one such plot. The decay time for the CC to fall to 1/e is 6.0 minutes. This is in very good agreement of Mehltretter (1978) who obtained characteristic decay times of 5.9 minutes.

A program has been initiated at the Swedish 50 cm solar vacuum telescope at La Palma to study the characteristics of granulation, such as the lifetime of granules, in photospheric regions close to sunspots.

Time evolution of prominences is being studied using timeseries of Ca II filtergrams of prominences observed by O. Engvold in 1978 at the solar vacuum tower telescope of National Solar Observatory, Sacramento Peak (Engvold 1981).

Large scale structures are separated from small scale structures by filtering techniques. The measurements of lifetime are thus done separately on large and small intensity structures in the prominences.

Low velocity flows in the solar atmosphere

A program to measure low velocity flows in the solar photosphere is initiated at the Swedish solar observatory, La Palma.

Spectroscopic observations (Beckers, 1978; Howard, 1978) have shown meridional flows of 20-40 m s^{-1} from pole to equator. Such flow speeds are observable with the CC method. The detection limit is reduced somewhat by the image distortion from seeing, which can however be largely eliminated when we average the measured shifts obtained from different frames recorded close in time.

Acknowledgements

I would like to express my warmest thanks to the staff of the Swedish Observatory at La Palma for their enthusiastic support and hospitality during my stay there in June/July 1986.

A special thanks to Dr. Oddbjørn Engvold, Institute of Theoretical Astrophysics, for his kind support and advices during my studies of the cross correlation method.

References

Andreassen, Ø., Engvold, O., and Muller, R.: 1984, "High Resolution in Solar Physics", Proceedings, Toulouse, France, Ed.: R. Muller, p. 91.
Beckers, J.M.: 1978, Proceedings Workshop on Solar Rotation, Publicazione Oss. Astrofisico de Cataria, 162, 166.
Bray, R.J., Loughhead, R.E., and Durant, C.J.: 1984, "The Solar Granulation", Cambridge University Press.
Darvann, T.A., Engvold, O., and Andreassen, Ø.,: 1986, (in preparation).
Engvold, O.: 1981, Solar Phys. 70, 315.

Howard, R.: 1978, Reviews of Geophys. and Space Phys. 16, 721.
Kusoffsky, U., Hosinsky, G., Darvann, T.A.: 1986, LEST site testing report, La Palma (in preparation).
von der Lühe, O.: 1983, Astron. Astrophys. 119, 85.
Mehltretter, J.P.: 1978, Astron. Astrophys. 62, 311.
Wyller, A.A. and Scharmer, G.B.: 1985, Vistas in Astronomy 28, 467.

THE SWEDISH FABRY-PEROT ECHELLE SCANNER AS A TWO-DIMENSIONAL UNIVERSAL FILTER AND STOKESMETER

A.A. Wyller
Grupo Sueco del Observatorio del Roque de los Muchachos
Royal Swedisch Academy of Sciences,
Stockholm, Sweden

In a recent review on trends in measurements of solar vector magnetic fields using the Zeeman effect Harvey (Harvey, 1985) stresses the trend away from the use of grating spectrographs to narrow-band filters to improve angular and temporal coverage. There is the problem of the accuracy with which line profiles can be reconstructed from filtergrams, although adequate techniques appear to be available (Caccin et al., 1983). In fact, in the same conference publication as Harvey's review there is a paper by Lites and Skumanich (Lites and Skumanich, 1985) which specifically addresses the problem of recovering vector magnetic fields and thermodynamic parameters from polarization measurements of photospheric line profiles measured with filtergraphs.

For a number of years the Swedish Station at Observatorio del Roque de los Muchachos (Wyller and Scharmer, 1985) has had operational an echelle-grating scanner with pinhole apertures. This unit is part of URSIES (Ultravariable Resolution Single Fabry Perot Echelle Scanner) spectrograph complex (Wyller and Fay, 1972). In its completed state the scanner includes a Fabry-Perot interferometer (Ramsay, 1966) which is placed in the collimated beam in front of the echelle grating (128 mm x 56 mm, 300 lines/mm, blaze angle $63°26'$) which has a dispersion of 1.2 Å/mm at NaD_1 (5890 Å). The interferometer may be moved on precision line bearings in and out of the collimated beam. It has a clear aperture of 55 millimeter in diameter. The spherical mirrors, the grating and the interferometer are encased in a cylindrical pressure vessel (length 1600 mm, diameter 700 mm) which can be filled with, e.g., freon gas. The spectral transmission can be finetuned in wavelength by pressure scanning (in freon, 10 mm Hg pressure change produces a wavelength change of 0.09 Å at the NaD_1). A Texas Instrument Pressure Controller and Regulator can provide pressure steps as small as 0.25 mm Hg.

Now, by placing the Fabry-Perot interferometer between the collimating mirror and the echelle grating the different transmission bands or orders of the interferometer, $\lambda_p = 2nt/p$, will be spatially separated by the grating in the camera focal plane. The plate separation is t, and n is the index of refraction. The angular factor, $\cos\beta=1$ for the small angular fields that we are concerned with. The resulting display of the transmission bands is called <u>a channel spectrum</u> (see Figure 1). The separation in Ångstroms between the adjacent channels is called <u>the free spectral range</u> $\Delta\lambda$. The separation of interference orders, λ_p, in units of transmission band width, $\delta\lambda$, is called the finesse, N_E, of the interferometer; typically the finesse may be 20 over a wide wavelength interval (4000 - 7000 Å). For a plate separation of 3 mm = t, λ = 6000 Å, $\Delta\lambda = \lambda/2nt$ = 0.6 Å, $\delta\lambda = \Delta\lambda/N_E$ = 0.03 Å.

If we use a pinhole entrance aperture, and with a perfect optical system for the spectrograph, the spectral channels will be spatially represented as pinholes of the same size in the exit focal plane since $f_{coll} = f_{cam}$ = 1500 mm. Let us now relax the restriction of a pinhole entrance aperture and adopt an entrance <u>slit</u> configuration perpendicular to the dispersion axis. Along such an axis each point on the entrance slit will generate a channel spectrum at the exit focal plane (see Figure 2). By selecting an exit <u>slot</u> (not slit) narrow enough to exclude the adjacent λ_{p-1} and λ_{p+1} channels or transmission bands we have <u>a one dimensional universal tuneable filter</u>. The pass band or channel, λ_p, can be coarse tuned by turning the grating and finetuned by changing the pressure in order to have a preselected wavelength fall in the λ_p channel and the exit slot aperture.

Placing a silicon diode array like a CCD at the exit focal plane, with pixel size 30 microns, in our existing telescope and spectrograph configuration each pixel will cover 0.03 mm x 9"/mm = 0.27 arcseconds along the slit image perpendicular to the wavelength axis, 0.03 mm x 1.2 Å/mm = 0.036 Å in wavelength perpendicular to the slit image. By pressure scanning (Δp = 1 mm Hg, $\Delta\lambda$ = 0.009 Å) we can fine-tune through Fraunhofer lines with halfwidths 0.10 Å, and with the photoelastic modulator system currently being developed by Hosinsky we can sample magnetic field in real time with diffraction limited resolution of 0.27 arcseconds.

In order to expand the application of this one-dimensional universal tuneable filter to a full two-dimensional filter let us make the following considerations. In our one-dimensional diagram let us consider <u>the channel spectrum of a spatial point displaced by one channel width along the dispersion axis</u> (see dotted channel spectrum in Figur 3). The same exit slot used in the one dimensional case to cut out λ_{p+1}^A and λ_{p-1}^A can also be used to cut out contamination by λ_{p+1}^B and λ_{p-1}^B. By continuing this argument of considering spatial points along the wavelength axis we see that the same exit slot width will cut out λ_{p+1}^i and λ_{p-1}^i for <u>all</u> spatial points which lie up to <u>half</u> the free spectral range, i.e. up to half the separation between adjacent channels λ_{p+1} and λ_{p-1} on either side of λ_p. Thus the same exit slot width (= $N_E \times \delta\lambda$ = free spectral range) will accomodate N_E spatial points at <u>entrance slot</u> (i.e. not a slit anymore). The exit slot will not allow any λ_{p+1}^i or λ_{p-1}^i channels to contaminate the λ_p^i channels within the exit slot.

Thus from this argument we see that URSIES in the interferometer echelle combination can be considered as a fully tuneable universal two-dimensional filter with <u>fully variable band width</u>. The band width of the filter is fully variable since our Fabry-Perot interferometer has a spacing between the plates which is continuously variable from 0.1 mm to 10 mm, or a band width variable from 0.9 Å to 0.01 Å. The plates adjusted to a separation of about 0.4 mm will give our URSIES filter a band width of 0.25 Å at Hα, i.e. equivalent to that of a Zeiss Hα filter.

What does the above mean in terms of spatial and spectral resolution and CCD's? If we go for diffraction limited performance with our Swedish solar telescope, each pixel will cover 0.03 mm x 9 "/mm = 0."27 in spatial resolution, and the spectral channel width should cover one pixel or 0.03 mm x 1.2 Å/mm = 0.036 Å. If the finesse is 20 (conservatively estimated) then we accomodate 20 channel widths à 0.036 Å along the dispersion axis, which is also 20 channels of 0."27 in spatial extension. That is about 5."4 in slot <u>width</u> along the dispersion axis. Perpendicular to the dispersion axis, i.e. along the slot <u>length</u> we can accomodate the full CCD pixel number, say 256, or 256 x 0."27 = 70 arcseconds. Thus we can sample some 5000 data points <u>simultaneously</u>, over an area of 5."4 x 70", where each data point is 0."27 (diffraction limited) and of spectral width 0.036 Å. We can step through a given spectral line with half width 0.10 = 3 channel widths by pressure steps of 4 mm Hg = 0.036 Å at NaD_1 in freon.

The only way of increasing the number of spatially sampled points is to <u>increase the finesse</u>. This can readily be done by incorporating special Fabry-Perot plates coated for a higher finesse over a more limited wavelength range, - a finesse of 50 can be readily attained. Thus the spatial slot width along the dispersion axis can be increased by a factor of 2.5 to cover an area of about 14 arcseconds by 70 arcseconds. The number of simultaneously spatially sampled points is increased to 12500 with or without the magnetic analyzer; each sampled point being 0."27 and 0.036 Å spatially and spectrally.

However, the limitation in the number of spatially simultaneously sampled points along the dispersion axis can be turned to a powerful advantage in the following way. The spatial sampling limitation was imposed by the requirement that the projected entrance slot width should not exceed the free spectral range at the exit focal plane. This was required so that a spatial point "i" along the dispersion axis would not contaminate a given λ_p^i channel with light from its λ_{p+1}^i or λ_{p-1}^i channels.

Now the same entrance slot will be imaged as a series of adjacent image slots in the light of the various spectral channels λ_{p-1}^i, λ_p^i, λ_{p+1}^i etc. If these images are allowed to fall side by side on a CCD device, the URSIES filter can sample the same entrance slot <u>simultaneously</u> in various portions of a broad spectral line profile like Hα or Na D with each imaged slot sampled in diffraction limited mode 0."27 and 0.036 Å. Referring to our earlier numerical example we had a slot width of 20 x 0.036 Å = 0.72 Å using 20 pixels in the direction of the dispersion axis of our CCD sized 256 x 256 pixels. This slot width allows 12 slots to be sampled <u>simultaneously</u> or an interval of 12 x 0.72 Å = 8.6 Å across a broad spectral line is sampled both non-magnetically and magnetically. In the case of the NaD lines both the temperature stratification and the magnetic field stratification could then be measured simultaneously in one CCD read out with no need for scanning. Figure 4 displays the Hα profile from the Jungfraujoch Atlas with 5 slots put in.

The free spectral range in our numerical example is fairly wide, 0.72 Å. This would be the slot width in the exit focal plane along the dispersion axis. This may be too coarse a sampling of even a fairly broad spectral line (except possibly in sunspots). In order to finegrain the sampling better, the free spectral range may be decreased by decreasing the channel width. As mentioned earlier our Fabry-Perot interferometer has a continuously variable spacing. The maximal spacing is 10 mm which gives a channel width of 0.01 Å at 6000 Å. The free spectral range is then 0.20 Å and the CCD (256 x 256 pixels) would then span 2.4 Å in 12 slot images, which is more appropriate for the photospheric Na D line profiles and the chromospheric Hα line profiles.

Instead of letting the adjacent image slots sample different parts of one and the same spectral line, one could just as well <u>let the slots sample two different but fairly adjacent spectral</u> lines. Recently Stenflo and Harvey (Stenflo and Harvey, 1985) have used the Fe I line pair 5247.06 and 5250.22 to investigate the properties of magnetic flux tubes using the Kitt Peak FTS instrument. Their spectral resolution was 20 m Å or 0.02 Å and they used a square entrance aperture of 5 x 5 arcseconds.

If we adopt a channel width of 0.04 Å, then with a finesse of 20 the free spectral range is 20 x 0.04 = 0.8 Å. So each image slot will be 0.8 Å wide along the dispersion axis, i.e. 5.4 arcseconds wide spatially and 70 arcseconds long perpendicular to the dispersion axis. The line separation between the aforementioned two Fe I lines is 3.16 Å or almost exactly 4 slot widths apart (see Figure 5). Since the spacing is continuously variable one can finetune the channel widths and the free spectral range so that the respective iron line profiles are centered within the first and the fifth spectral channels in their respective image slots. Thus one can <u>simultaneously</u> in the two iron lines sample 5000 data points which are each 0".27 and 0.04 Å in respectively spatial and spectral extent. The spectral channel of width 0.04 Å can then be stepped through the iron line profiles by stepping with the grating or by pressure scanning.

REFERENCES

B. Caccin, R. Falciani, G. Roberti, A.M. Sambuco, L.A. Smaldone, Solar Physics, <u>89</u>, 323, 1983.
J. Harvey, NASA Conference Publication no. 2374, Measurements of Solar Vector Magnetic Fields, p. 109, 1985.
B.W. Lites and A. Skumanich, NASA Conference Publication no. 2374, Measurements of Solar Vector Magnetic Fields, p. 342, 1985.
J.V. Ramsay, Applied Optics <u>5</u>, 1297, 1966
J.O. Stenflo and J. Harvey, Solar Physics, <u>95</u>, 99, 1985.
A.A. Wyller and T. Fay, Applied Optics <u>11</u>, 1152, 1972.
A.A. Wyller and G. Scharmer, Vistas in Astronomy, vol <u>28</u>, 467, 1985.

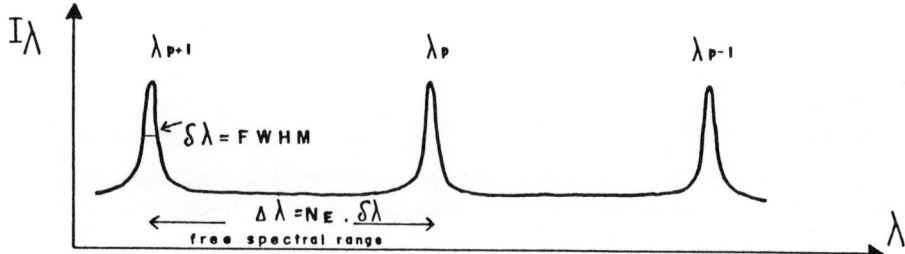

Figure 1. Intensity Distribution of Entrance Pinhole Image = Channel Spectrum.

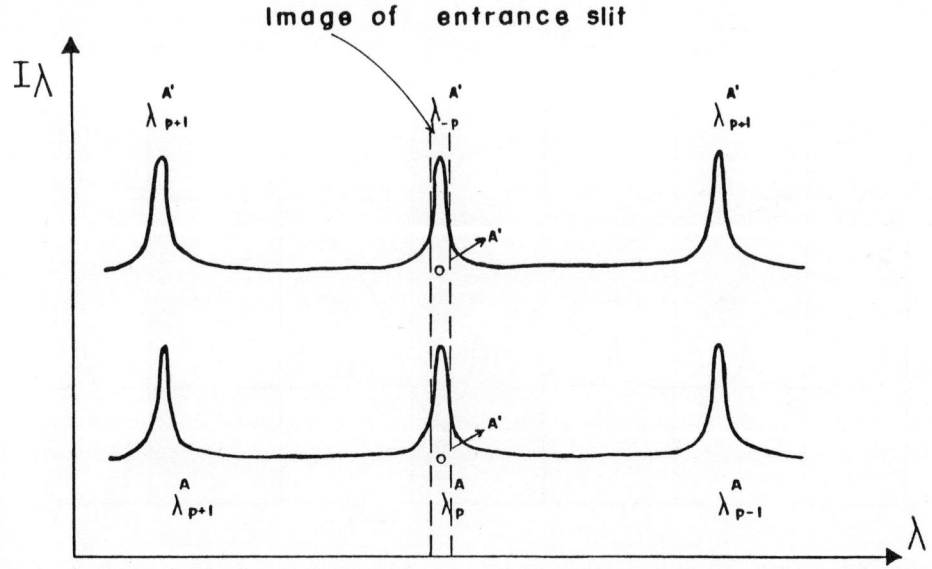

Figure 2. Channel Spectrum of "pinhole" points A and A' along Slit Length. Both Intensity Distributions are perpendicular to Slit Length.

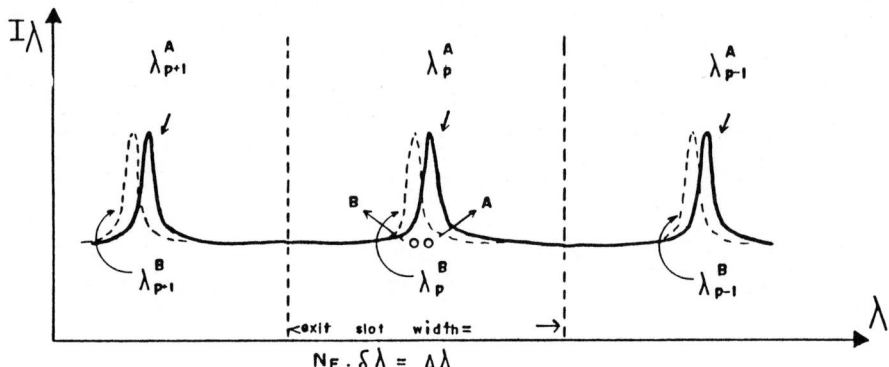

Figure 3. Full Line: Channel Spectrum of "pinhole" A on Slit
Dotted Line: Channel Spectrum of "pinhole" B
displaced perpendicular to Slit Length.

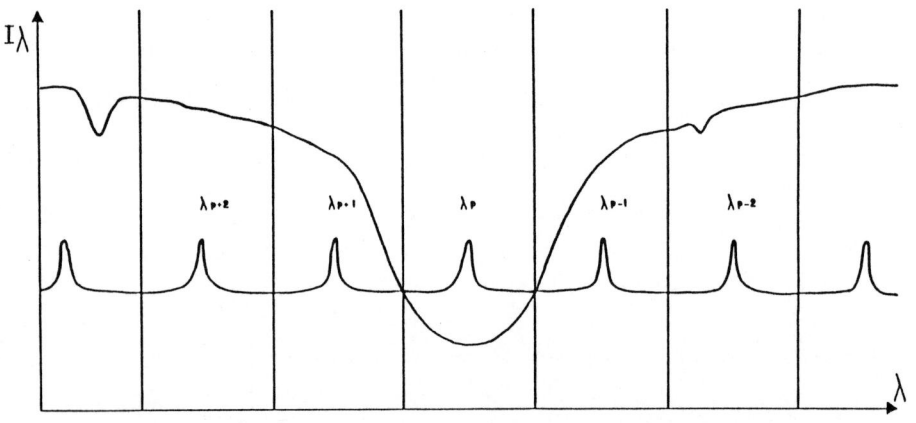

Figure 4. Exit Slots p+2, p+1, .. all with widths 0.8 A to
free spectral range. Superimposed H-alpha profile
from Jungfraujoch Atlas to same wavelength scale.

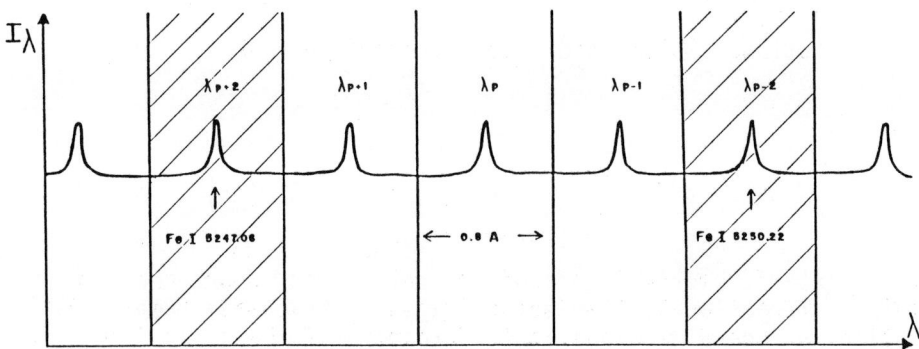

Figure 5. Simultaneous Dual Line Filtergramming.

IMAGING INTERFEROMETRY WITH NON-REDUNDANT ARRAYS

J. B. Zirker
National Solar Observatory/Sacramento Peak

Interferometry is one of several current approaches toward obtaining diffraction-limited solar images. I have made numerical simulations of one-dimensional interferometry of the solar limb, using two (or three) non-redundant arrays (e.g., masks with suitably spaced holes) at an image of the objetive of a meter-class telescope. Random phase aberrations that are linear (but arbitrarily large) across each hole were considered. Results of these simulations will be shown, with emphasis on the dependence of phase errors on fringe noise. Diffraction-limited resolution in the restored intensity profile of the solar limb seems feasible if the fringe noise is less than about 3%. Examples of real fringes, obtained with the Sacramento Peak Tower Telescope will also be shown.

MAGNETIC FIELD MEASUREMENTS WITH THE SWEDISH

SOLAR TELESCOPE ON LA PALMA

Henrik Lundstedt
Lund Observatory, Box 43, S-221 00 Lund, Sweden

A Consortium for solar magnetic field studies at Holger Crafoord Laboratory for solar physics was formed in Zürich, August 1986. Members of the consortium are,

The Holger Crafoord Laboratory for Solar Physics, La Palma	Lund Observatory, Lund University, Lund	Inst. für Astronomie, ETH, Zürich
U. Kusoffsky	H. Lundstedt	J.O. Stenflo
A. Wyller	B. Larsson	S. Solanki
G. Scharmer		
G. Hosinsky		

Our planned magnetograph is illustrated in Figure 1 and discussed below. In an early stage only Stokes I and V will be measured, but at a final stage the full Stokes vector is planned to be measured.

I. The Littrow Spectrograph

The spectrograph is described in detail in G. Scharmer et al., 1985. Excellent videofilms recorded with the Littrow spectrograph were shown during the meeting.

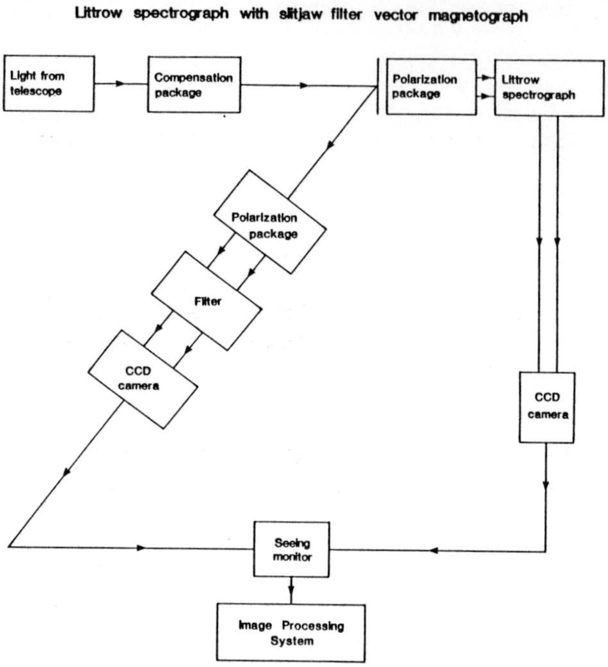

Littrow spectrograph with slitjaw filter vector magnetograph

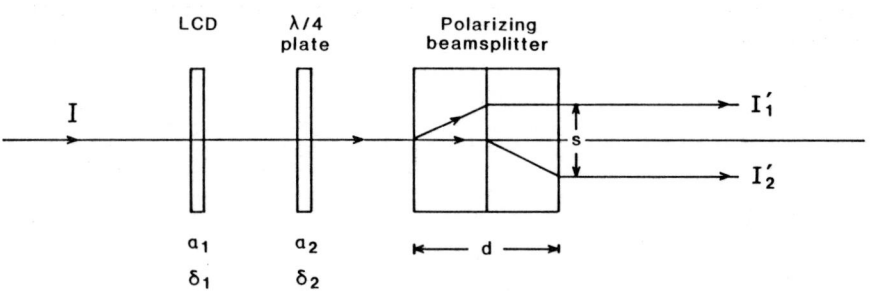

Fig. 1. The planned magnetograph for solar telescope on La Palma.

II. The Polarization Package

POLARIZATION PACKAGE

Fig. 2. The polarization package consists of a liquid cristal (LCD), a quarter wave plate and a polarizing beamsplitter. The positions angles are α_1 and α_2, and the retardations are δ_1 and δ_2.

The Mueller matrix for the polarization package is given by

$$\begin{pmatrix} I' \\ Q' \\ U' \\ V' \end{pmatrix} = \frac{1}{2} \begin{pmatrix} 1 & \pm 1 & 0 & 0 \\ \pm 1 & 1 & 0 & 0 \\ 0 & 0 & 0 & 0 \\ 0 & 0 & 0 & 0 \end{pmatrix} \begin{pmatrix} 1 & 0 & 0 & 0 \\ 0 & \cos 2\alpha_2 & -\sin 2\alpha_2 & 0 \\ 0 & \sin 2\alpha_2 & \cos 2\alpha_2 & 0 \\ 0 & 0 & 0 & 1 \end{pmatrix} \begin{pmatrix} 1 & 0 & 0 & 0 \\ 0 & 1 & 0 & 0 \\ 0 & 0 & \cos\delta_2 & \sin\delta_2 \\ 0 & 0 & -\sin\delta_2 & \cos\delta_2 \end{pmatrix} \begin{pmatrix} 1 & 0 & 0 & 0 \\ 0 & \cos 2\alpha_2 & \sin 2\alpha_2 & 0 \\ 0 & -\sin 2\alpha_2 & \cos 2\alpha_2 & 0 \\ 0 & 0 & 0 & 1 \end{pmatrix}$$

(2.1)

$$\begin{pmatrix} 1 & 0 & 0 & 0 \\ 0 & \cos 2\alpha_1 & -\sin 2\alpha_1 & 0 \\ 0 & \sin 2\alpha_1 & \cos 2\alpha_1 & 0 \\ 0 & 0 & 0 & 1 \end{pmatrix} \begin{pmatrix} 1 & 0 & 0 & 0 \\ 0 & 1 & 0 & 0 \\ 0 & 0 & \cos\delta_1 & \sin\delta_1 \\ 0 & 0 & -\sin\delta_1 & \cos\delta_1 \end{pmatrix} \begin{pmatrix} 1 & 0 & 0 & 0 \\ 0 & \cos 2\alpha_1 & \sin 2\alpha_1 & 0 \\ 0 & -\sin 2\alpha_1 & \cos 2\alpha_1 & 0 \\ 0 & 0 & 0 & 1 \end{pmatrix} \begin{pmatrix} I \\ Q \\ U \\ V \end{pmatrix}$$

where α_1, δ_1 are the position angle and retardation for the LCD and α_2 and δ_2 for the wave plate. If we choose $\alpha_1=0°$, $\delta_1=0$, $\alpha_2=45°$ and $\delta_2=\pi/2$ then

$$I'_1 + I'_2 = I$$
$$I'_2 - I'_1 = V$$

(2.2)

If we on the other hand choose $\alpha_1=0°$, $\delta_1=\pi$, $\alpha_2=45°$ and $\delta_2=\pi/2$ then

$$I'_1 + I'_2 = I$$
$$I'_2 - I'_1 = -V$$

(2.3)

Our two images from the beamsplitter fall on two separate CCD-detector areas with efficiencies e_1 and e_2, respectively. $e_{1,2}$ fluctuates from pixel to pixel. The signal in the two images is (Stenflo, 1985)

$$S_{1,0} = \frac{1}{2} e_1 (I+V)_0$$
$$S_{2,0} = \frac{1}{2} e_2 (I-V)_0$$

(2.4)

If we now shift the phase for the LCD to $\delta_1=\pi$, then

$$S_{1,\pi} = \frac{1}{2} e_1 (I-V)_\pi$$
$$S_{2,\pi} = \frac{1}{2} e_2 (I+V)_\pi$$

(2.5)

With the definition

$$S_V = (S_1 - S_2)_0 - (S_1 - S_2)_\pi$$
$$S_I = (S_1 + S_2)_0 + (S_1 + S_2)_\pi$$
(2.6)

After some manipulation (J.O. Stenflo, private communication) we obtain

$$V/I \approx S_V/S_I ,$$
(2.7)

while $\left|\dfrac{e_1 - e_2}{e_1 + e_2}\right| \ll 1$ and $(S_1+S_2)_0 - (S_1+S_2)_\pi$ is expected to be small.

This polarization value is free from seeing and pixel effects.

III. The Filter

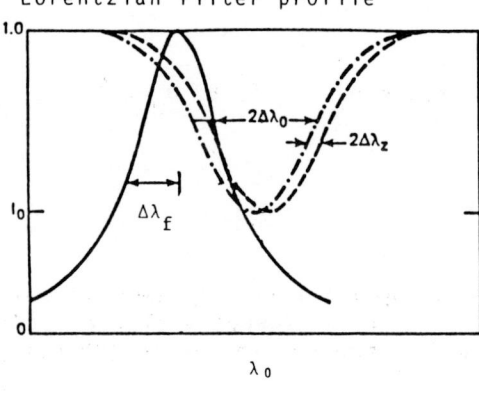

Zeeman shifted Gaussian line profile
and
Lorentzian filter profile

5250.22 Å

Fig. 3. Profiles of the two circularly polarized components (dashed curves) and a filter profile (solid curve).

Our old Zeiss H-alpha filter, with a 0.25 Å passband, will be rebuilt for the Fe 5250.22 line, with a passband of 0.15 Å, ±16 Å. The minimal detectable B with the magnetograph may be estimated the following way; Assume a Gaussian line profile, with a half width $\Delta\lambda_0 = 0.07$ Å and a central intensity I of 0.4. The Fe 5250.22 line is shifted $\Delta\lambda_Z = 4.227 \cdot 10^{-5}$ B Å. The Lorenzian filter profile is characterized by a

HWHM of $\Delta\lambda_f=0.075$ Å. If R=(the product of the intensity between the two polarizations and the filter profile)/(the product of the Gaussian profile and the filter profile) then the magnetic sensitivity becomes

$$\Delta B = \frac{1}{R \cdot S/N} \approx \frac{1}{6.5 \cdot 10^{-4} \cdot S/N} \quad (3.1)$$

For an ordinary CCD-detector S/N=50, then $B=31/\sqrt{n}$ gauss, where n is the number of differences averaged.

IV. The Image Processing System

A Zeiss Kontron DEC/IPS has recently been installed (Scharmer, 1986) at the Swedish solar station. Only the best CCD-images of opposite polarization, selected due to sharpness in real time, will be subtracted and stored. In order to further enhance the S/N ratio, a large number of subtracted images will be averaged.

References

Scharmer, G.B., Brown, D.S., Pettersson, L. and Rehn, J.: 1985, Appl. Opt. 24, No. 16, 2558
Scharmer, G.B.: 1986, The Swedish 50 cm solar telescope: Concepts, performance and auxillary instrumentation, this volume.
Stenflo, J.O.: 1985, LEST Foundation Tech. Rep. No. 12.

SUMMARY OF JOINT DISCUSSION ON INSTRUMENTATION, SESSION 12

J.B.Zirker
National Solar Observatory
Sunspot, N.M., USA

1. Most of the discussion concerned the paper by A. Wyller on his concept for a Fabry-Perot Echelle Scanner, to be used as a universal filter and stokesmeter. Among the points made were:

A. Righini reported severe stray-light problems with a Fabry-Perot, universal filter combination at Arcetri Observatory, and suggested the Swedish instrument would also encounter such problems if a wide entrance slot were used. Wyller said he will rely on established techniques for deconvolving line profiles from a stray light background.

A. Title suggested that solid Michelson interferometers were far less expensive as monochromators than the proposed Swedish instrument. Wyller pointed out that much of the development costs for the Fabry-Perot system had already been invested by the group at CSIRO in Sydney, Australia. Additionally, Title thought the proposed instrument would have too small a field of view to be useful. R. Dunn and A. Wyller agreed, however, that even a 15 arcsecond field would be satisfactory, if the design specifications for spectral and spatial resolution were met.

D. Bonacini suggested using piezoelectric modulation for the Fabry-Perot instead of the slower pressure scanning method.

2. Some discussion followed the paper by T.A. Darvann on cross correlation of solar images. O. von der Luehe asked questions concerning the application to seeing evaluation at the Swedish tower telescope. Darvann quoted resolution of about 0.3", image motion of 4" to 5", and, average distortion of .12" to 0.2". P. Brandt cautioned the audience not to use these preliminary figures to compare LaPalma with other sites. Some participants expressed skepticism at Darvann's figures, since they imply a Fried parameter (r_0) of about 50 cm. Darvann also said his correlation technique works as well with only one bit digital data (black or white levels), as with 8 bits.

3. After A. Title's invited review, F. Kneer asked how far a 5"x5" area moved horizontally on the Sun in 4 minutes. Title said .27" (i.e. 800 meteres/sec) for a typical element. However the structures are of such a size that a 5"x5" box moves at a rate of only 0.1" per hour.

THE SWEDISH 50 CM VACUUM SOLAR TELESCOPE: CONCEPTS AND AUXILIARY INSTRUMENTATION.

G.B. Scharmer

Swedish Research Station for Astrophysics, Royal Swedish Academy of Sciences, c/o Stockholm Observatory, S-13300 Saltsjöbaden, Sweden.

Abstract. The paper reviews the concepts of the Swedish 50 cm vacuum solar telescope. The auxiliary instrumentation, including the Littrow spectrograph, the Image Sharpness Selector and the image acquisition system, are also discussed briefly. The strengths and weaknesses of the entire system are pointed out in order to guide future users towards optimized observing programs and procedures.

CONCEPTS

The design goals of the 50 cm vacuum solar telescope are described in some detail in a paper by Scharmer et al. (1985). Here only the most important properties will be reviewed.

The telescope was installed in December 1985 replacing the previous 25 cm telescope. Mechanically the new telescope uses a turret design (Figure 1) similar to that of the Sacramento Peak vacuum telescope (Dunn 1964). The image forming element is a doublet lens with the crown component also constituting the vacuum seal. The vacuum does not introduce a major optical deformation of the crown component but a minor change in the optical design was necessary to eliminate a weak spherical aberration.

The image quality obtained with this relatively small telescope is remarkably high. This was achieved by paying particular attention to factors influencing the image quality. Among these are **i)** a simple optical design which uses components that are easy to manufacture and adjust, **ii)** very high optical quality corresponding to a total wave front aberration through the nine optical surfaces of approximately 1/15 wave ptp (Brown 1987), **iii)** carefully designed mirror cells which do not introduce deformations of the mirror surfaces, **iv)** a lens cell that does not cause heating of the edges of the lens, which would lead to spherical aberration **v)** a turret design (see Figure 1) that eliminates the disturbances usually caused by domes and heated mirror surfaces and **vi)** a separate guiding telescope, outside the aperture of the main telescope, which does not degrade the point spread function by causing scatter of energy into the outer diffraction rings. Tests on stars with a shearing interferometer (Brown 1987) and knife edge tests on the limb of the sun (Dunn 1986, private communication) show that the telescope *optical* quality is nearly diffraction

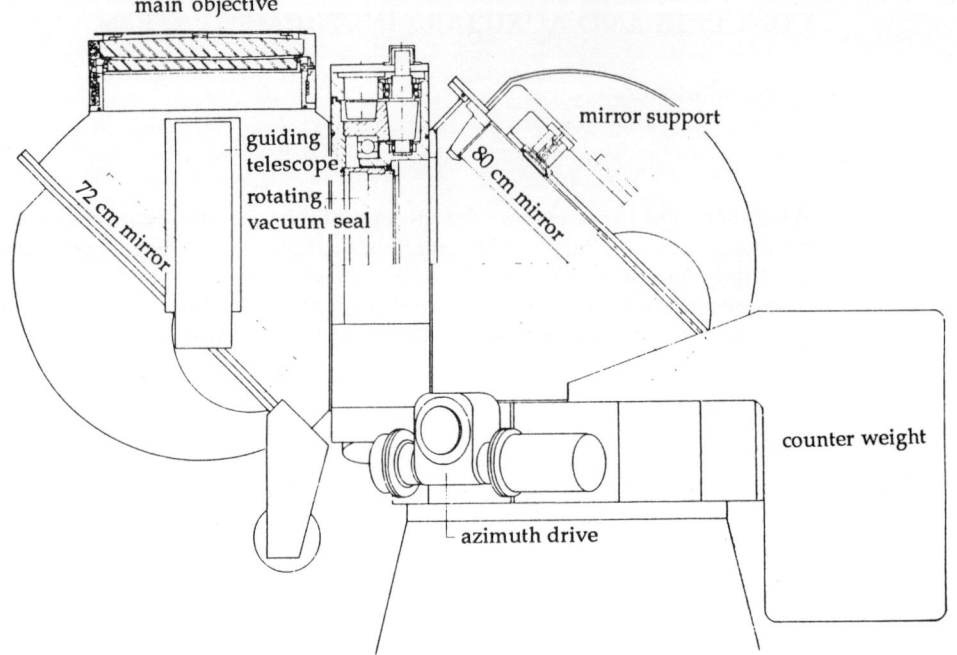

Figure 1. Cross-section of the turret showing the annular diaphragm that prevents heating of the lens cell and the lens, the two flat mirrors, one mirror support, the rotating vacuum seals and the separate guiding telescope.

limited also when in operation.

The high optical quality of the telescope and the excellent seeing at La Palma indicates that auxiliary instrumentation and research programmes should be designed primarily for high spatial resolution rather than for maximum flexibility. Potential observers should recognize that the telescope can be used with advantage only for certain types of observations and a succesful application for observing time must take full advantage of the valuable properties of the telescope and instrumentation without being limited by its drawbacks.

LITTROW SPECTROGRAPH

A short Littrow spectrograph for work requiring very high *spatial* resolution has been built and is located on one of the optical benches in the spectrograph room. The spectrograph is equipped with a holographic grating (2400 grooves/mm, efficiency 50%, length 220 mm, resolution 480,000) which is used in the first order. The focal length of the spectrograph is 3 meters and the dispersion is approximately 1 mm/Å. This corresponds to 10 mÅ/pixel if a CCD with 10 μm pixel size is used. For larger pixel sizes magnification optics is used. With a 25 μm slit, which corresponds to 0"22 on the sun, the spectral resolution is 147,000 at 4100 Å, 168,000 at 4800 Å, 201,000 at 5300 Å, 240,000 at 5900 Å and 286,000 at 6400 Å. With a 15 μm slit the spectral resolution is increased by 67% but the efficiency is lowered by 40% as compared to when the slit width is matched to the telescope spatial resolution. The advantages of the spectrograph are: high optical quality, a small number of optical surfaces, high spectral purity, absence of overlapping orders and short exposure times enabling nearly diffraction limited spatial resolution. The disadvantages are: relatively low spectral resolution, a weak contribution of *additive* scattered light from the surfaces of the Littrow lens, poor mechanical stability and that it is impossible to make simultaneous observations at two or more different wavelengths.

IMAGE ACQUISITION SYSTEM

After much consideration we have decided not to buy a conventional CCD-system. The advantage of such systems is primarily a large dynamic range and high photometric accuracy (12 bits A/D conversion). However, for photometric or spectrographic studies of solar fine structures down to the resolution limit of earth-based solar telescopes, the main observational errrors come from atmospheric image degradation rather than from the number of bits in the A/D converter. It therefore seemed appropriate to buy an image acquisition system that minimizes atmospheric seeing effects and enables new, powerful observing techniques that can provide qualitatively new information on solar fine structures.

We have recently purchased the image processing system DEC/IPS from Kontron in Munich. This system is characterized by the following properties: **i)** The system allows synchronized digitization of images from two CCD-cameras with 8 bits accuracy at video rates (in February 1987 only one camera input has been installed), **ii)** The system has a large freely definable image memory which can be expanded up to 64 Mbytes (we presently have 8 Mbytes), **iii)** The system has a pipe-lined image processor and a large subroutine library for standard reductions of the images, **iv)** the DEC/IPS is connected to a MicroVAX II via the Q-bus giving possibilities for reductions also with software developed for VAX/VMS, **v)** the system is flexible and Kontron's policy seems to be that future products will be compatible with the now existing DEC/IPS, enabling upgrading rather than replacement of the system as new products become available and. The 10 Mhz A/D convertor allows up to 30 768x560 images to be digitized per second and the large image memory allows bursts of such images to be stored in real time.

At La Palma the best seeing usually occurs in the middle of the day. However, even during good conditions the seeing can be very intermittent with rapid variations in the image quality occuring on time scales much shorter than a second.

To take advantage of the best moments of seeing an Image Sharpness Selector (ISS) has been constructed at the Stockholm Observatory. The ISS measures the contrast of small-scale structures in real time using the video signal from a CCD-camera. By selecting a window of appropriate size around the target of interest the ISS can descriminate even strongly differential seeing (unlike seeing monitors that work on the limb of the sun). The sharpness values are read out 50 (PAL) or 60 (NTSC) times per second and are analyzed by a Z-80 which sends commands to the DEC/IPS via the MicroVAX RS-232 interface. The ISS is programmed to operate in two modes: either the ISS sends commands to the DEC/IPS which allows the *sharpest* image to be stored in a given time interval or it orders the DEC/IPS to take bursts of images whenever the sharpnes value exceeds a specified *threshold*. For many applications only one image out of 250 or 300 (corresponding to one image every ten seconds) is needed to fully cover the time evolution of dynamic phenomena. The ISS systematically selects the best of these images and significantly enhances the quality of the data.

The DEC/IPS and ISS is particularily well suited for *time studies* of solar fine structures which seems to be a neglected field in observational solar physics. The Micro VAX 560 Mbyte Winchester can store over 2000 512x512 images, corresponding to six hours of uninterrupted observations if one frame is obtained every ten seconds.

The possibility to connect two *synchronized* CCD cameras allows two different images of the same object to be obtained simultaneously so that the *seeing* is exactly the same (apart from wavelength variations) for the two images. This greatly simplifies the interpretation of such pairs of images. Particular applications which require this involves: simultaneous filtergrams in two different wavelengths, spectra and corresponding slit-jaw image (the Littrow spectrograph can be used with the DEC/IPS and the ISS) and video-magnetograms (see below and Lundstedt, these proceedings).

We point out that the DEC/IPS is limited by its 8-bit A/D converter and is not suitable for research programmes requiring very high photometric accuracy or long exposure times. It is possible that we will install a 12 bit slow-scan (3.5 Mhz) interface for such programmes.

VIDEO MAGNETOGRAPH

In january 1987 the Crafoord foundation donated 140.000 $ for the construction of a video-magnetograph similar to that now in operation at Big Bear. The good image quality and several important improvements in the design should enable magnetograms with a spatial resolution better than 0.5 arc secs. The video magnetograph is built around the DEC/IPS and is described in more detail by Lundstedt (these proceedings).

CONCLUDING REMARKS

The excellent image quality of the Swedish 50 cm solar telescope makes it an attractive choice for many projects requiring high spatial resolution. The image acquisition system DEC/IPS and the Image Sharpness Selector ISS give unique opportunities for following the evolution of small-scale solar structures. Many of the 'standard' programmes carried out at other solar observatories can, however, not be carried out with the now existing or planned equipment. I hope that observers interested in applying for observing time will carefully consider programmmes that are not only important but also *ideal* for the available equipment. Special equipment or software cannot be provided by us.

The spectrometer URSIES is described in a separate article by Wyller (these proceedings).

ACKNOWLEDGEMENTS

The succes of the Swedish solar telescope was the result of a team-work in which many individuals contributed. Lennart Pettersson, who made the mechanical design and Dr. David S. Brown, who made the optical designs and the optical tests, deserve special credits. We have also benefited repeatedly from informative discussions with Dr. Dick Dunn.

The telescope and the computing facilities were financed by several donations from the Wallenberg foundation. Similarily, the observatory could not have been built without the generous donations from the Crafoord foundation.

REFERENCES

Brown, D.S. (1987). Technical Report on Optical Tests of the Swedish 50 cm Vacuum Solar Telescope.
Dunn, R,B. (1964). An evacuated tower telescope. Appl. Opt. 3, 1353.
Scharmer, G.B., Brown, D.S., Pettersson, L., Rehn, J. (1985). Concepts for the Swedish 50-cm vacuum solar telescope. Appl. Opt. **24**, 2558.

PRESENT AND FUTURE FACILITIES FOR THE VACUUM GREGORY COUDÉ TELESCOPE AT IZAÑA

E. Wiehr
Universitäts-Sternwarte
D-3400 Göttingen

INTRODUCTION

The idea to use a Gregory-Coudé type telescope for solar observations stems from P.ten Bruggencate (1958) who faced the enormous advantage of reducing heat, straylight and instrumental polarization. In contrast to a Cassegrain type with its hyperbolical secondary mirror in front of the prime focus, the elliptical "Gregory"-secondary allows to place a (water cooled) field stop in the prime focus. Using a 200 arcsec aperture the illumination on all mirrors behind the prime focus is reduced by a factor of about 100. Since straylight originates almost entirely from mirror surfaces rather than from the sky (Stellmacher and Wiehr, 1970; the earth's atmosphere contributing only in case of significant sky-dust and freshly aluminized mirrors), the reduced illumination on most telescope mirrors is the essential reason for the very low straylight in the Gregory telescope. A consequence of the parallactic (Coudé) mounting is that the relative angle between the two folding flat mirrors varies only with declination, all other angles remaining constant. Therefore, the instrumental linear polarization and phase retardation are sufficiently constant during one observing day (Wiehr, 1971). They can thus be compensated by means of a tilted glass plate followed by a Bowen compensator, both fixed at the ALPHA axis (Wiehr and Roßbach, 1974). A fine-adjustment of the Bowen compensator can be done by controlling the Stokes-V profile from a magnetic fluxtube without transverse field component (Scholiers and Wiehr, 1985).

These particular advantages of the Gregory-Coudé telescope indicate its conceptional field of application: Observation of solar objects sensitive to scattered light, and measurements of polarization. Consequently, the first observations with that new telescope in the early sixties, after its installation at the Locarno solar station of the Göttingen observatory in 1959, concerned resonance polarization (c.f. Brückner, 1963), prominence spectroscopy (c.f. Stellmacher, 1969) and umbra spectroscopy (c.f. Stellmacher and Wiehr, 1970; Wöhl et al., 1970). However, the main disadvantage of that telescope was still its large sensitivity to internal seeing. The tube of the telescope had thus been evacuated to 0.1 torr since 1970. New ZERODUR optics and a new grating without astigmatism further improved the instrument thus allowing to obtain spectra with good spatial resolution up to 1.5 arcsec (c.f. Wiehr et al., 1987).

The whole instrumentation had been operated at the Locarno station until August 1984 when it was dismantled for a transfer to the Izaña site at Tenerife (Fig. 1a). In the course of that re-installation the telescope had been completely overhauled at the workshop of the

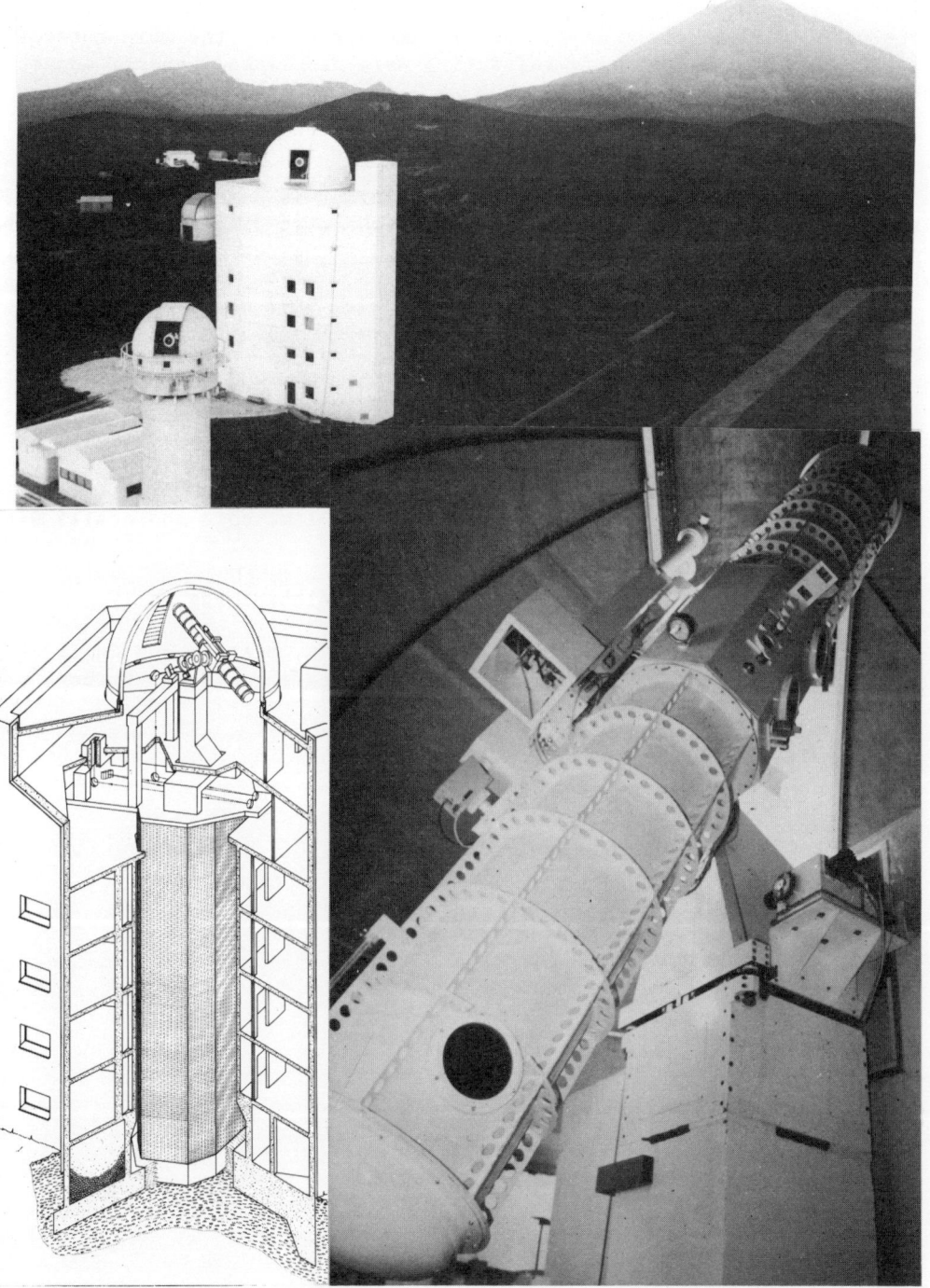

Fig.1: a)Gregory building (together with Newton) as seen from the VTT at sunrise. b)Cross-cut view of the Gregory building showing the inner tower of bricks supporting the instrumentation. c)The evacuated Gregory-Coudé telescope with its two small refractors.

Göttingen observatory; among other improvements, the main tube was stiffened, the two auxiliary refractors were displaced, several adjustment screws were vacuum fed-through.

INSTRUMENT DESCRIPTION

Since the general instrumentation has been preserved, only its basic data shall be summarized (c.f. schematic diagram Fig. 2); for more details see Brückner et al. (1967), Wiehr et al. (1981), Schröter et al. (1985):

Entrance aperture $\emptyset$ = 45 cm, d = 8 cm BK-7 glass window.
Primary mirror of ZERODUR $\emptyset$ = 45 cm, f = 240 cm.
Water-cooled field-stop in the prime-focus; Dawe-pinhole with $\emptyset \simeq$ 200 arcsec.
Secondary mirror, elliptical 1 : 10.4.
Two folding flats under 45°.
Coudé focus at f = 25 m $\simeq$ 8.25 arcsec/mm, 25 mm field.

Guiding refractor imaging the solar limb on 2 photocells via control of ALPHA- and DELTA- drives.

Auxiliary refractor for whole-disc H-Alpha image on TV.

Slit jaw imaging simultaneously in Ca^+K, H-Alpha and white light partly simultaneously on 35 mm camera, TV and screen.

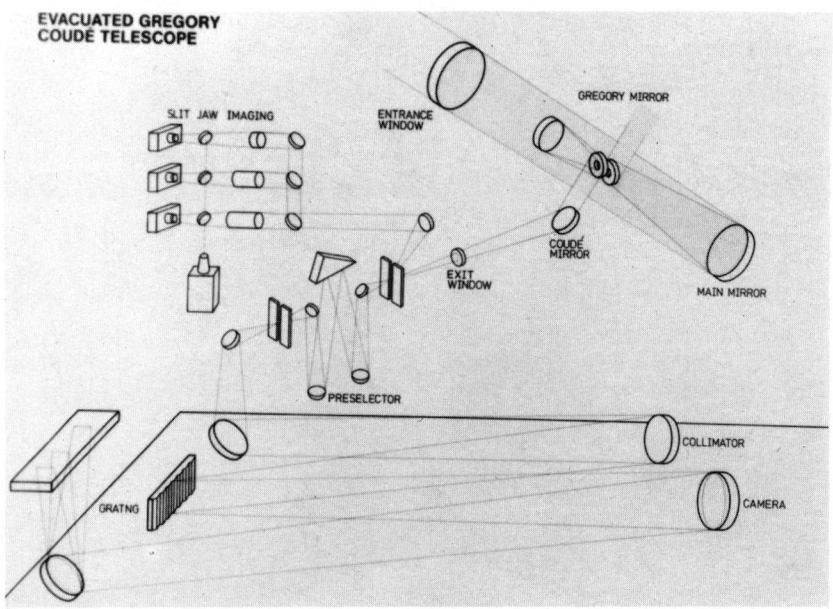

Fig. 2: Optical light pass of Gregory telescope, slit jaw imaging device, predisperser and spectrograph.

Prism predisperser f = 1 m removable for use of filters.

Czerny type spectrograph f = 10 m with echelle grating, 300 grooves/mm, blaze angle = $60°$, Ca^+K in the 15th, Ca^+IR in the 7th order with 110 resp. 225 mÅ/mm.

Spectrum 2.4 * 40 cm, 8.25 arcsec/mm.
Various 35 mm cameras; 9*12 plates.
Diode array 100*100 (c.f. Küveler and Wiehr, 1985).
Stokes vector polarimeter (Wiehr, 1974).
Image intensifiers (c.f. Hellwig, Stellmacher, Wiehr, 1984).
Spectrum scanner with two photomultipliers (under construction).

The computer system of the Gregory is shown in Fig. 3: Telescope, spectrograph and arrays are controlled by a TEXAS microprocessor, host computer of which is a 64 bit COMPUTER-AUTOMATION NM 4/30 mini-computer (Wittmann, 1987); data pre-processing and image-processing will be possible with a PDP 11/73 common with the VTT.

The building (Fig. 1) consists of a 9.6 m * 15.6 m outer tower made of concrete and a 4 * 4 m inner tower made of bricks. These two different materials assure a vibrational discoupling; the proper frequencies of 4 Hz and, resp., 2.5 Hz being different from the 3.2 Hz of the telescope.

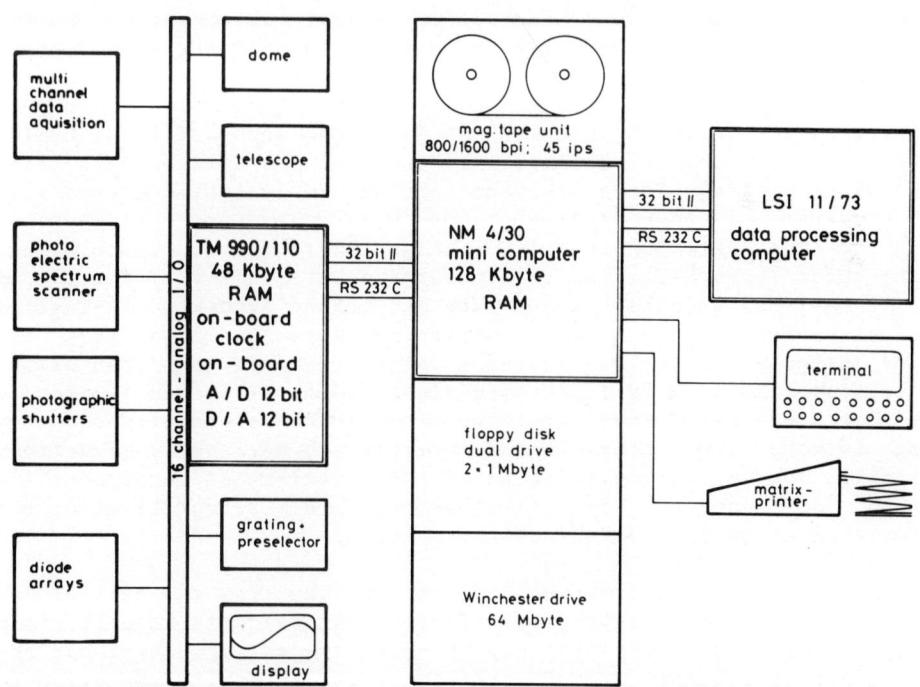

Fig. 3: Schematic diagram of the Gregory's computer system.

A concrete plate on top of the inner tower supports the spectrograph as well as three `bridges` supporting telescope, slit-jaw devices and, resp., spectrum camera devices. The outer walls are insulated by 10 cm glass wool and covered by 4 mm thick aluminum plates painted with TiO-white colour. This minimizes both, heat lost from the building and radiative heating by the sun. The height of the telescope entrance 25 m above ground minimizes disturbing influences from local turbulence. The outer tower protects the inner one against wind impact and supports the $\emptyset$ = 8 m fiber-glass dome, which automatically follows the telescope by computer control via ALPHA- and DELTA- encoders.

ADJUSTMENT AND OPTICAL TESTS

After the telescope was mounted in May 1985 at Izaña (Fig. 1c), a new adjustment procedure developped by the author has been applied for an optimum alignment of the two axes of the paraboloid (primary) and the ellipsoid (secondary): A He-Ne-Laser is fixed at the primary's backside and then laterally moved until its beam exactly passes the center of the mirror (non-aluminized area $\emptyset$ = 1 cm). The Laser is then tilted until its beam intersects the declination axis. In a next step the secondary (Gregory) mirror is laterally shifted until it is exactly centered with respect to the Laser beam. In a last step primary and secondary mirrors are tilted until a multiple (back and forth) reflection of the Laser beam occurs. For an optimum adjustment, the various circles of light visible on the primary must be perfectly concentric. This adjustment procedure has been proved to be sufficiently sensitive to take care of the rather delicate alignment of the two mirror axes: Under very good seeing conditions, a star showed more than five diffraction rings!

The spectrograph had been installed in May 1986. The particular building for the Gregory required some alterations of the optical path (Fig. 1b) as compared to the former Locarno concept. The slight (convex) surface curvature of the grating is compensated by the convergence of grooves if the grating normal is oriented toward the direction of the outgoing beam (here N-E). The then remaining astigmatism of about 4 mm (f = 10 m!) is slightly larger than at Locarno since the reflection angles at collimator and camera mirror are somewhat larger (2.4° instead of formerly 1.8°). This residual astigmation had then been diminished to less than 1 mm by a backward-displacement of the collimator by 25 mm ($\approx$ 0.0025*f). The resulting slightly convergent light on the grating affects the focus in the vertical and horizontal directions differently, due to the modification of the interference condition. The resulting focus curve (Fig. 4) shows negligible astigmatism for the whole working range (50 to 70°) of the grating.

After elimination of interference fringes arising from the exit window (30 cm in front of the spectrograph slit), first reliable spectra could be obtained in September 1986. Fig. 5 shows the selected best one obtained under excellent seeing conditions. Its intensity variation in the continuum at 6495.3 Å can be compared with that of the selected

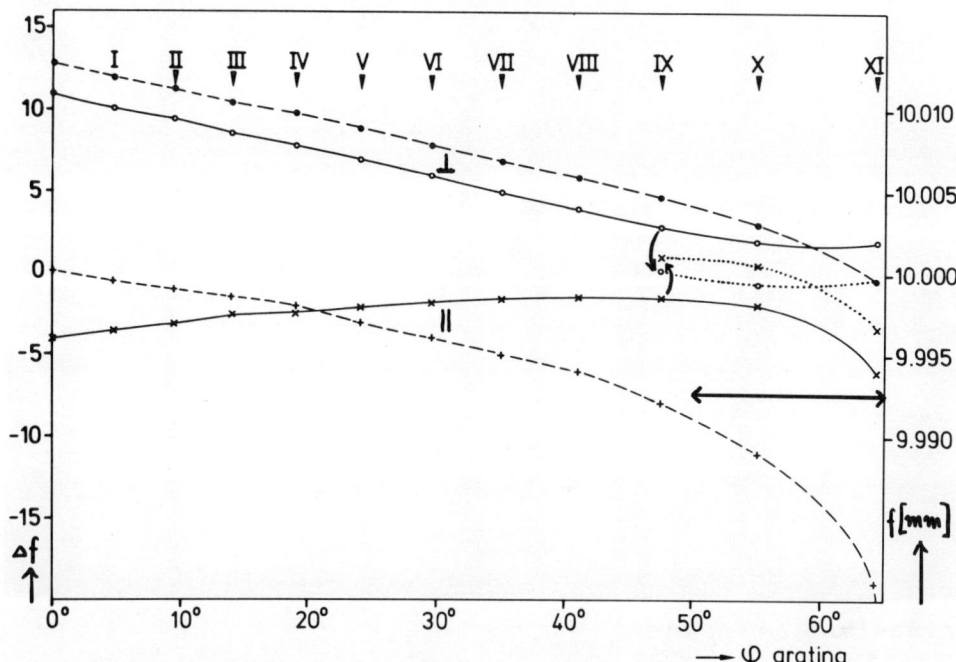

Fig. 4: Spectrum focus as a function of grating angle for Hg5461 in the zero through 11th order (upper scale); differences of spectral line focus (II) and solar structure focus (I) measure the astigmatism which is essentially reduced for reversed grating(--------); further reduction with non-parallel light on the grating (........).

best spectrum so far obtained at the former Locarno station with the same instrument under the same apparative conditions (slit, prefilter, exposure time, emulsion, development, photometry): The spectrum in Fig. 5a yields roughly twice as much continuum modulation. In addition, one finds an about two times higher spatial resolution which agrees with the empirical MTF deterrmined by Wiehr (1987) and indicates that instrumental limits are reached in case of best seeing moments at the Tenerife site.

Fig. 5a (next page): Spectrum 6493.5<λ[Å]<6499.8 at disc center obtained on Sept. 24, 1986 with the evacuated Gregory-Coudé telescope at Tenerife immediately after final elimination of astigmatism and interference fringes in the f=10m spectrograph.

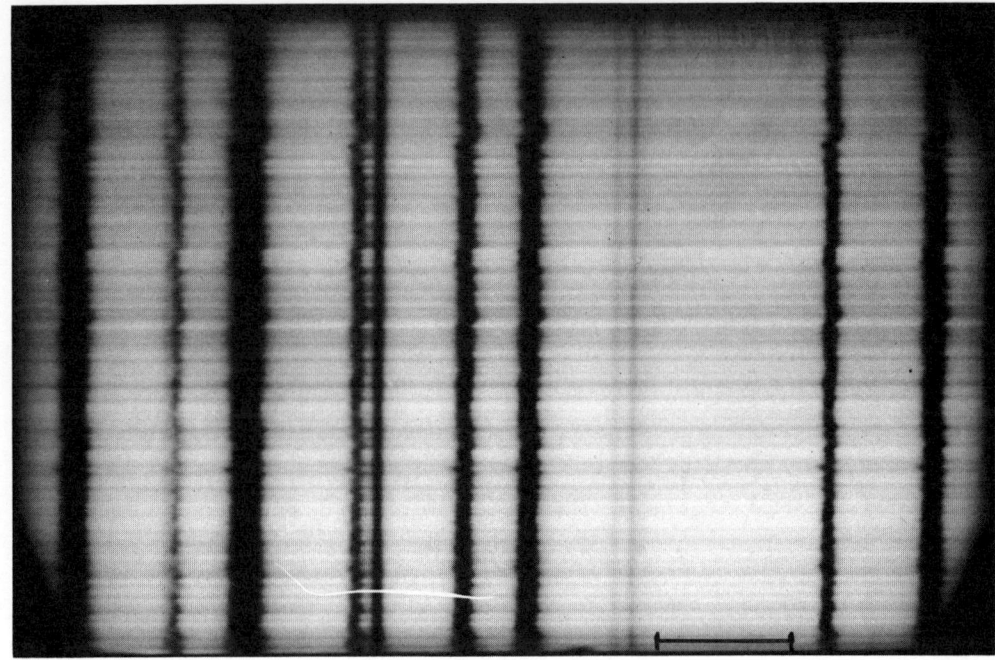

Fig.5a (see prec.page)

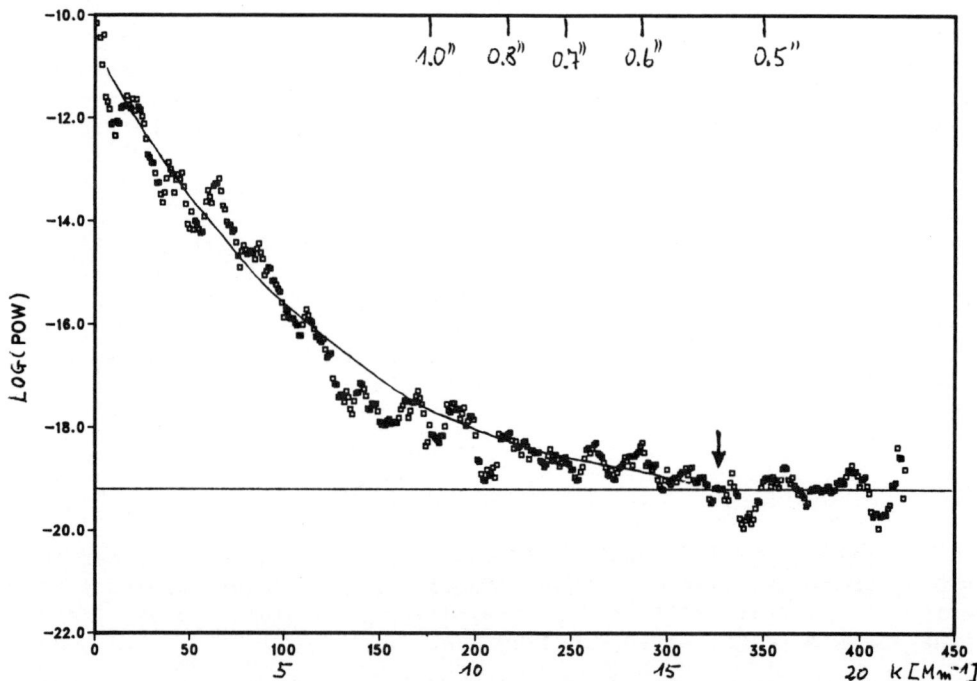

Fig.5b: Power spectrum of the continuum intensity fluctuations at 6498.3 ± 0.3 Å (⊢──┤ in Fig.5a); the spatial resolution of about 0.53 arcsec exceeds the theoretical one (0.48" given by the grating height) by only 10%.

REFERENCES

Bruggencate, P.ten: 1958, Publ. Göttingen Obs., No 122
Brückner, G.: 1963, Z.f.Astrophys. 58, 73
Brückner, G., Schröter, E.H., Voigt, H.H.: 1967, Sol.Phys. 1, 487
Hellwig, H., Stellmacher, G., Wiehr, E.: 1984, Sol.Phys. 94, 285
Kayser, H.: 1900, Handbuch d. Spektroskopie Nr. 1, 441
Küveler, G., Wiehr, E.: 1985, Astr.Astrophys. 142, 205
Scholiers, W., Wiehr, E.: 1985, Sol.Phys. 99, 349
Schröter, E.H., Soltau, D., Wiehr, E.: 1985, Vistas in Astron. 28, 519
Stellmacher, G.: 1969, Astron.Astrophys. 1, 62
Stellmacher, G., Wiehr, E.: 1970, Astron.Astrophys. 7, 432
Wiehr, E.: 1987, invited lecture at this workshop
Wiehr, E.: 1971, Sol.Phys. 18, 226
Wiehr, E.: 1974, Sol.Phys. 35, 351
Wiehr, E., Roßbach, M.: 1974, Sol.Phys. 35, 343
Wiehr, E., Wittmann, A., Wöhl, H.: 1981, Sol.Phys. 68, 207
Wiehr, E., Grosser, H., Knölker, M., Stellmacher, G.: 1987, poster at this workshop
Wittmann, A.: 1987, manual "GRECOS operating system"
Wöhl, H., Wittmann, A., Schröter, E.H.: 1970, Sol.Phys. 13, 104

Present and Future Solar Observational Facilities of the German Vacuum Tower Telescope

by Dirk Soltau

Kiepenheuer-Institut für Sonnenphysik
Schöneckstraße 6
D-7800 Freiburg
Federal Republic of Germany

1. Introduction and present status

After many years of preparation, the work on the VTT-project has now entered in its final phase. As long ago as 1974 the late Dr Johannes Peter Mehltretter started the work on the design. In November 1985 the rotatable spectrograph tank has been assembled. In August 1986 the vacuum tank of the telescope has been successfully assembled and rests now in its final position. At the present time the removable dome is finished and now being tested. By the end of this year the coelostat and many parts of the electronics will be installed. This paper shall give an overview about what an observer can expect to find in the observatory within the near future.

2. Description of the telescope

The optical system of the Vacuum Tower Telescope (VTT) consists of a spherical mirror (D-0.7 m, f-46 m) which rests at the bottom of an evacuated steel tank.

The light is fed into this tank by a classical coelostat system of two plane mirrors (D-0.8 m) which are mounted on top of the building at 35 m above ground. In order to have the coelostat working in the free airmass the cylindrical dome has been constructed completely removable. The two parts of it may be moved independently and may also serve as a windscreen if desired. Since the length of the vacuum tank is about half of the focal length, the converging beam is folded by an additional plane mirror mounted near the top of the tank.

In the observing room at ground level the telescope produces a solar image approximately 1 m behind the exit window. The 15cm image covers about 600 arcseconds (scale 4."6/mm). Because of the coelostat mounting the image is practically nonrotating. It can also be transferred in one of three optical laboratories by the use of a 45°-mirror. A small fraction (<1%) of the incoming light is used for an auxiliary light path which produces an image of the whole solar disc. It is used to stabilize the main image by sensoring the position of the disc and correcting it by tilting the coelostat's secondary mirror. By moving the sensors on a x,y,z-table the solar image may be scanned with a stepwidth less then the diffraction limited resolution (0."2). Technical information about the telescope and some of its subsystems may be found in the following papers: Mehltretter (1975) who did the original design., Soltau (1980, 1983), , Schröter et al. (1985), Schröter and Wiehr (1985) and Schmidt and Soltau (1986).

3. Postfocus equipment which will be available from the beginning of operation

3.1. The Echelle Spectrograph

The main postfocus installation is the 15m vertical Echelle spectrograph. It is mounted inside a steel tank situated in an underground shaft. The spectrograph uses a large blazed Bausch&Lomb grating (420mm x 220mm) with 79 rules per mm and a blaze angle of 63°. One observes in high orders (typically 57 for CaK, resp. 22 for 10380Å) and a predisperser is used to select the wavelength regions of interest. The linear dispersion varies from 14.2 mm/Å for CaK to 6.4mm/Å for 10380Å.

Many possibilities exist to observe different wavelengths simultanously. There is also the possibility to avoid the predisperser and to enter the spectrograph directly. In this case, order selection may be done by using appropriate filters and also a different grating may be used (e.g. 600 rules/mm).

The extraordinarily large slit height of 120mm allows for the covering of nearly the whole field of view. The curvature of the spectral lines will then be compensated by curved slits.

Another important feature of the spectrograph is the possibility to rotate the whole spectrograph tank and thus giving the entrance slit an arbitrary orientation with respect to the solar image. The rotation axis will be in the middle of the slit within 0.5mm.

3.2. The Slit Jaw Camera System

A three channel slit jaw camera will provide three monochromatic images of the aluminized slit jaws simultanously. A Hα-image, a Ca-K-image and a white light image (six selectable interference filters) can be transmitted via an internal TV-System and may be displayed on three monitors. Photography of the slit jaws is also possible where even the whole entrance slit may be imaged.

3.3 The photographic equipment

At the exit, the spectrograph produces a 500 mm wide spectrum with a typical linear dispersion of 80mÅ/mm at 5000Å. In this plane a number of housings for photographic film or plates in different formats (including 70 mm) may be mounted. Exposure time and film transport is controlled by computer.

3.4. The Two Channel Spectrum Scanner

A two channel spectrum scanner allows for high resolution spectral scanning. A mechanical support carries two photomultipliers, one of which is mounted on a scanning table. The other one can be fixed in an arbitrary position. The scanning table can be moved with a positioning accuracy of better than 2 microns ($\approx$ 1.5 mÅ). The total scanning range is 150 mm ($\approx$ 12 Å). An analog electronic circuit may form the ratio of the two signals thus yielding an intensity independent signal. In this mode the system is a stand alone system. It also may be connected with the anolog to digital converter of the computer.

3.5. The Single Detector Doppler Compensator

The integral line shift may be measured with a new Doppler compensator which avoids some of the disadvantages of other types. It uses only one detector (photomultiplier) which alternately measures the intensities of the blue and the red wing of the line profile. A normalized error signal is calculated. A microprocessor system uses this intensity-independent value to control a tilting plane parallel glass plate which compensates the line shift. The device does not suffer from the problems which arise from the use of two different detectors for both line wings (unequal sensitivities, unequal nonlinearities). Moreover, it is possible to optimize the system for different line profiles by adjusting certain system parameters via software. The accuracy of the system is expected to be better than 5 m/s.

3.6. The Linear Reticon Array System

A system consisting of three linear Reticon diode arrays (1024 pixels each) which can be positioned independently in the spectrum will be available. The arrays can be mounted on the precision scanning unit which is mentioned above. The distance between two pixels is 25μ.

3.7. The CCD Camera

Our CCD Camera uses the thinned version of the RCA SID 53612X0 chip with 320 x 512 pixels (30μ each). Tests show that in monochromatic use the fringe pattern has a contrast of 5 to 10% depending on wavelength. The pattern is stable, and correcting for a photometric accuracy of 1% should be possible. The controller (Photometrics Ltd., Tucson, USA) allows for different readout modes including variable gain, binning and readout of selected areas. For data acquisition we use a DEC PDP11/73 with 2 Megabyte of RAM and 320 Megabytes of disc storage. This equals approximately 1000 full frames.

3.8. Computer facilities and Operating System

A fully equipped Computer Automation Naked Mini 4 computer is used to control the instrument's motions and data acquisition in most cases. A comfortable Operating System has been written. It covers scanning possibilities in many different ways, data acquisition and analog to digital conversion of up to eight channels, control of shutters and film transport, file handling and visualizing of data.

The above mentioned PDP 11/73 computer will mainly be used for data compression and image processing of the CCD-data. It will be a powerful tool for data reduction at the site.

4. Postfocus equipment which will be available soon after the beginning of operation

4.1. The Multichannel Subtractive Double Pass Device

The diameter of the spectrograph's collimator and camera mirror has been chosen large enough so that a double pass configuration is possible without overlapping of the incoming with the outgoing light bundle. To take advantage of this, there is an agreement that the Meudon solar group may install their Multichannel Subractive Double Pass unit

(MSDP) in the spectrograph. It uses a two-dimensional window instead of the normal entrance slit. The field is 270" x 35". In the spectral plane nine wavelength bands are selected and spatially seperated. The light returns and the nine wavelength channels recombine into nine quasi monochromatic images of the spectrograph's entrance window. The wavelength resolution is about 150 mÅ which is sufficient to cover the profiles of broad chromospheric lines. One example of the many possible applications is the construction of two-dimensional velocity maps of prominences where all data have been obtained simultanously. For details see (Mein, 1981)

4.2. The Zeiss Universal Birefringent Filter

Another instrument which will reside in the VTT-building as a national contribution according to phase B of the JOSO project, is a Zeiss universal birefringent filter. The filter was purchased by the Italian National Research Council in 1982 und is runned by the Osservatorio Astrofisico di Arcetri in Florence (Cavallini et al., 1982; Cavallini et al., 1986).

The filter consists of nine crystal groups which can be rotated in order to adjust the transmission peaks und thus selecting the wavelength characteristics of the filter. The mechanical and electronic parts have been altered and control is now done by computer. For 5890 Å (Na D) the filter has a measured FWHB of 180 mÅ.

5. Future Facilities

There will be many possibilities to expand the system once it is in operation. The large building contains three optical laboratories which are temperature controlled and allow for many applications including the assembly of horizontal spectrographs of ten meter focal length.

At the moment there are several considerations and activities at the Kiepenheuer-Institut to expand the equipment in different ways:

• A polarimetric device using the CCD camera as a detector for the V+ and V- Stokes parameter simultanously is being investigated
• Active optics to compensate for image motion is being investigated using the correlation tracking method
• The purchase of a commercially available Fourier Transform Spectro- meter is considered
• The assembly of the existing 4m Rowland spectrograph which worked successfully in Locarno is considered

6. Conclusion

The above mentioned postfocus equipment make the VTT a very versatile instrument and it hopefully will serve the solar community as a valuable research tool for many years to come. If we can continue our work according to schedule we hope to have »first light« in spring 1987.

7. Acknowledgements

It should be emphasized that this instrument is the result of the combined effort of many people in the Kiepenheuer-Institut and the Instituto de Astrofisica de Canarias over a time span of many years. It is not possible to give justice to all of them at this place. Instead I would just like to mention the late K.O. Kiepenheuer who initiated the project and the late J.P. Mehltretter who did the optical and mechanical design.

REFERENCES
Cavallini, F., Ceppatelli, G., Falciani, R., Righini, A.: 1982, JOSO Annual Report 1982
Cavallini, F., Ceppatelli, G., Righini, A., Paloschi, S., Tantulli, F.: 1986, JOSO Annual Report 1982
Mehltretter, J. P.: 1975, JOSO Annual Report 1975
Mein, P.: 1981, Proceedings of the Japan-France Seminar on Solar Physics, p. 285
Schmidt, W. and Soltau, D.: 1986, Geowissenschaften in unserer Zeit Vol. 4, p.87
Schröter, E. H., Soltau, D., Wiehr, E.: 1985, Vistas in Astronomy, Vol. 28, p. 519
Schröter, E.H. and Wiehr, E.:1985, Sterne und Weltraum 6/85, p. 319
Soltau, D.: 1980, Proceedings of the Sacramento Peak National Observatory Conference »Solar Instrumentation: What's next?«, p.600
Soltau, D.: 1983, Proceedings of the Kunming workshop on Solar Physics and Interplanetary Travelling Phenomena, p.1191

Table: The Vacuum Tower Telescope in Numbers

Telescope:
Diameter of the Coelostat-mirrors	80 cm
Diameter of the entrance window	75 cm
Diameter of the primary mirror	70 cm
Focal length of the telescope	46 m
f-ratio	1 : 77
Image scale	4."6 / mm
Diffraction limited resolution	0."2
Diameter of the Airy disc	90 µm
Field of view	13'.8

Auxiliary telescope:
Aperture	5 cm
focal length	20 m
image scale	10"/mm

Spectrograph
Entrance slit height	12.5 cm
Focal length of predisperser	4.2 m
Focal length of main spectrograph	15 m
Echelle grating 220 x 420 mm, 79 grooves per mm	
blaze angle	63º

At a wavelength of 5000Å one observes in the 45th order, the separation of two neighbouring orders is about 110Å and the linear dispersion is 12.5 mm/Å

THE FRENCH POLARIZATION - FREE TELESCOPE THEMIS

J. RAYROLE
Observatoire de Paris, section d'Astrophysique de Meudon
5, place Janssen, 92 195 MEUDON Principal, Cedex

THEMIS introduces a new concept of Solar magnetograph. It is supported by all the French solar group. It has been designed for high accurary measurements of magnetic and velocity fields inside the fine structures of the solar atmosphere. In modern solar magnetometry we have to solve two different problems :
1/ High spatial, spectral and possibly tempororal resolution measurements of the state of polarization inside line profiles
2/ Interpretation of the observed values in term of the magnetic field vector components

To solve these problems, we need a large aperture telescope free of instrumental polarization up to the polarimeter package and associated with a high resolution spectrograph, working simultaneously in a spectral range as large as possible.

Large spectral range is needed to allow the simultaneous observations of several spectral lines formed at different levels as well as the separation of the magnetic field measurement of the temperature and density variations.

The optical scheme has been optimized for that purposes, and a special attention has been brought to solve the following points :
- Instrumental polarization introduced by the entrance window
- Cooling of the secondary mirror
- Control of the spatial positioning of the telescope
- Precise scanning mechanism.

A 90cm evacuated Ritchey - Chretien telescope focus a 6 minutes field in front of the analyzer of polarization. A flat holed mirror deflects the non used light for real time evaluation of the atmospheric disturbances and low frequency limb tracking guider. Then, a collimator mirror focuses a pupil image on the actif mirror used for real time correction of the wave front tilt and transfer optics focus a F/57 beam on the spectrographs entrance slit. With 3 interchangeable gratings the predisperser spectrograph associated with an echelle spectrograph allows numerous combinations for classical spectroscopy as well as for multichannel subtractive double pass process with or without analysis of the polarization.

The telescope is mounted as an altazimuth at the top of a 18 meters concrate pilar (Elevation axis at 22 meters). The vertical spectrograph tank is rigidly connected to the fork base. The outer part of the tower supports a 9 meters Dome to protect instruments for sun heating and wind. The analyzer of polarization (fig. 1a and 1b) has been computed to cancel astigmatism introduced by anisotropic media and to be achromatic in the range 3800 A to 11 000 A. The I,Q,U,V polarized incident beam travels through the phase retardation optics and then is divided by mean of a polarizing beam splitter into 3 planed polarized beams coming from the same portion of the solar surface. Two of them (B1 and B2) are used for magnetic field measurement, the third one for the granulation correlation tracker. The principle of a granulation correlation tracker is well known. Live images detected by a two-dimensional diode array detector are continuously compared with a previously stored reference. Image displacement is measured by detecting the location of the global maximum of the correlation function. The computation at a minimum rate of 100 Hz requires complex modern hardware. The use of two crossed linear detectors considerably reduces the problem complexity. A prototype of an image motion detector has been built with two 128 S elongated pixels RETICON linear detectors. The acquisition of the reference image and live images as well as the computation of images correlation functions are performed with a fast numerical processor TMS 32010. The error signals for the two directions (X,Y) are calculated in 5 milliseconds allowing the system to perform a correction 100 times per second. An example of an one axis correlator signal is showned on the figure 1c.

The tilting mirror has two main functions :

- Accurate scan of the field of view without speed up the inert mass of the telescope
- Real time correction of the wave front tilt due to atmospheric disturbances.

The lightened sandwich mirror and its support has been computed by method of finite elements. The principle of the device is showned on the figure 1d. The mirror is fixed in 3 points (at twice third of the radius) on a light rigid metallic plate. The mirror and its supporting structure are connected with an elastic stay to the fork frame. The 400 Hz resonance frequency of such a device is high enough to allow corrections every 10 milliseconds.

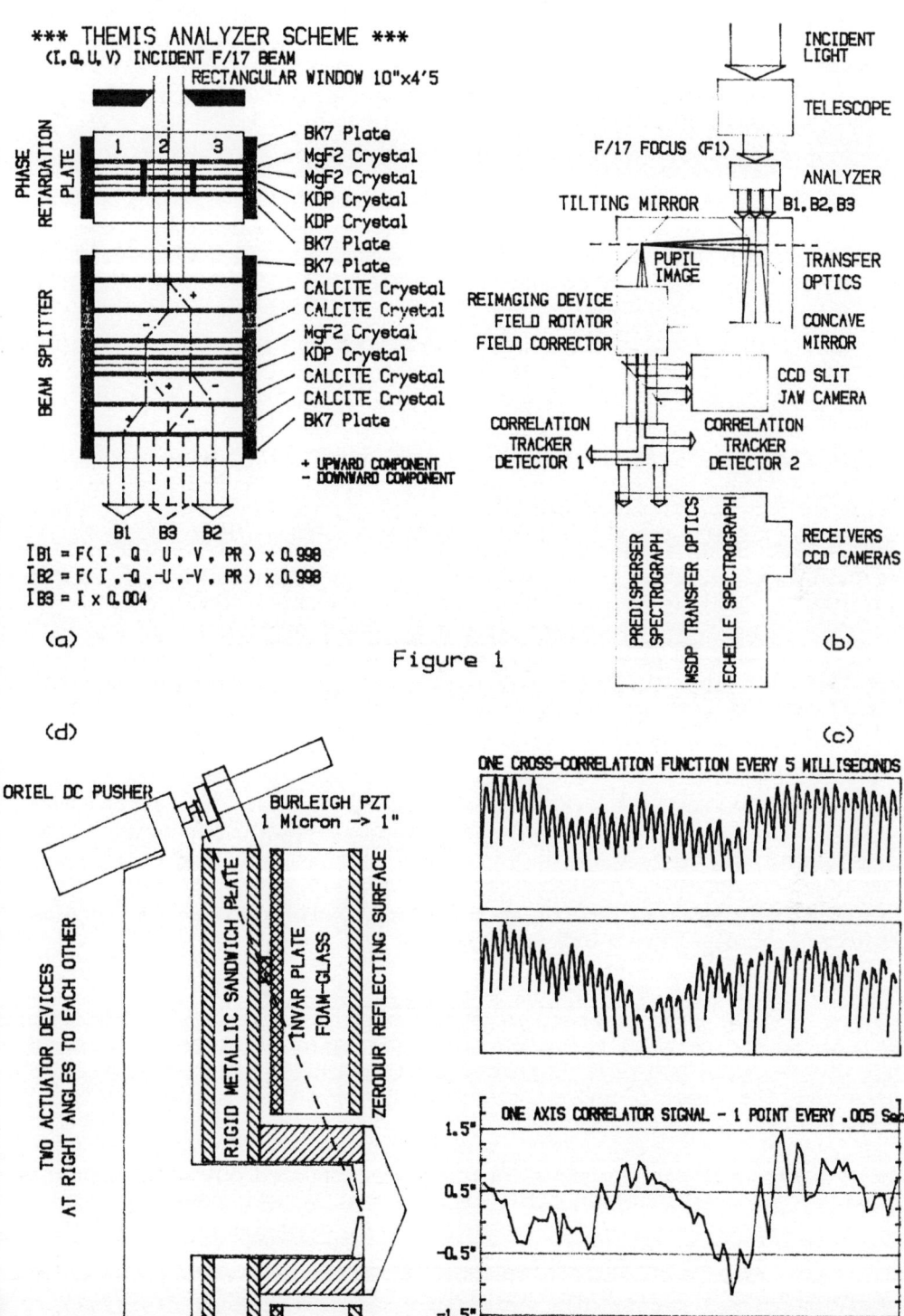

Figure 1

OTHER SOLAR FACILITIES AT THE OBSERVATORIO DEL TEIDE

M. Vázquez
Instituto de Astrofísica de Canarias, 38200 - La Laguna,
Tenerife, Spain.

1 INTRODUCTION

In the following will be described the main characteristics of other Solar Telescopes at the Observatorio del Teide, where the Spanish observing time is, at least, 50%. For a complete list of telescopes see Schröter (1984).

2 THE SOLAR LABORATORY

As such is defined the ensemble of solar instruments which look at the Sun in integrated light or with low spatial resolution. We are planning to locate these in a new building, pyramid-shaped, whose construction has been finished recently (Roca-Cortés 1984).

2.1 The K Resonant scattering spectrometer

Since 1976 a resonant scattering spectrometer has been in operation, working in the KI 7699 resonance line, to measure the radial velocity of the Sun, producing well known results on global oscillations and the internal Solar structure (see Brookes et al. 1978 for a description).

Resonant scattering intensities in the red and blue wings of the K 769.9 nm line, together with time and total incident light, are logged, every second, on a BBC microcomputer.

This work has been carried out in close collaboration with the Physics Department of Birmingham University.

2.2 The Na Resonant scattering spectrometer

This spectrometer, which has been in operation at Izaña since 1985, works in the Na D line (589 nm). The main difference with the K instrument is the use of a liquid crystal modulator (Eccles et al. 1985) instead of the electro-optical light modulator (Pockels cell with KDP crystal).

The comparison of simultaneous observations, carried out with both instruments, has been recently published by Isaak et al. (1986).

2.3 Two-dimensional Spectrometer

In a first version (Brookes et al. 1981) this instrument used a two-dimensional positioning device to study the spatial distribution of solar velocity fields. Later, in 1985, spatial filters specially designed to study oscillation modes with l=m were inserted in the optical path. A description and first results is given by Perez-Hernandez (1986).

2.4 S.L.O.E. (Solar Luminosity Observations Equipment)

The design of this four-channel photometer was carried out at ESTEC (European Space Research and Technology Centre, Noordwijk) under the direction of Dr. V. Domingo. The main aim of the project was to test the performance of excellent ground-based sites for luminosity variations. In August 1984 it was installed at Izaña and since March 1986 a twin model is in operation at Sierra Nevada (Granada, Spain).

In short, it consists of 4 channels centered at 500, 680 (I) and 1060 nms with a half-width of 10 nm. Four photovoltaic detectors record the signals stored in an HW computer. The first results have been recently published by Jimenez et al. (1986 a,b).

3 THE RAZDOW TELESCOPE

This type of Solar Telescope was designed by NASA, in the sixties, for a patrol network and manufactured by the Razdow Laboratories. It is a 25cm refractor (f=1 m) equipped with an H-α birrefringent filter. It was installed in 1969 in the present tower of the VNT, and was dismantled in 1972. After a complete refurbishment, carried out at the IAC workshops, it was re-mounted in 1980 in a small tower at Izaña.

A Motorola microprocessor controls the following observational parameters: Frequency of pictures with a 35cm camera, filter band width (0.5 Å or 1.0 Å), shift from the line center (±1 Å), control of the filter's temperature and focal length given full disk picture or only 1/3 of it.

An RCA camera TC 1004 equipped with an Ultricon tube enables the display of H-α images on a TV monitor.

For white-light observations a secondary lens (O=15 cms, f=234 cms) can be used. Automatic control of dome rotation is at present being considered.

4. THE VAKUUM NEWTON TELESCOPE (VNT)

This instrument belongs to the Kiepenheuer Institut (Freiburg) and was installed at Izaña in June 1972. During the first few years it was used for site-testing campaignes, which culminated in 1979 with the comparison between the sites of Roque de los Muchachos and Izaña (Brandt & Wöhl 1982). Later it has been used mainly for scientific observations.

The optical scheme of this 40 cm Newtonian reflector has been described by Mattig & Casanovas (1975). The light enters throught a window of 50cm diameter, inciding in the main mirror, a f/7.5 parabola of 40 cms, made of zerodur. A secondary magnifying lens system produces a 35 cm image (scale 5.5"/mm) giving a f_{eff} = 37.5 m . In the following shall be described the main existing instrumentation:

a) <u>Interference filters</u> (central wavelength; half width in nms). (416.3; 5.7) (500.0; 8.9) (550.0; 18.0) (574.0; 14.0) and (679.0; 16.2). Neutral filters set.

b) <u>Calibration system:</u> Step-wedge with 10 different values.

c) <u>Photographic cameras:</u> 35mms film (type Nikon F-2 and F-3) 16mms film (Bolex).

d) <u>Birrefringent filters:</u> H-α (0.5 A); Ca II K (0.3 A).

e) <u>Pinhole photometer</u>
It consists mainly of an EMI 9658 B photomultiplier (spectral response S20). The filters available are (413.8; 16.9) (493.1; 16.9) (568.4; 14.4) (673.9; 16.3) and (776.4; 15.8). The measurements can be made with 0.5"; 1" and 2" holes. The data are stored in a BBC microcomputer.

Other characteristics : Integration time to be selected from 0.03 to 10 sec. The dark current is substracted automatically.

f) <u>Image Scanner</u> : It is engaged in the tracking system of the telescope (α and β). By selecting a time (in units of 10 msec) and speed it is possible to travel a certain distance in one of both axes.

g) In the past year another secondary optical path has been installed, where the KIS image stabilizer is being tested.

5 REFERENCES

Brandt, P.N., Wöhl, H. (1982). Solar site testing campaign of JOSO on the Canary Islands in 1979. Astron. Astrophys. 109, 77.
Brookes, J.R., Isaak, G., Van der Raay, H.B. (1978). A resonant scattering solar spectrometer. Mon. Not. R. Astr. Soc 185, 1.
Brookes, J.R., Isaak, G.R., Van der Raay, H.B. (1981). A two-dimensional solar spectrometer. Solar Phys. 74, 503.
Eccles, D.G., Elworth, Y., Van der Raay, H.B., Palle, P.L., Roca-Cortés, T. (1986). Comparison of solar oscillation data obtained from a study of the Na and K Fraunhofer absorption lines. In IAU Symp. No. 123 "Advances in Helio and asteroseismology". Aarhus.
Jimenez, A., Palle, P.L., Perez-Hernandez, F., Regulo, C., Roca-Cortés, T., Domingo, V., Korzennick, S. (1986 a). Earth based observations of solar luminosity oscillation. In IAU Symp. No. 123 "Advances in Helio and asteroseismology". Aarhus.
Jimenez, A., Palle, P.L., Roca-Cortés, T., Domingo, V., Korzennick, C. (1986 b). Ground based measurements of solar intensity oscillations. Astron. Astrophys. (In press).
Mattig, W., Casanovas, J. (1975). The 40 cm Vacuum test Telescope of the Fraunhofer Institut. In JOSO Annual Report, pg. 18

Perez-Hernandez, F. (1986). Rotacion del Interior del Sol. Tesina. Universidad de La Laguna.
Roca-Cortés, T. (1984). The Solar Laboratory. _In_ JOSO Annual Report pg.
Schröter, E.H. (1984). Solar Observational facilities on the Canary Islands - operational, under construction, planned. _In_ JOSO Annual Report. pg. 29.

Round Table Discussion

Richard B. Dunn
National Solar Observatory, Sunspot, NM 88349, U.S.A.

The following is an impression on the round table discussion at the end of the conference. It is not intended as a complete summary of the days' proceedings, and I have reorganized the comments. The discussion was lively, and I cannot possibly give proper credit to all!

Two broad areas were to be addressed: "What can be done best, with what restriction, and what cannot be done with the presently foreseen facilities on the Canary Islands? What extensions of these facilities are desirable?" and "What are the candidate problems for future integrated observational experiments, for which all available solar observing facilities could be used simultaneously?"

The idea was to uncover any weaknesses in the instrumentation that existed or that was planned, and to see if there was a need for cooperative programs, especially ones that involved more than one instrument or more than one observatory or different countries. Please understand that there were plenty of ideas and plans of individuals and groups to use their instruments, but could the authors of the papers, or anyone in the group suggest that something had been overlooked?

First of all the list of instruments planned at La Palma and Izaña is formidable, and clearly represents the highest concentration of solar instrumentation in the world. The sites are good, and may be the best in the world, and they are easily accessible to European astronomers.

The instruments include the 1-meter aperture, low-polarization THEMIS under construction, the 70-cm aperture Vacuum Tower Telescope (VTT), the 45-cm Vacuum Gregorian Telescope (VGT), the 50-cm aperture Swedish evacuated telescope on La Palma, the 40-cm Vacuum Newtonian Telescope (VNT), the Solar Lab (which includes a K and Na resonant scattering spectrometer, solar luminosity instrument and a 25-cm Razdow), possibly the 45-cm Utrecht telescope, and more.

I thought Peter Brandt and Dirk Soltau's suggestion to compile a list of the instruments and their capabilities a good one. I believe there is a list, published by the Instituto de Astrofosica de Canarias a few years ago, for all the telescopes including stellar; but a much more detailed one is needed for solar instruments - perhaps a compilation

Round table discussion

more like the "NSO User's Manual". This is a rather large job!

<u>ACTION ITEM:</u> Compile a list of the instruments along with a detailed description of the focal plane instruments, computers, video processors, etc.

The strongest contender for additional focal plane instruments was clearly the magnetograph. J.O. Stenflo brought up the <u>video</u> magnetograph on several occasions. C. Zwaan thought an up-to-date magnetograph observing several spectral lines simultaneously worthwhile. W. Livingston pointed out the contribution of the NSO Vacuum Telescope magnetogram/10830 Å monitor and the need to augment that. Soltau suggested that NSO magnetograms might be obtained via the Space Environmental Solar Image System (SELSIS), accepting the time delay. THEMIS is aimed directly at spectrographic measurements of all the Stokes vectors. The Lund group is working actively on a video magnetograph for the Swedish Vacuum Telescope.

<u>ACTION ITEM</u>: Are additional magnetographs needed? What should they look like?

Other instrumentation candidates include a medium dispersion spectrograph for flares (Falciani), a coronagraph (Schröter), Fourier Transform Spectrograph (Schröter), 4-dimensional spectrograph (x,y,v,t) (Kneer) and more instruments for the IR. Better guiding was worthwhile (Falciani).

Oskar von der Lühe thought there should be more on-site computer capability. The Swedish group now has a Microvax, and additional computers are planned for the VTT.

If there were to be cooperative programs between observatories, then better communications were needed, starting with the telephone system. Networking the computers ("E mail") and a fiber link was discussed, but the latter was considered too expensive at the moment. "Walkie Talkies" were suggested.

Alan Title thought that video processors and video techniques would play an important role and that software standards are important and need to be developed. He stresses that agreements need to be made from the onset.

The observatories might benefit from a dedicated complex of instruments aimed at activity.

For comparing the two Canary Island sites, especially in the context of LEST, I thought it important to add a CCD TV to the 40-cm evacuated telescope on Izana. The scale should be the same as that used by Scharmer - i.e., 25 arcsec across the TV picture.

There was considerable discussion, often sparked by C. Zwaan, H. Spruit, H. Schmidt and R. Falciani, throughout the session on whether or not to formally organize the observatories for cooperative programs and for possible interfacing with space experiments. Targets such as flares or a specific science program could form the basis of a cooperation. The problems of archiving and data bases were discussed. Standards, especially for magnetic field measurements, were an issue. There was a general feeling that there should be some level of formal cooperation, but at the end it was agreed to leave the organization of cooperative programs to individual scientists. In the past this has been the most natural way and has been productive.

The need for a "User Group" did come up (C. Zwaan) and I think it is very important. Stenflo pointed out that JOSO could possible play this role, or at least organize such a group.

I pointed out that perhaps one of the most important contributions of Sac Peak was to bring together solar physicists from all over the world. A one-year or longer stay at Sac Peak was an opportunity to concentrate on some scientific area that was lost in the noise at home. This worked at Sac Peak because there was considerable resident scientific staff, together with the instruments, computers and large library, all in a relatively isolated environment. In the long run it contributed to the stability of the observatory. Since most of the European observatories in the Canaries are considered by their sponsors to be "observing stations", where individuals will come for just a few weeks, this role must be played by the Instituto de Astrofisica de Canarias. I encourage them to do this and to build up their staff and facilities there (especially computer and library) until the Canaries is the most important center for solar physics in the world. Other participants are encouraged to provide stipends for long-term visits to the Canaries.

LIST OF PARTICIPANTS

Ambastha, A.	Udaipur Solar Observatory, INDIA
Anguera Gubau, M.	IAC, La Laguna, Tenerife, SPAIN
Anton, V.	Kiepenheuer-Institut, Freiburg, GERMANY
Anzer, U.	Max-Planck-Inst.f.Astroph., Garching, GERMANY
Balasubramaniam, K.S.	Indian Inst. of Astrophysics, Bangalore, INDIA
Ballester Mortes, J.L.	Dept. de Fisica, Univ. Palma de Mallorca, SPAIN
Bonaccini, D.	Oss. Astrofisico Arcetri, Firenze, ITALY
Bonander, K.	Stockholm Observatory, Saltsjöbaden, SWEDEN
Bonet Navarro, J.A.	IAC, La Laguna, Tenerife, SPAIN
Brandt, P.	Kiepenheuer-Institu , Freiburg, GERMANY
Busse, F.	Lehrstuhl Theoret. ıysik, Bayreuth, GERMANY
Cacciani, A.	Jet Prop.Lab.Cal.In: . of Technology, Pasadena, Ca., USA
Carlsson, M.	Inst. of Theoretica: Astrophysics, Blindern, Oslo 3, NORWAY
Cavallini, F.	Oss. Astrofisico Arcetri, Firenze, ITALY
Ceppatelli, G.	Oss. Astrofisico Arcetri, Firenze, ITALY
Collados Vera, M.	IAC, La Laguna, Tenerife, SPAIN
Darvann, T.A.	Inst. of Theoretical Astrophyscis, Blindern, Oslo 3, NORWAY
Degenhardt, D.	Universitäts-Sterr ırte, Göttingen, GERMANY
Deinzer, W.	Universitäts-Stern\ te, Göttingen, GERMANY
Del Toro Iniesta, J.C.	IAC, La Laguna, Ten fe, SPAIN
Deubner, F.L.	Inst.f.Astr.u. Astrc Univ. Würzburg, GERMANY
Dunn, R.B.	National Solar Obser ory, Sunspot, N.M., USA
Edwin, P.M.	Dept. of Appl. Math. Univ. of St. Andrews, Scotland, UNITED KINGDOM
Eriksson, K.	Astronomiska Observatoriet, Uppsala, SWEDEN
Falciani, R.	Oss. Astrofiscio Arcetri, Firenze, ITAlY
Ferriz Mas, A.	IAC, La Laguna, Tenerife, SPAIN
Garcia de la Rosa, J.I.	IAC, La Laguna, Tenerife, SPAIN
Godoli, G.	Inst. de Astronomia de l'Universita, Firenze, ITALY
Goossens, M.	Astron. Inst. Kath. Univ. Leuven, Heverlee, BELGIUM
Grosser, H.	Universitäts-Sternwarte, Göttingen, GERMANY
Hammer, R.	Kiepenheuer-Institut, Freiburg, GERMANY
Hammarbäck, G.	Astronomiska Observatoriet, Uppsala, SWEDEN
Holweger, H.	Inst.f.Theoret. Physik u. Sternw.,Kiel, GERMANY
Hosinsky, G.	Obs. del Roque de los Muchachos, La Palma, Canary Islands, SPAIN
Hurlburt, N.	DAMTP, Univ. Cambridge, UNITED KINGDOM
Jensen, E.	Inst. of Theoret. Astrophy., Blindern, Oslo 3, NORWAY
Jimenez Mancebo, A.J.	IAC, La Laguna, Tenerife, SPAIN
Kneer, F.	Universitäts-Sternwarte, Göttingen, GERMANY
Knölker, M.	Universitäts-Sternwarte, Göttingen, GERMANY
Komm, R.	Kiepenheuer-Institut, Freiburg, GERMANY

Participants

Koutchmy, S.	Inst. d'Astrophysique CNRS, Paris, FRANCE
Kusoffsky, U.	Obs. del Roque de los Muchachos, La Palma, Canary Islands, SPAIN
Larsson, B.	Lund Observatory, Lund, SWEDEN
Laufer, J.	Inst.f.Astr.& Astroph. Univ. Würzburg, GERMANY
Livingston, W.	National Solar Observatory, Tucson, Arizona, USA
Lühe, O. von der	National Solar Observatory, Sunspot, N.M., USA
Lundstedt, H.	Lund Observatory, Lund, SWEDEN
Lustig, G.	Inst. f. Astronomie, Universität Graz, AUSTRIA
Malmort, M.	Stockholm Observatory, Saltsjöbaden, SWEDEN
Maltby, P.	Inst. of Theoretical Astroph., Blindern, Oslo 3, NORWAY
Marmolino, C.	National Solar Observatory/S.P., Sunspot, N.M., USA
Marquez Rodriguez, I.	IAC, La Laguna, Tenerife, SPAIN
Mattig, W.	Kiepenheuer-Institut, Freiburg, GERMANY
Mein, P.	Observatoire de Meudon, Meudon, FRANCE
Moreno Insertis, F.	IAC, La Laguna, Tenerife, SPAIN
Nesis, A.	Kiepenheuer-Institut, Freiburg, GERMANY
Oekten, A.	University Observatory, Istanbul, TURKEY
Pahlke, K.D.	Universitäts-Sternwarte, Göttingen, GERMANY
Pallé Manzano, P.L.	IAC, La Laguna, Tenerife, SPAIN
Priest, E.	Appl. Math. Dept., Univ. of St. Andrews, Scotland, UNITED KINGDOM
Rayrole, J.	Observatoire de Meudon, Meudon, FRANCE
Regulo, C.	IAC, La Laguna, Tenerife, SPAIN
Righini, A.	Oss. Astrofisico Arcetri, Firenze, ITALY
Roberti, G.	Oss. Astron. di Capodimonte, Napoli, ITALY
Roberts, B.	Dept. Appl. Maths., Univ. of St. Andrews, Scotland, UNITED KINGDOM
Roca Cortes, T.	IAC, La Laguna, Tenerife, SPAIN
Rodriguez Hidalgo, I.	IAC, La Laguna, Tenerife, SPAIN
Rosner, R.	Center for Astrophysics, Harvard University, Cambridge, MA, USA
Roudier, T.	Observ. du Pic du Midi, Bagnères de Bigorre, FRANCE
Ruiz Cobo, B.	IAC, La Laguna, Tenerife, SPAIN
Sànchez Almeida, J.	IAC, La Laguna, Tenerife, SPAIN
Scharmer, G.	Stockholm Observatory, Saltsjöbaden, SWEDEN
Schmidt, H.U.	Max-Planck-Inst. f. Astroph., Garching, GERMANY
Schmidt, W.	Kiepenheuer-Institut, Freiburg, GERMANY
Schmitt, D.	Universitäts-Sternwarte, Göttingen, GERMANY
Schröter, E.H.	Kiepenheuer-Institut, Freiburg, GERMANY
Schüßler, M.	Kiepenheuer-Institut, Freiburg, GERMANY
Severino, G.	Oss. Astron. di Capodimonte, Napoli, ITALY
Solanki, S.K.	Inst.f.Astron., ETH-Zentrum, Zürich, SWITZERLAND
Soltau, D.	Kiepenheuer-Institut, Freiburg, GERMANY
Spruit, H.C.	Max-Planck-Inst. f. Astroph., Garching, GERMANY
Steffen, M.	Inst.f. Theor. Physik u. Sternwarte Univ. Kiel, GERMANY
Stellmacher, G.	Inst. d'Astrophysique Paris, FRANCE
Stenflo, J.O.	Inst.f. Astron., ETH-Zentrum, Zürich, SWITZERLAND

Title, A.	Lockheed Research Lab., Palo Alto, CA, USA
Trujillo Bueno, J.	Universitäts-Sternwarte, Göttingen, GERMANY
Ulmschneider, P.	Inst. f. Theor. Astrophysik, Heidelberg, GERMANY
Vàzquez, M.	IAC, La Laguna, Tenerife, SPAIN
Vince, I.	Astronomical Observatory, Belgrad E, YUGOSLAVIA
Weiss, N.	DAMTP, Cambridge, UNITED KINGDOM
Weisshaar, E.	Theoret. Physik, Universität Bayreuth, GERMANY
Wiehr, E.	Universitäts-Sternwarte, Göttingen, GERMANY
Wittmann, A.	Universitäts-Sternwarte, Göttingen, GERMANY
Wöhl, H.	Kiepenheuer-Institut, Freiburg, GERMANY
Wyller, A.	Obs. del Roque de los Muchachos, La Palma, Canary Islands, SPAIN
Zirker, J.B.	National Solar Observatory, Sunspot, N.M., USA
Zwaan, C.	The Astronomical Inst., Utrecht, NETHERLANDS